# RADIO ASTRONOMY

# RADIO
# ASTRONOMY

## Shubhendu Joardar, PhD
(Tata Institute of Fundamental Research)

and

## James R. Claycomb, PhD
(Houston Baptist University)

**MERCURY LEARNING AND INFORMATION**

*Dulles, Virginia*
*Boston, Massachusetts*
*New Delhi*

Publisher: David Pallai
MERCURY LEARNING AND INFORMATION
22841 Quicksilver Drive
Dulles, VA 20166
info@merclearning.com
www.merclearning.com
1-800-758-3756

This book is printed on acid-free paper.

S. Joardar and J. R. Claycomb. *Radio Astronomy.*
ISBN:978-1-936420-35-3

Library of Congress Control Number: 2013944480

141516321

# CONTENTS

# PREFACE

## PREFACE TO THIS EDITION

This edition of *Basic Techniques of Radio Astronomy* has been supplemented with context-connection boxes, including NASA's Deep Space Network (DSN), the South Pole Telescope (SPT), the Low-Frequency Array (LOFAR), Space Very Long Baseline Interferometry (VLBI), pulsar dispersion and distance, and plane waves in conducting and dielectric media. Three appendices have been added on radiation potential formalism, the physics of radio spectral lines, and a table of world radio observatories. Several problems have been added to emphasize the physical aspects of radio astronomy. Supplementary materials and all textbook figures are included on the companion disc.

*James R. Claycomb*
*Houston, TX*
*March, 2014*

## PREFACE TO THE ORIGINAL EDITION

This book is based on my lectures on the subject of radio astronomy and techniques delivered at various places including the GMRT. It is designed from the point of view of a large audience consisting of persons who are associated with radio astronomy. They include (i) students, (ii) astronomers, (iii) engineers, and (iv) academicians. The book begins with the foundation of radio astronomy and then goes on to explain the fundamentals, polarization, designing radio telescopes, understanding radio arrays, interferometers, receiving systems, mapping techniques, image processing, and propagation effects, all in relation to radio astronomy. A special chapter at the end presents the GMRT radio array as an example of these techniques. For the benefit of students, two additional chapters have been added as appendices. They describe the coordinate systems used in radio astronomy and the essential aspects of antenna theory. It is advisable to go through the first two appendices together with the first chapter. The review questions at the end of each chapter contain direct hints for easy solutions. For constructing radio images from interferometer data, readers are advised to freely download the relevant software from the NRAO website. Although AIPS is popularly used, I encourage students to use CASAPY, which is based on python scripts. Tutorials for software operation are also available for free download. In either case, the system requirement is Linux, preferably Fedora Core, Ubuntu, or Debian. The bibliography at the end lists only published materials. Materials that are available free on the Internet have also been cited. I am thankful to the entire radio astronomical community, whose direct or indirect contributions have helped in the completion of this work. I express my special thanks towards them. Finally, I give my best wishes to the audience. If anyone benefits from this work, I shall consider my efforts as both successful and rewarded. The inspiration behind this work came from my Gurudev Ananta Sri Baba, and I dedicate this work to Him.

*S. Joardar*
*Pune, India*
*March, 2012*

# FOUNDATIONS OF RADIO ASTRONOMY

## 1.1 INTRODUCTION

Radio astronomy studies the Universe at radio frequencies. It analyzes the radio signals received from distant objects in the Universe.

### 1.1.1 Location of the Signals in the Universe

The radio astronomical signals may be categorized regionally as follows:

(*i*) Our solar system consists of the Sun, planets and their satellites, comets, asteroids, and the interplanetary medium. Strong transient radio signals reach us from the Sun and Jupiter. Relatively weak continuum radio signals from the Sun as a black body also reach us. The distance from the Sun to Earth is 150,000,000 km, which is 1 AU (astronomical unit). The solar system is a part of the galaxy in which we live, known as the *Milky Way* or the *Galaxy*. The Sun is at a distance of 8.5 kpc (kiloparsec) from the center of the Galaxy; 1 pc = 3.26 light years. Solar radiations are usually measured in units of SFU (solar flux unit); $1 \, \text{Jansky} = 10^{-26} \, \text{W/m}^2\text{/Hz}$.

(*ii*) The stars are objects like our Sun but vary in size, mass, and temperature. They, too, emit as black bodies and possibly also have stronger transient emissions. However, they are far away from the solar system but within the Galaxy, and their distances are usually measured in kpc. Stars are also constituents of other galaxies, which are not distinctly visible as individuals. A large number of the stars form binary systems, where one star revolves around the other. Because the distance is much larger than that to the Sun, the flux density received on Earth is much less. Thus the unit of measurement varies from a milli-Jansky to a Jansky, where $1 \, \text{Jansky} = 10^{-26} \, \text{W/m}^2\text{/Hz}$.

(*iii*) Stars are also seen in the form of small stellar systems known as *star clusters*. They are also seen in larger systems known as *galaxies*. The distances between the galaxies are measured in Mpc (megaparsec). Because galaxies contain stars, they also produce radio emissions as a part of their spectra. The central region of any galaxy usually contains a super massive black hole surrounded by a bulge of stars. Hence, intense radio emission is very likely to originate from the central region.

(*iv*) Materials in the space between the stars is known as the *interstellar medium* (ISM). It contains dust- and gas-forming giant clouds. They appear as bright or dark nebulae, depending on the state of emission. Hot nebulae radiate at frequencies corresponding to their elementary composition, which is known as *spectral-line emission*. They also absorb power at certain frequencies when wide-band signals pass through them. This is known as *spectral-line absorption*.

(v) Special types of stellar remnants, like pulsars and supernovae, may also radiate at radio frequencies. Objects like planetary nebulae and quasars may also radiate in the radio spectrum. We also receive 2.7 K black-body microwave radiation coming from the Universe in all directions. This radiation is known as the CMB (cosmic microwave background). Galaxy clusters composed of many galaxies exist at extremely large distances and also radiate in radio waves.

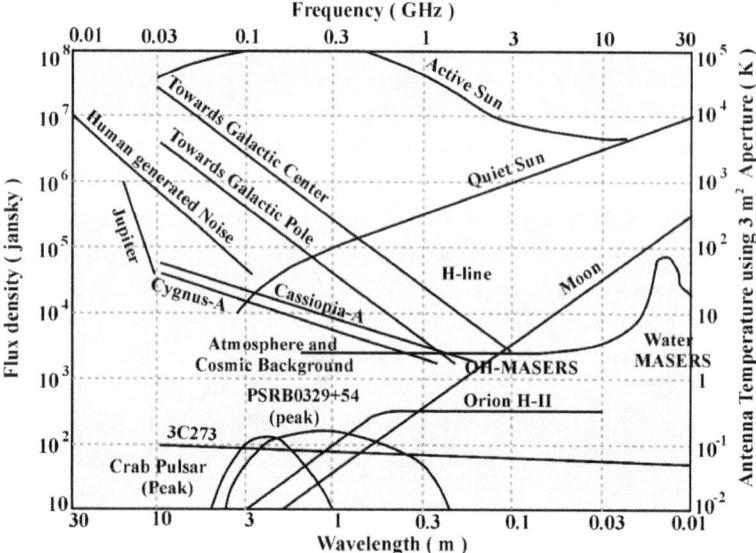

**FIGURE 1.1** Some measured flux densities near the surface of the Earth and corresponding antenna temperatures, using a 3m² aperture antenna. These are produced from different astronomical sources at different radio frequencies.

Fig. 1.1 shows some radio flux densities measured near the Earth as a result of various radio emissions from different astronomical objects. The corresponding antenna temperatures (see Section 1.4.1) using a 3m$^2$ aperture are shown. Observe that at lower frequencies, the human-generated noise dominates over most of the signals obtained from distant astronomical objects, especially from those outside the solar system. The unit Jansky is used here. It is named after the discoverer of radio astronomy, Karl Jansky.

### 1.1.2 Transparency of the Atmosphere

The Earth's atmosphere is not completely transparent to electromagnetic radiation at all frequencies. This is shown in Fig. 1.2. Hence, only a limited portion of the actual astronomical spectrum is available on Earth. The sites of radio observatories are chosen based on the intended observation frequency and are validated based on the graph of Fig. 1.2. Ground-based radio astronomy is limited by our atmosphere within a range of 30 MHz to 30,000 MHz.

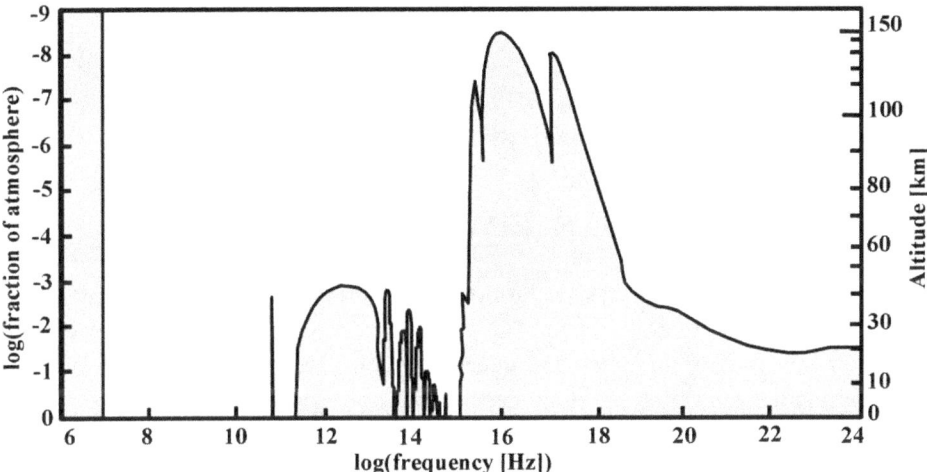

**FIGURE 1.2** Atmospheric transparency to electromagnetic radiation as a function of frequency. The gray regions imply that the atmosphere is opaque. The dark line above the gray regions shows the height above sea level where the atmosphere becomes completely transparent.

### 1.1.3 Processes of Radio Emissions

Basically, there are three major processes responsible for astronomical emissions. These are (*i*) thermal black-body emission, (*ii*) spectral-line emissions, and (*iii*) synchrotron radiation. Thermal Bremsstrahlung radiation is also seen when radiation takes place from hot plasma. Thermal black-body radiation comes from hot and warm objects like the Sun, planets, or the CMB. This radiation is continuous, and the relation between the intensity and wavelength is obtained by Planck's law, which will be explained later. The spectral-line radiations are due to transitions within atoms and molecules. They originate in clouds of gas in space and can be detected both in emissions and absorptions. The synchrotron radiation comes from spiraling charged particles at relativistic velocities in magnetic fields. Acceleration and deceleration of charged particles due to interaction with heavier charged particles results in Bremsstrahlung radiation.

### 1.1.4 Purpose of Radio Astronomy

The basic purpose of radio astronomy is to examine the received radio radiations and attempt to deduce the conditions under which these were emitted. The material composition, temperature, and the distances (in some cases) can be determined. The size and brightness distribution of the source can be mapped when the radiation comes from an extended source.

In the following sections, we begin with the discovery of radio astronomy and early work in the field. We then introduce the basic mechanisms responsible for generation of the radio spectrum by astronomical objects. Finally, we shall move into the radio techniques for data analysis and imaging. In various places, different coordinate systems have been used, which are described in Appendix A.

## 1.2   DISCOVERY AND EARLY WORKS

Karl Jansky, a radio engineer, was the first to discover the celestial nature of certain radio noises. While working at the Bell Telephone Laboratories, he was trying to locate the source of static noise that interfered with radio reception at wavelengths of 14.6 m and 10 m. He found that some of the noise was originating from nearby thunderstorms, while other noise was from very distant sources. He also listened to a faint hiss that was not related to the thunderstorms at all. He recorded this data for over a year and noted the crests and troughs in the intensity of the weak signal. Fig. 1.3 shows a portion of this record. The peaks in the record appear at an interval of nearly 22 minutes. Note that the peak slowly shifts from south to southwest after two hours. Because the azimuth of a source located on the Earth does not change, this implied that the signal was of extraterrestrial origin. In December 1932, he published his results. Following this, he published a few more papers. He called his discovery *star static*. He showed that the signal originated from the center of the Milky Way in the direction of Sagittarius. He suggested a narrow-beam antenna for resolving the signal more clearly.

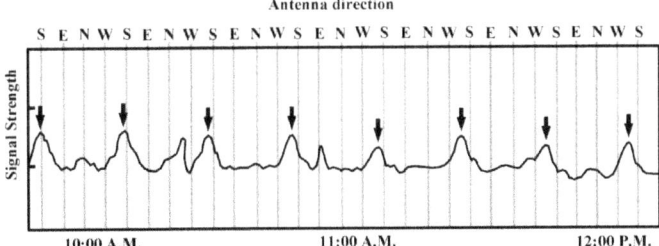

**FIGURE 1.3** A small portion of Karl Jansky's records from February 1932. The changing direction of the peak from south to southwest indicates that the source was not stationary with respect to Earth and hence had an extraterrestrial origin.

Another radio engineer, Grote Reber, was attracted by the discovery of radio noise from the Milky Way. He built the first radio telescope using a paraboloid reflector having a diameter of 9.5 m. It was a meridian telescope and only the declination could be changed. Reber worked at shorter wavelengths. In 1939, he found radiation at 1.87 m that was strongest in the direction of the Milky Way. It confirmed Jansky's discovery. Some of his work was published in 1940 and 1942. He further published a radio map in 1944, showing equal-intensity contours of radio noise on the celestial sphere (see Fig. 1.4). Reber was the first to interpret the antenna temperature as a measure of the equivalent temperature of the distant radio sources. He is officially the first radio astronomer. In 1944, Jan Hendrik Oort noted Reber's paper. He suggested that radio line emissions could be used to investigate the motion of the gas from which the emissions originated. Later, van de Hulst suggested that when the spin of an electron spontaneously reversed itself in a hydrogen atom, the photon wavelength would be 21.106 cm (1420.406 MHz), which is in the radio frequency part of the spectrum. This line would be detected provided that interstellar space contained a large amount of hydrogen.

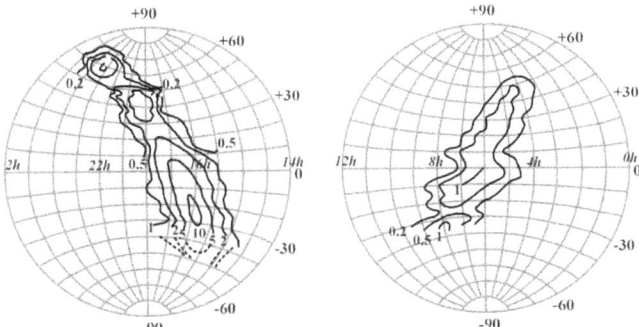

**FIGURE 1.4** Reber's map of radio emissions from the Milky Way at 160 MHz (1.875 m). The contours are plotted with equatorial coordinates. They more or less coincide with the Milky Way.

During World War II, radio interference sources were identified that confused radar location detectors. This resulted in extensive research efforts to design powerful radio transmitters and sensitive receivers. This major technological boost was a boon for radio telescopes. At the end of World War II, a number or radar engineers systematically studied the new discoveries of radio astronomy using these instruments. They discovered discrete radio sources. The angular precision of telescopes remarkably improved with the introduction of radio interferometry. During 1949, Bolton, Stanley, and Slee showed that the *Crab Nebula* (a supernova remnant), *Centaurus A* (associated with galaxy *NGC 5128*), and *Virgo A* (associated with galaxy *M87*) were emitting at radio frequencies. The diffuse radio emissions from the Milky Way and these findings together established that discrete radio sources exist in concentration in the Galaxy or beyond. In 1951, Graham Smith measured the positions of *Cygnus A* and *Cassiopeia A* with an accuracy of about 1 minute of arc. These are two bright sources of radio waves. Cassiopeia A is from a supernova remnant within our Milky Way. Cygnus A is associated with a distant galaxy, and this proved that radio sources could be used for cosmological studies. In 1960, another bright radio source, *3C295,* was discovered. It was associated with the brightest galaxy in a cluster of galaxies that had a redshift of 0.461.

## 1.3  NATURE OF COSMIC RADIO EMISSIONS

Cosmic radio signals are generated by natural processes and have the form of Gaussian noise. When received, the voltage output of the antenna resembles a series of extremely short-duration pulses that occur randomly. These are like noise from a communication receiver and not like communication signals. Across a bandwidth $\Delta v$, where $v$ stands for frequency, the RF (radio-frequency) envelope of the waveform varies randomly proportional to $1/\Delta v$. The radiation is continuous over a very wide spectral range. On a spectrum analyzer over a limited bandwidth, continuum radiation appears the same as noise.

### 1.3.1 Properties of Random Signals

A random or stochastic process has an inate indeterminacy in its future evolution that can be described by a probability distribution. Even though the initial condition is known, the evolution of the process has many possibilities. However, some evolutionary paths are more probable and others less so. A random process can be continuous or discrete. Because cosmic emissions are continuous, they behave as continuous random processes. Their probability density functions $p(x)$ can be Gaussian, as expressed in Eq. (1.1), where $\sigma$ is the standard deviation (square-root of the variance), $\mu$ is the mean, and $x$ is the random variable. Fig. 1.5 shows a Gaussian (normal) distribution with zero mean and three different variances $\sigma^2$.

$$p(x) = \frac{1}{\sigma\sqrt{2\pi}} \exp\left(\frac{-(x-\mu)^2}{2\sigma^2}\right) \qquad (1.1)$$

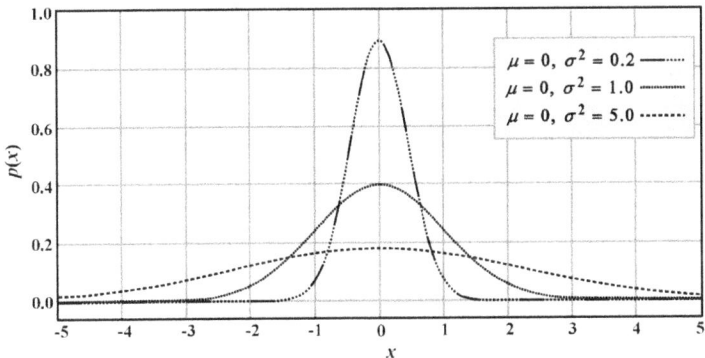

**FIGURE 1.5:** The Gaussian probability distribution $p(x)$ for zero mean $\mu = 0$ and three different values of the variance $\sigma^2$ for the random variable $x$.

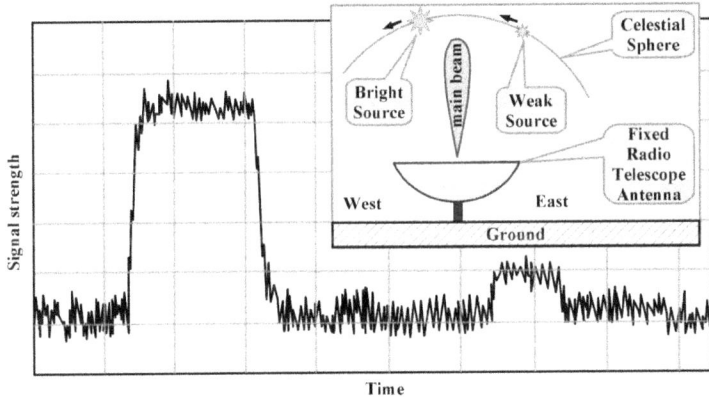

**FIGURE 1.6:** A drift signal curve produced from two radio sources (strong and weak) when they cross the main beam of the antenna.

As an example, consider two cosmic sources (one strong and one weak) drifting across the main beam of a static radio telescope, as shown in Fig. 1.6. The type of signal that would be recorded is one in which the noise sources are similar but of different magnitudes. The $x$-axis represents time, increasing from left to right, and the $y$-axis represents the corresponding signal strength. On the left we find evidence of a strong source passing through the beam, then after a short time there appears a second source. Throughout the duration of this record there are rapid, small-scale fluctuations of random intensity, or *noise*. The noise is always present in the background. The signal is visible only if it is greater than the noise. In other words, the SNR (signal-to-noise ratio) should be larger than unity. The SNR is about five for a strong source and slightly larger than unity for a weak source.

### 1.3.2 Basic Emission Mechanisms

Radio astronomical sources can be broadly classified into two categories: (*i*) thermal, and (*ii*) nonthermal. Thermal radiation is obtained when a black body is heated and is governed by Planck's law. The peak emission frequency depends on the temperature. Most of the nonthermal radiation is either synchrotron radiation or Bremsstrahlung radiation. These are explained below.

#### *1.3.2.1 Thermal Radiation*

A hot object radiates a unique spectral flux depending on its temperature. A black body absorbs all the electromagnetic energy incident on it, but does not reflect it or let it pass through. Conversely, if a black body has a nonzero temperature, it radiates energy at frequencies governed by Planck's law, as shown in Fig. 1.7. An example of a black body is a star.

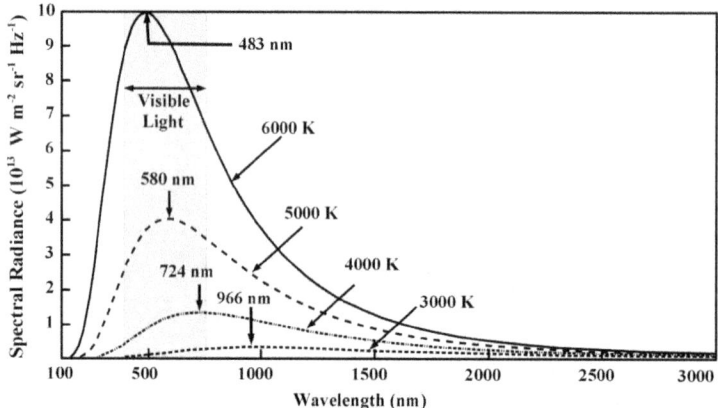

**FIGURE 1.7:** Approximate black body models of stars at different temperatures, determined by Planck's law. Note that the wavelength at peak emission is highly temperature dependent.

A star is not strictly a black body, but may be approximated as one within a limited frequency range. This is because a star contains various elements that produce absorption spectral lines. These wavelengths correspond to the elemental composition of its outer layers, such as the photosphere and chromosphere. Hence, when a star's luminosity is plotted against the wavelength, the absorbing lines (glitches) appear on the curve. Sometimes, the stars also produce sudden emissions, like radio bursts, which vary in frequency and power. The star is said to be *active* during this period. The intensity of these bursts is much higher than when the star is in a normal state or *quiet*. If we ignore the absorption lines and if the star is quiet, then within a small frequency range the star may be approximated by a black body. This is shown in Fig. 1.7. As long as the star is hot, there will always be some emission of radiation that falls within the radio band, though with considerably less intensity than at the peak.

### 1.3.2.2 Nonthermal Radiation

Radio nonthermal radiation is generally either (*i*) synchrotron, or (*ii*) Bremsstrahlung. These as described below.

(*i*) **Synchrotron Radiation:** Synchrotron electromagnetic radiation occurs in the Sun, Jupiter's magnetosphere, active galaxies, and pulsars. In 1953, it was proposed by Shklovsky that the white light obtained from the Crab Nebula was a result of synchrotron radiation. This radiation is produced from electrons at relativistic speeds spiraling around magnetic field lines. The force $\vec{F}$ on a moving charge in a magnetic field is the cross product of the particle velocity $\vec{v}$ with the magnetic field $\vec{B}$ times the charge $q$, as expressed in Eq. (1.2). This is shown in Fig. 1.8a. In Fig. 1.8b, an electron is trapped by a magnetic field and spins around it. As the electron progresses, it emits radiation along the direction of its motion. The radiation is strongly plane polarized, with its electric field lying parallel to the orbital radius of the electron about the magnetic field.

$$\vec{F} = q\vec{v} \times \vec{B} \tag{1.2}$$

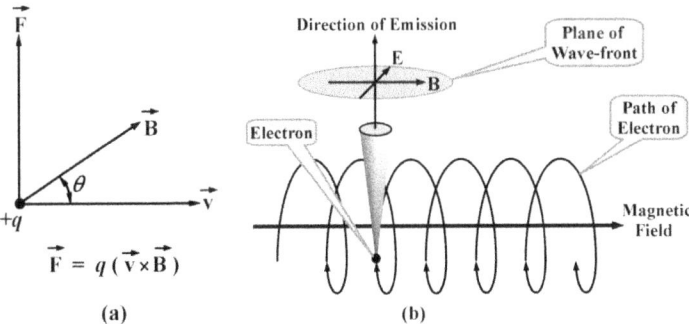

(a)  (b)

**FIGURE 1.8:** (a) The force on a moving charge in a magnetic field is equal to the cross product of the particles velocity with the magnetic field times the charge. (b) An electron at relativistic speed spirals around a magnetic field line resulting in synchrotron emission. It is polarized, with its electric field lying parallel to the orbital radius of the electron about the magnetic field.

An electron possesses an energy $E$, as shown in Eq. (1.3), where $m_e$ is the electron mass, $c$ the speed of light, and $\gamma$ is the *Lorentz factor. The Lorentz factor $\gamma$ is* given by Eq. (1.4), where is $v$ the speed of the particle. The characteristic synchrotron frequency of emission $v_c$ is given by Eq. (1.5), where $B$ is the magnitude of magnetic field in units of Gauss, and $m_e = 511$ keV/$c^2$. In actuality, the emission spectra is broad, ranging essentially from zero to the synchrotron frequency. The width of the spectral distribution for a single electron is small compared to the characteristic emission frequency $v_c$, and its polarization can be easily found. However, in the case of emission from an ensemble of electrons, the degree of polarization depends on the complexity of the magnetic field. If the magnetic field distribution in the spatial region from which the radiation is observed is homogeneous, then the observed polarization could be about 75%. On the other hand, if the spatial region within the radio telescope's view contains a complex magnetic field for which the direction is changing rapidly with time, the degree of the observed effective polarization is significantly reduced.

$$E = \gamma m_e c^2 \tag{1.3}$$

$$\gamma = \frac{1}{\sqrt{1 - \left(\dfrac{v}{c}\right)^2}} \tag{1.4}$$

$$v_c = \frac{3\gamma^2 eB}{4\pi m_e c} \approx 4.2 \times 10^6 \gamma^2 B \text{ Hz} \tag{1.5}$$

Shklovsky predicted that the white-light emission from the Crab Nebula was likely to be linearly polarized. It was later confirmed that the emissions from some regions of the Crab Nebula are nearly 60% linearly polarized. The spectral characteristics of synchrotron radiation are shown in Fig. 1.9.

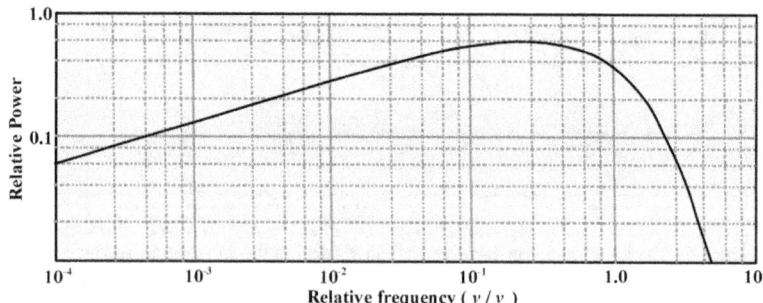

FIGURE 1.9: Spectral characteristics of synchrotron radiation.

(*ii*) **Bremsstrahlung Radiation:** The German word *Bremsstrahlung* means braking or deceleration. Bremsstrahlung radiation is produced by the acceleration of any charged particle when deflected by another charged particle. It has a continuous spectrum. This phenomenon was discovered by Nikola Tesla between 1888 and 1897.

Let an electron $e^-$ with velocity $v$ pass by a charge $Ze^+$, where $Z$ is an integer equal to the number of protons in a bare nucleus. As shown in Fig. 1.10a, the electron's path is altered from a straight line to a curve. Let the shortest distance between the electron and the positive charge be $b$. The electron accelerates until it reaches the shortest distance $b$, after which deceleration takes place. The acceleration and deceleration are not uniform. The process emits photons within a range of frequencies.

Bremsstrahlung radiation is continuously emitted in a plasma due to the mechanical interaction of free electrons with the ions. This is also known as *thermal Bremsstrahlung,* as it depends on the temperature of the plasma. In a uniform plasma with electrons, the power-spectrum density of the Bremsstrahlung radiation is shown in Eq. (1.6), where $T_e$ is the absolute temperature of the electrons (Maxwell-Boltzmann distribution), $n_e$ is the number density of the electrons, $r_e$ is the classical radius of an electron, $m_e$ is its mass, $k_B$ is the Boltzmann constant, and $c$ is the speed of light. The effective state of the ion charge $Z_{\text{eff}}$ is the average over the charge states of the ions and is defined in Eq. (1.7), where $n_Z$ is the number density of the ions. The special function $E_1$ is defined as the exponential integral in a complex plane and is shown in Eq. (1.8). The quantity $w_m$ is expressed in Eq. (1.9), where $k_m$ is the maximum or cutoff wave number.

$$\frac{dP_{Br}}{d\omega} = \frac{4\sqrt{2}}{3\sqrt{\pi}} \left(n_e r_e^3\right)^2 \left(\frac{m_e c^2}{k_B T_e}\right)^{1/2} \left(\frac{m_e c^2}{r_e^3}\right) Z_{\text{eff}} E_1\left(w_m\right) \tag{1.6}$$

$$Z_{\text{eff}} = \sum_Z Z^2 \frac{n_Z}{n_e} \tag{1.7}$$

$$E_1\left(z\right) = \int_z^\infty \frac{e^{-t}}{t}\, dt, \ |Arg(z)| < \pi \tag{1.8}$$

$$w_m = \frac{\omega^2 m_e}{2 k_m^2 k_B T_e} \tag{1.9}$$

A Bremsstrahlung spectrum is plotted in Fig. 1.10b in arbitrary units. The spectrum rapidly decreases as a function of angular frequency, going to zero as $\omega \to \infty$. Fig. 1.10c shows the Bremsstrahlung and synchrotron radiation from the spiral galaxy NGC925.

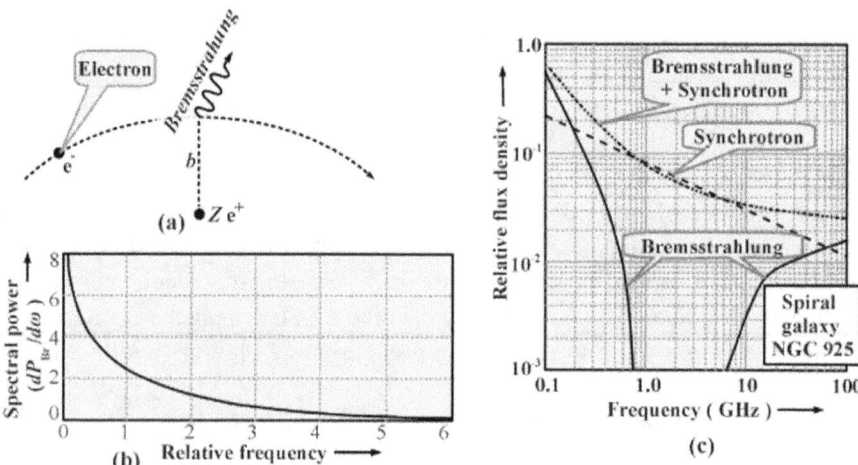

**FIGURE 1.10:** (a) An electron $e-$ crossing an ion with charge $Ze+$ undergoes acceleration and deceleration, resulting in Bremsstrahlung photon emission. (b) A tentative plot of the Bremsstrahlung spectrum using arbitrary units. (c) Bremsstrahlung and Synchrotron radiation from the spiral galaxy NGC925.

## 1.4  RECEIVING PRINCIPLES AND SIGNAL PROCESSING

The pillars of radio astronomy are radio engineering principles, as is evident from the history of radio astronomy. Understanding radio data processing and imaging requires a prior knowledge of engineering concepts such as antenna temperature, receiver temperature, signal-to-noise ratio, polarization, correlations, etc. It also requires good knowledge of digital signal processing and image processing.

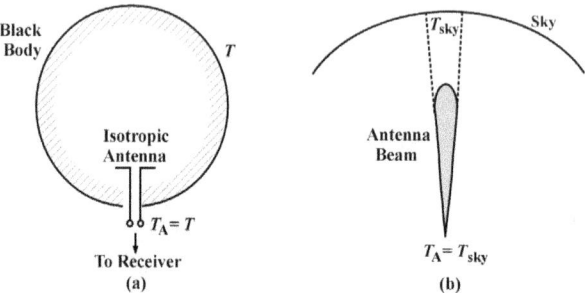

**FIGURE 1.11:** (a) An isotropic antenna is enclosed inside a black body kept at temperature $T$. The antenna temperature $T_A$ is equal to the black-body temperature $T$. (b) The sky brightness temperature $T_{sky}$ is gathered by an antenna because the main beam of the antenna is pointing towards the radiation.

### 1.4.1 Concept of Antenna Temperature

Let an isotropic antenna (see Appendix B.3) be placed inside a spherical cavity that radiates as a perfect radiator and behaves like a black body. This is shown in Fig. 1.11a. Let the antenna be connected to a receiver. The enclosed antenna will be at cavity temperature and will add an appropriate level of noise to the receiver circuit. Changing the temperature of the cavity will change this noise. In this way, one can calibrate the output of the receiver directly in terms of the temperature. The isotropic antenna is then replaced with one that has a typical pattern of those found in radio telescopes, as shown in Fig. 1.11b. Although the radiation is now received from only a small part of the inner wall of the cavity, one can still relate the strength of the signal to the temperature. With this reasoning, the sky could replace the cavity, and so the measured signal is now a measure of the temperature of a portion of the sky $T_{\mathrm{sky}}$. This kind of temperature is called the *antenna temperature* $T_A$.

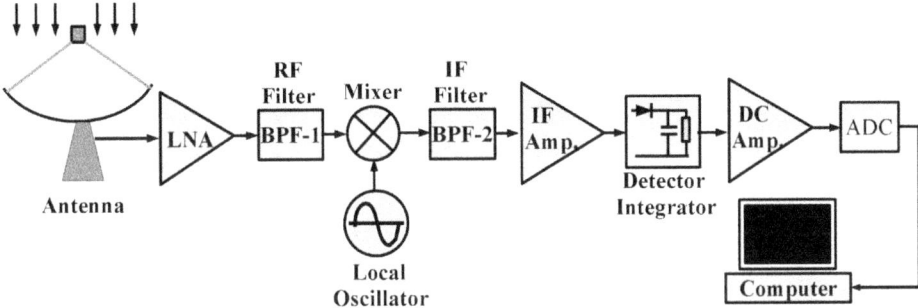

**FIGURE 1.12:** A basic heterodyne radio telescope receiver. The antenna, low-noise RF amplifier (LNA), RF filter, mixer, local oscillator, IF filter, IF amplifier, detector, DC amplifier, and data recording computer are shown.

### 1.4.2 Basic Radio Receiving System

A complete radio telescope in its simplest configuration is shown Fig. 1.12. An extremely narrow antenna beam (high directivity and very low side lobes) is produced by the dish at the operating frequency centered at $\nu_0$. The gain of the antenna is assumed to be fixed over the frequency band of observation. The beam is pointed to a radio source in the sky. A low noise RF amplifier (LNA) operates, covering the frequency band of observation. This is followed by a band-pass filter (BPF-1) with a center frequency $\nu_0$ and a bandwidth $\Delta\nu$. The signal is then converted to an intermediate frequency $\nu_{\mathrm{IF}}$ by mixing it with the local oscillator frequency $\nu_{\mathrm{LO}}$ such that $\nu_{\mathrm{IF}} = \nu_0 - \nu_{\mathrm{LO}}$. An IF filter having a bandwidth $\Delta\nu_{\mathrm{IF}} = \Delta\nu$ is used before sending the signal to the IF amplifier after conversion. A square-law detector followed by an integrator is used to detect the power of the signal over a fixed integration time. This is then converted to a digital signal by using the ADC (analog to digital converter) and sent to a computer for recording as a function of time.

### 1.4.3 Improving the Signal-to-Noise Ratios

Continuum radio signals are extremely weak near the Earth's surface. Hence the SNR (signal-to-noise ratio) of any receiving system (antenna with receiver) is poor. In order to recover the signal information, certain mathematical operations are performed at different stages of the radio telescope receiver system. The most important of these are the correlation functions (especially cross-correlation and the Wiener-Khinchin theorem), which provide more information about the spectra, as explained in the following sections. These are often used in interferometer-type radio telescope arrays.

#### 1.4.3.1 Correlation Functions

The cross-correlation is a measure of the similarity between two waveforms as a function of a time-lag $\tau$ applied to one of them. For two continuous functions of time $t$ denoted as $f(t)$ and $g(t)$, the cross-correlation $r(\tau)$ is shown in Eq. (1.10), where $f^*(t)$ represents the complex conjugate of $f(t)$.

$$r(\tau) = \int_{-\infty}^{\infty} f^*(t)\,g(t+\tau)\,dt \qquad (1.10)$$

A special case of cross-correlation is autocorrelation, where the signal is correlated with itself. The autocorrelation $R(\tau)$ of a continuous function of time $t$ is denoted as $f(t)$ is shown in Eq. (1.11).

$$R(\tau) = \int_{-\infty}^{\infty} f^*(t)\,f(t+\tau)\,dt \qquad (1.11)$$

#### 1.4.3.2 Wiener-Khinchin Theorem

The *Wiener-Khinchin* theorem, also known as the *Wiener-Khintchine* theorem, states that *the power spectral density of a wide-sense-stationary random process is the Fourier transform of the corresponding autocorrelation function*. Mathematically, if $S(v)$ is the power spectral density of a continuous random signal $x(t)$, then it may be expressed as the Fourier transform of the autocorrelation function $R(t)$. This is shown in Eq. (1.12).

$$S(v) = \int_{-\infty}^{\infty} R(\tau)e^{-j2\pi v\tau}\,d\tau \qquad (1.12)$$

### 1.4.4 Radio Arrays

Radio astronomical signals are extremely weak near the Earth's surface. To overcome this, the required collecting area of an aperture antenna is huge and sometimes beyond the scope of construction. However, a large aperture can be synthesized using a certain number of small antennas. We give here a taste of the basic principles using simple models, such as the source at the zenith. We shall explain these in more detail in later chapters.

#### 1.4.4.1 Filled Array

Consider an aperture strip having an area $A_e$ as show in Fig. 1.13a. An approximate aperture strip may be constructed using four parabolic dish antennas placed side by side. We may express $A_e$ as the sum of the individual aperture areas of the dishes as shown in Eq. (1.13).

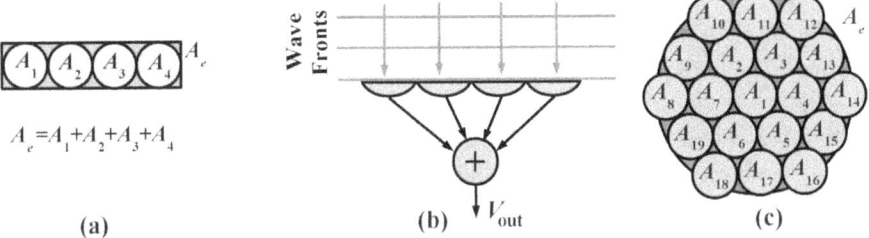

FIGURE 1.13: (a) A rectangular aperture is synthesized using four small circular apertures. (b) Four dish antennas are pointed to a source at the zenith. At any instant of time, the same wave front reaches all the individual antenna apertures. (c) Synthesis of a large circular aperture using many small circular apertures.

$$A \approx A_1 + A_2 + A_3 + A_4 = \sum_{i=1}^{4} A_i \qquad (1.13)$$

This arrangement works well as long as the source is at the zenith. For any other source position, the shadows of the dishes fall on each other, and the same wave front does not reach all the antennas simultaneously. This effectively reduces the synthesized aperture. Fig. 1.13c shows an example of this technique, where a large circular aperture is synthesized by using many small circular apertures. Due to the problem of a reduced synthesized aperture, these type of arrangements are seldom used. We shall talk about these in detail in Chapter 4, Section 4.8.1.

#### 1.4.4.2 Phased Grating Array

For a phased grating array, the antennas are separated over a distance large enough to prevent shadowing. The beam shape has several crests and troughs. A one-dimensional phased array pointed to a radio source at the zenith of the sky is shown

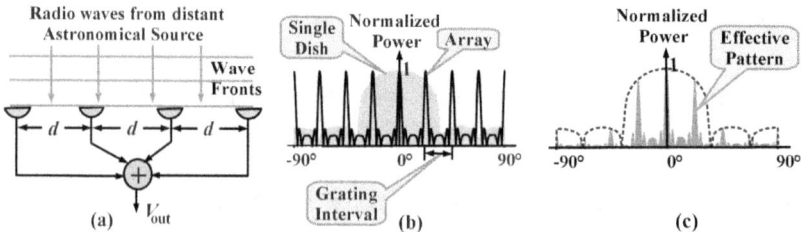

**FIGURE 1.14:** (a) Signals from a source at the zenith are added because they are in phase. (b) Normalized power patterns of a dish and an array of isotropic antennas. (c) Effective system response.

in Fig. 1.14a. The antennas are separated by a distance $d$ from each other and simultaneously receive the same wave front from the source at the zenith. These signals are therefore in phase. Fig. 1.14b shows the normalized power pattern of (*i*) a single dish, and (*ii*) an array of isotropic antennas. Only half of the pattern −90° to +90° in one plane is shown. Note that the array response resembles a grating. The angular distance between two adjacent crests is known as the *grating interval*. The grating interval depends on $d$ and the wavelength. The effective pattern of the phased array is shown in Fig. 1.14c. One of the major drawbacks of a grating array is that there could be more than one crest under the major lobe, which can significantly pick up unwanted signals. We shall talk about these in detail in Chapter 4, Section 4.8.2.

### 1.4.4.3 Correlator Arrays

Correlator arrays are popular for continuum observations over a limited bandwidth. Antenna output voltages are multiplied in pairs followed by a time integration. Fig. 1.15a shows a correlator array using four dish antennas looking towards a radio astronomical source at the zenith. The correlated signals are saved in a computer. Cross-correlations originate between any two different antennas, while autocorrelations come from every single antenna. Fig. 1.15b shows a tentative plot of the radiation pattern formed by the cross-correlation product of a pair of antennas. Note that the beam splits, forming a fringe-like pattern. The number of lobes in the beam pattern increases if the distance $d$ of separation between the two antennas is increased or the wavelength of observation is reduced. The correlator output comes after integration in time, which improves the SNR. The final output of the system contains both $r(\tau)$ and $R(\tau)$ as defined by Eqs. (1.10) and (1.11). The cross products $r(\tau)$ are greater in number. If $n_a$ is the number of antennas, the number of cross-correlations $n_{cc}$ and autocorrelations $n_{ac}$, respectively, given by Eqs. (1.14) and (1.15). These values are saved along with time, frequency, bandwidth and the antenna coordinates (see the $u$, $v$, *w* coordinate system in Appendix A.9). These are essential for constructing radio images of the source.

$$n_{cc} = n_a(n_a - 1)/2 \qquad (1.14)$$

$$n_{ac} = n_a \qquad (1.15)$$

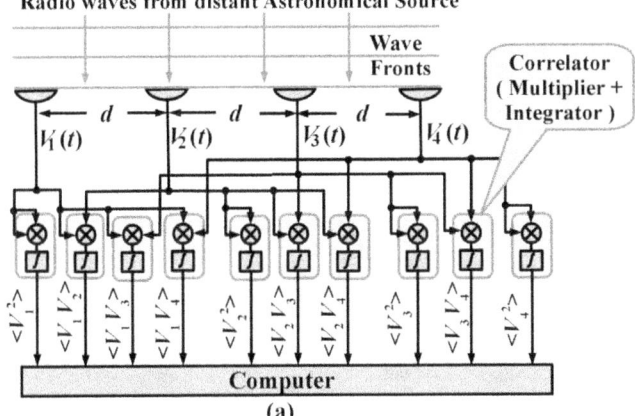

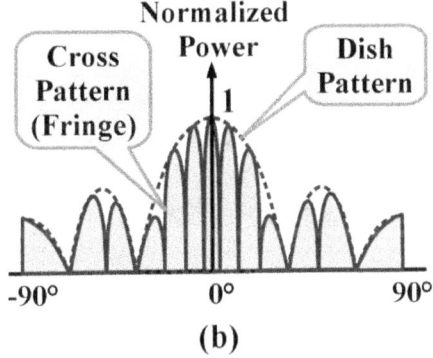

**FIGURE 1.15:** (a) A four-antenna correlator array observing a source at the zenith. Signals from each antenna pair are correlated and saved in a computer. (b) A tentative plot of the fringe pattern formed by the cross-correlation product of a pair of antennas.

The GMRT (Giant Meter-wave Radio Telescope) is one of the largest correlator arrays at meter wavelengths. It is located near Narayangaon (80 km north of Pune), India. An aerial picture of several central square antennas of the GMRT is shown in Fig. 1.16. The entire array consists of 30 steerable antennas, each with a 45 meter diameter. These are spread in a *Y*-configuration across a radius of 15 km, with the central square antennas close to the center. There are 5 or 6 antennas in each arm, and 14 antennas in the central square. The operating frequencies range roughly from 120 MHz to 1500 MHz, with a bandwidth of 32 MHz, and having two polarizations.

**FIGURE 1.16:** Aerial view of some of the GMRT central square antennas. The photograph is from Google Maps.

## 1.5 CONSTRUCTING RADIO MAPS

Radio intensity-images of observed sources are prepared after processing the data obtained from the radio telescopes. In principle, the image can be formed by scanning the source. However, an interferometer radio array consisting of several antennas distributed over a large area on the ground is preferred, for this will give a good SNR and a high angular/pixel resolution to the image. The basic principles are explained below.

### 1.5.1 Scanning to a Radio Image

Similar to an optical image formation when using a telescope where lenses or concave mirrors are used, the formation of a radio image depends on the concentration of the waves coming from a radio source into a single point using a paraboloid dish antenna. If the beam-shape of the dish antenna is extremely narrow, such that it is narrower than the source extent, it may be used to scan the radio source. The signal intensity may be plotted as a function of the angular scan, and a one dimensional image can be formed. If the antenna is scanned in two dimensions (along two orthogonal axes), a small area will be covered, and the constructed image will be two dimensional.

### 1.5.2 The Basic Mapping Technique

Assume that a highly directional single-dish radio telescope is pointed towards a distant radio astronomical source, as shown in Fig. 1.17a. Parallel rays illuminate the dish aperture. Consider the aperture plane as the origin of a $u$, $v$, $w$ coordinate system (see Appendix A.9) with $w$ pointing towards the source, as shown in Fig. 1.17b. Let the two-dimensional electric field distribution on the sky be centered on an $l$, $m$ coordinate system (see Appendix A.10). Let the antenna beam-width be negligible. The two-dimensional electric field distribution $\mathcal{E}_{\mathrm{dish}}(u, v)$ is the result of the two-dimensional electric field distribution $I_{\mathrm{sky}}(l, m)$ on the sky.

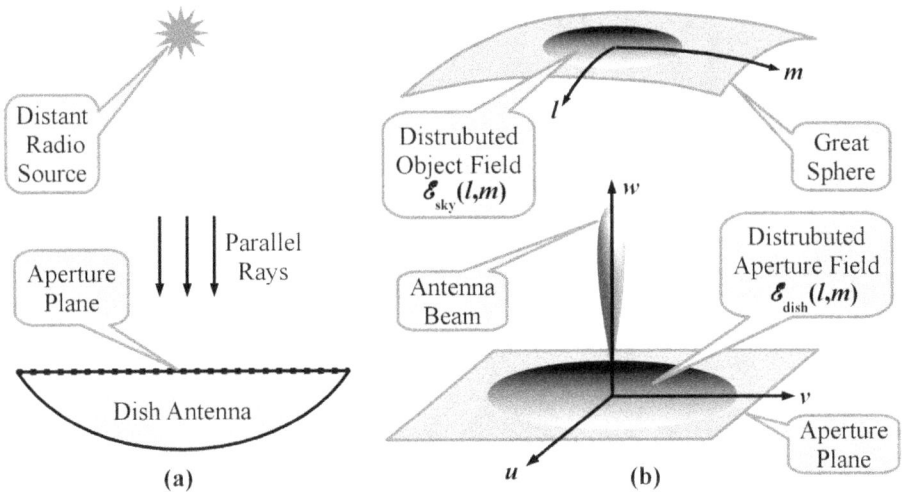

**FIGURE 1.17:** (a) A distant radio source illuminates the aperture of a dish antenna with parallel rays. (b) Electric field distributions of the source (on the great sphere) and antenna aperture (on the aperture plane).

Let us take the autocorrelation of each of the electric field points on the $u - v$ plane and represent them as $W(u, v)$. Applying a two-dimensional spatial Fourier transform to $W(u, v)$, one obtains the intensity field distribution $I(l,m)$ of the source in $l$-$m$ space. This is a very important result because it represents the intensity map of the extended radio source on the sky.

Single-dish radio telescopes are rare mainly due to the limitation of large dish construction. Fig. 1.18 shows the world's largest steerable single-dish radio telescope, the Green Bank Telescope (GBT) located at NRAO, West Virginia. Prior to the present GBT, there was an old GBT that was destroyed.

Fig. 1.19 shows a radio image of the Cygnus-X region at a frequency of 790 MHz obtained by the GBT. The antenna beam width is approximately 16 arc minutes. This observation was made by the GBT commissioning team led by Ron

**FIGURE 1.18:** World's largest steerable single-dish radio telescope known as the GBT (Green Bank Telescope) located at NRAO, West Virginia.

Maddalena. For the coordinate system used, refer to the universal equatorial system in Appendix A.4.

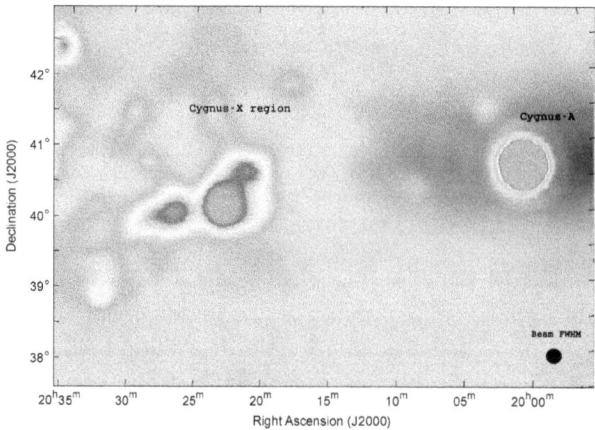

**FIGURE 1.19:** An image of the Cygnus-X region at 790 MHz, made during the commissioning of the GBT. The signals from Cygnus-A are completely saturated. Copyright NRAO/AUI/NSF.

### 1.5.3 The Basic Super-Synthesis Technique

An antenna array aided by the Earth's rotation can synthesize a very large antenna aperture. This overcomes the difficulty of constructing a single large antenna. This technique was named by Martin Ryle as *super-synthesis*.

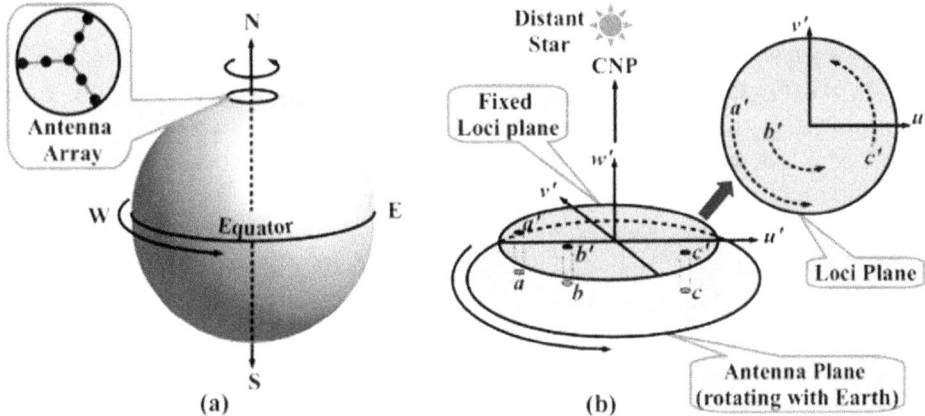

**FIGURE 1.20:** (a) An array of antennas located at the north pole is rotating with the Earth. (b) It is observing a radio source in the direction of the CNP (celestial north pole).

Assume an antenna array to exist at the north pole over a plane region as shown in Fig. 1.20a. Let these antennas track a distant radio source located at the celestial north pole (exactly along the rotational axis of Earth). Due to Earth's rotation, the entire antenna array rotates as seen from the radio source. We use a rectangular coordinate system $u'$, $v'$, $w'$ which is a special case of the $u$, $v$, $w$ coordinate system when the source under observation is located on the north celestial pole. Let the origin of $u'$, $v'$, $w'$ be located near the center of the plane containing the antennas, as shown Fig. 1.20b. Let the $w'$-axis point towards the celestial north pole (towards the source) and let the $u'$-$v'$ plane be stationary with respect to the source. As seen from the source towards the antennas, due to Earth's rotation, the position of the antennas will appear to be moving over the $u'$-$v'$ plane. For simplicity we consider three antennas ($a$, $b$, and $c$). The loci $a'$, $b'$, and $c'$, respectively, of the antennas $a$, $b$, and $c$ on the $u'$-$v'$ plane will be circles over a period of 24 hours. At very short intervals of time, we may record the voltage outputs of individual antennas and place these values on the $u'$-$v'$ plane. If there are many antennas at different radial distances from the origin of the $u'$,$v'$,$w'$ coordinates, then over a period of 24 hours, we shall find the entire $u'$-$v'$ plane highly populated with the voltage values resulting from the radio source. We may consider the populated area of the $u'$-$v'$ plane as an aperture of a large antenna. Hence, after proper calibration of the data in time and amplitude with respect to each other (antennas), we may apply a two-dimensional spatial Fourier transform of the $u$-$v$ plane data to obtain the the intensity field distribution $I(l,m)$ of the source.

### 1.5.4 Super-synthesis using Correlator Arrays

The basic idea behind super-synthesis is the same as described above. However, instead of using the voltages obtained from each antenna, as in the previous case, here the correlation products between any pair of antennas are used to fill the $u - v$ plane. The $u$, $v$, $w$ coordinates have dimensions in number of wavelengths. For example, the baseline distance $d$ between any two antennas in $u$, $v$, $w$ coordinates is $d_\lambda = d/\lambda$, where $\lambda$ is the wavelength. As shown in Eq. (1.14), if there are $n_a$ antennas in the array, the number of cross-correlation terms $n_{cc}$ is $n_a(n_a - 1)/2$. Thus the $u$-$v$ plane is well populated when the number of antennas is large. In a two-antenna interferometer, if we assume the position of one of the antennas to be fixed at the origin of the $u$, $v$, $w$ coordinate system, then a line joining the two antennas on the $u$-$v$ plane will rotate through 180° in 12 hours. A similar observation results if the other antenna is considered as the origin of the $u$, $v$, $w$ coordinates. Thus, in principle it is possible to cover 360° of baseline rotation in only 12 hours.

#### 1.5.4.1 Relation Between Visibility and Correlation

For making a radio image, we need to compute the spatial coherence function known as *visibility* (see Section 5.2.2) from the values of the cross-correlations $r(\tau_g)$ on the $u$-$v$ plane, and then apply a spatial Fourier transformation. The visibility values are nothing more than scaled values of the cross-correlations. Visibility is a function of the $u,v,w$ coordinates and is represented as $\mathcal{V}(u,v,w)$.

#### 1.5.4.2 The van Cittert-Zernike Equation

We find the values and positions of the visibilities on the $u,v,w$ coordinate system and construct the image by relating them to the $l,m$ coordinate system on the celestial sphere. The direction cosines ($l$, $m$) are measured with respect to the $u$- and $v$-axes. Hence, a synthesized image on the $l$-$m$ surface represents a flat projection of the spherical celestial region.

Let $P_n(l, m)$ be the normalized power pattern of a single antenna in $l,m$ coordinates. It is sometimes called the *primary beam* pattern. It is assumed that both the antennas forming the interferometer have identical beam shapes. Let the intensity distribution of the radio source (sky) in $l,m$ coordinates be $I(l, m)$. Let the visibilities in the $u,v,w$ coordinates be represented as $\mathcal{V}(u,v,w)$. The van Cittert-Zernike equation is shown in Eq. (1.16).

$$\mathcal{V}(u,v,w) = \iint \frac{P_n(l,m)I(l,m)}{\sqrt{1-l^2-m^2}} e^{-j2\pi\left[ul+vm+w\left(\sqrt{1-l^2-m^2}-1\right)\right]} \, dl \, dm \qquad (1.16)$$

It indicates that the complex-valued visibility function, $\mathcal{V}(u,v,w)$, is a Fourier-like integral of the sky brightness, $I(l, m)$, multiplied by the primary beam response of an interferometer, $P_n(l, m)$, and $1/\sqrt{1-l^2-m^2}$. The $u,v,w$ coordinates have been defined in such a way that the $w$-axis always points to the direction of the radio source. If the extent of the radio source is small, the values of $l$ and $m$ are sufficiently small, so that the term $w\left(\sqrt{1-l^2-m^2}-1\right)$ in the exponent is $-\frac{1}{2}(l^2+m^2)w \approx 0$ and can be neglected. We may thus express Eq. (1.16) as shown in Eq. (1.17).

$$\mathcal{V}(u,v,w) \times \mathcal{V}(u,v,0) = \iint \frac{P_n(l,m)I(l,m)}{\sqrt{1-l^2-m^2}}e^{-j2\pi(ul+vm)}dldm \qquad (1.17)$$

Thus for a restricted small range of $l$ and $m$, the visibility $\mathcal{V}(u,v,w)$ is approximately independent of $w$.

### 1.5.4.3 Filling the **u-v** Plane with Visibilities

The visibilities in the $u$-$v$ plane serve as the synthesized aperture of a huge antenna formed from the array antennas. The visibilities are used to populate the $u$-$v$ plane as densely as possible. Depending on the number of antennas in the array, at any instant of time a number of visibilities are generated. For example, the VLA (Very Large Array, New Mexico), consisting of twenty seven antennas positioned in a Y-configuration, can produce 351 visibilities for one polarization in one frequency band. Fig. 1.21a shows the $u$-$v$ plane coverage for a set of visibilities obtained from the antenna correlations at any instant of time when the radio source under observation is located at the zenith. With Earth's rotation, the star-like structure rotates in the $u$-$v$ plane. Hence with time, more and more data fill the $u$-$v$ plane. The position of the antennas are adjusted such that the baselines (distances between any two antennas) are unique, because each of the visibility values marked on the $u$-$v$ plane are at a radial distance equal to the baseline length expressed in number of wavelengths. If there are baselines of equal length, the visibilities corresponding to these will overlap at some or more points with Earth's rotation. The VLA antennas are set on rail tracks so the baselines are adjustable. If the $u$-$v$ plane is not densely populated to the user's satisfaction over a 12-hour observation period, the antenna positions can be readjusted to form new unique baselines, and further data can be obtained to increase the population density.

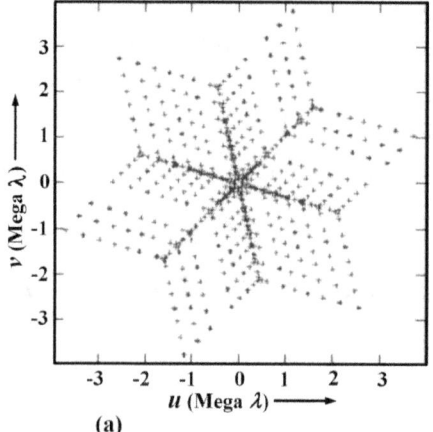
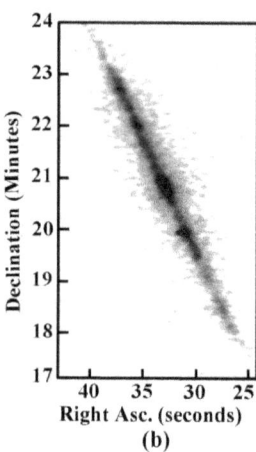

**(a)**                                        **(b)**

**FIGURE 1.21:** (a) The *u-v* plane coverage at any instant of time obtained from a source at the zenith using the VLA. (b) A radio picture of the galaxy NGC 891 taken at 5 GHz using the VLA.

After populating the *u-v* plane with sufficient data, Eq. (1.17) is applied to obtain the intensity distribution $I(l,m)$, which is the radio image of the source. Before such an operation, data inspection and calibration are essential to avoid errors from the instrument, radio-frequency interferences, and scintillations. Fig. 1.21b shows a radio picture of the galaxy NGC 891 at 5 GHz using the VLA.

## 1.6   RADIO INTERFERENCE

Cosmic radio signals are extremely weak compared to man-made radio signals and noise. For example, the radio flux received from the Sun could be only a few SFU, whereas a distant radio station may produce several thousand SFU. The power flux densities received from distant radio galaxies could vary from several micro-Jansky to few milli-Jansky. We have seen that astronomical sources produce signals resembling Gaussian noise, generally in both polarizations over a wide frequency range. A locally generated wide-band radio noise source like that from electrical arc welding may be sufficiently high to obscure the astronomical signal. Radio noise or a signal of any kind interfering or obscuring the astronomical signals is termed RFI (radio-frequency interference) by radio astronomers.

The unwanted radio signals interfere with the radio data through the side lobes of the antenna feed (*i*) directly, or (*ii*) by dish mesh leakage, or (*iii*) by dish spillovers. RFI may also leak into the system through the shielding of cables and the electronic receiver system. This is one of the major reasons for constructing radio telescopes at very remote places (less human population). Fig. 1.22 shows a few possible ways RFI is picked up in VHF/UHF radio telescopes. Interference can also occur due to the side lobes of the dish antenna.

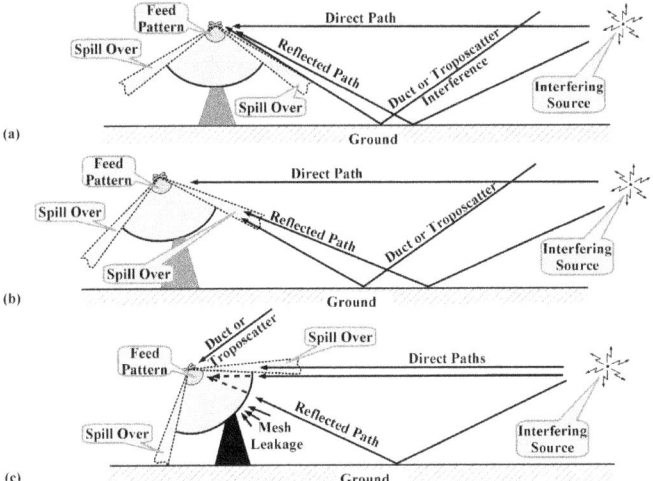

**FIGURE 1.22:** RFI in the VHF/UHF bands. Radio noise sources interfere through antenna-feed side-lobes (a) directly, or (b) through dish spillover, or (c) by spillover, dish mesh leakage, duct- and troposcatter, when the antenna points in different directions.

| BOX 1.1 | Deep Space Network |
|---|---|

The NASA Deep Space Network (DSN) operates three tracking stations, one each at locations in California, Spain, and Australia. The antenna facilities are separated by roughly 120 degrees of longitude around the globe, enabling continuous two-way communication with spacecraft at distances greater than 30,000 km from the Earth. Each of the DSN tracking stations consist of a 70 m antenna, one or more 34 m beam waveguide (BWG) antennas, and a 26 m antenna, as depicted in Fig. 1.23. The 70 meter antenna located at Goldstone, California is shown in Fig. 1.24. The 70 m reflectors are the largest antennas in the DSN, with high surface accuracy, and are capable of tracking spacecraft at distances up to ~100 A.U. The 34 m BWG antennas feature five reflecting radio frequency mirrors to guide signals into the pedestal room, located near the antenna base for easier access. The Cassegrain design of the BWG antennas is shown in Figure 1.25. Three of the BWG antennas are located in Goldstone CA, two in Madrid, and one in Canberra Australia. The 26 m antennas at all three DSN locations are used for tracking Earth satellite spacecraft. Earth satellites are used to transfer data between DSN sites and the command center located at the NASA Jet Propulsion Laboratory (JPL).

The DSN has provided two-way communication to spacecraft during the Apollo missions to the Moon, to Voyager 1 and Voyager 2 spacecraft exploring the outer solar system and beyond, as well as to the Galileo and Cassini spacecraft to Jupiter and Saturn. Numerous other missions have been supported by the DSN. More recently, the DSN has provided communication with the New Horizons space probe scheduled to rendezvous with Pluto in 2015 as well as with Mars' satellites and rovers.

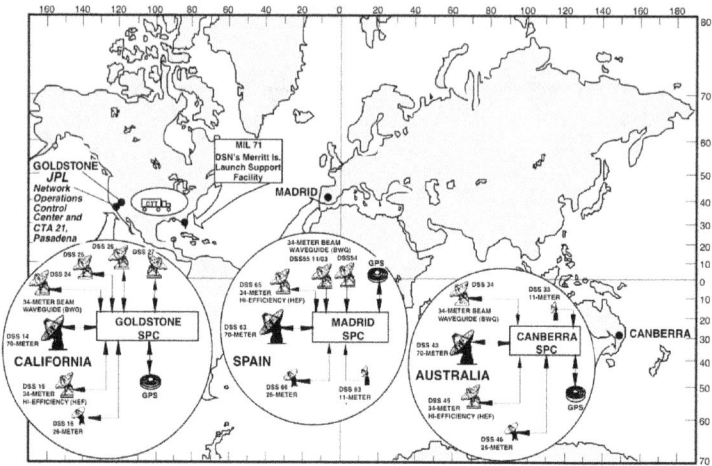

**FIGURE 1.23:** Location of the NASA Deep Space Network (DSN) telescopes. Image Credit: NASA/JPL.

**FIGURE 1.24:** 70 meter DSN antenna located at Goldstone, California. Image Credit: NASA

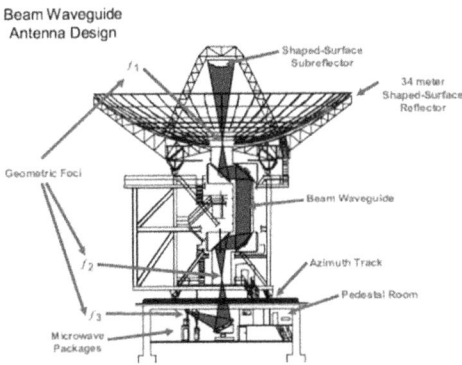

Beam Waveguide
Antenna Design

Shaped-Surface Subreflector

34 meter
Shaped-Surface
Reflector

$f_1$

Geometric Foci

Beam Waveguide

$f_2$

Azimuth Track

Pedestal Room

$f_3$

Microwave
Packages

**FIGURE 1.25: Schematic of the DSN** 34 m beam waveguide antenna. Image Credit: NASA

The Mars rover Curiosity landed in Gale Crater on Mars on August 6, 2012. The rover hosts a suite of instrumentation for sample acquisition, preparation, and analysis. It is nearly twice the size of the Mars Exploration Rovers Spirit and Opportunity that landed on Mars in 2004. One of Curiosities mission objectives is to determine if Mars could have supported life at some time in the past.

Curiosity is equipped with one short-range and two long-range antennas. The rover may transmit long-range signals directly to the DSN telescopes using either the low-gain antenna (LGA) or the high-gain antenna (HGA) depicted in Fig. 1.26. The HGA transmits a narrow beam that must be directed to a specific DSN telescope. The HGA data rate is up to 32 kbit/ s sent in the X-band frequency range between 7 GHz and 8 GHz. The omnidirectional LGA does not require steering but transmits at a much lower data rate, up to 15 bit/s. The LGA is primarily used for receiving signals from Earth. Curiosity may also communicate with the Mars Odyssey or Mars Reconnaissance Orbiter using the short-range UHF-band antenna at 400 MHz. Signals are then relayed to Earth by one of these Mars satellites.

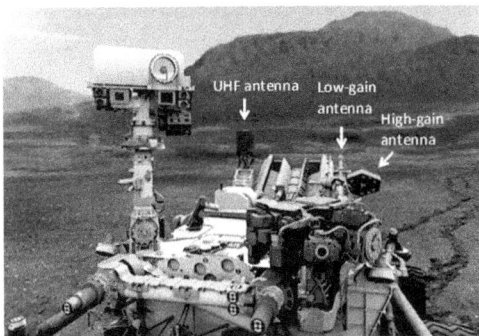

UHF antenna    Low-gain antenna

High-gain antenna

**FIGURE 1.26:** Depiction of the Mars rover Curiosity showing locations of the UHF, high-, and low-gain antennas. The UHF antenna sends signals to Mars orbiters that are then relayed to Earth. The high- and low-gain antennas transmit directly to DSN telescopes. Image credit: NASA/JPL-Caltech

## REVIEW QUESTIONS

1. About every eleven years the solar activity peaks. For a radio source located very close to the Sun during this period, what should be its minimum flux density for detection when using a 1.4 GHz radio telescope? [Hint: See Fig. 1.1]

2. Convert 1 SFU to units of Jansky.

3. How many AU are there in 1 pc?

4. What is a spectral-line emission? What is the difference between spectral-line emission and spectral-line absorption?

5. What is a random process? How do cosmic radio signals behave over a period of time?

6. Distinguish between thermal and nonthermal emissions.

7. Explain the mechanisms of the two emission processes: (*i*) synchrotron, and (*ii*) Bremsstrahlung.

8. From the signals received, when can we say that the Sun or a star is active? When is it said to be quiet?

9. Explain the meaning of antenna temperature.

10. The same signal is received using two receivers. Each of these receivers generates its own noise. Correlations are produced between the receiver outputs. How many unique correlation outputs will be produced?

11. In problem number 10, it is observed that the cross-correlation products have better SNR than the autocorrelation products. Explain.

12. How does time integration improve the SNR in auto- and cross-correlations? Explain.

13. What are the major drawbacks of a filled array?

14. What is the major drawback of a grating array?

15. For a correlator array consisting of 30 antennas with single polarization, how many auto- and cross-correlations can be formed? How many if dual polarized antennas are used? [Hint: Use Eqs. (1.14) and (1.15)]

16. Explain the Wiener-Khinchin theorem.

17. How are visibility and cross-correlation related?

18. Explain the van Cittert-Zernike equation.

19. What is the purpose of data inspection and calibration before constructing a radio image?

20. What kinds of interference signals and noise are termed as RFI by radio astronomers?

# FUNDAMENTALS OF RADIO ASTRONOMY

## 2.1   INTRODUCTION

It is necessary to understand the fundamental concepts used in radio astronomy, especially for those who are new to the subject. Many of these concepts require a prior knowledge of astronomical coordinate systems (Appendix A) and antenna principles (Appendix B). Concepts such as brightness flux density, effect of the antenna beam pattern on observations, black-body radiation, absorption and emission, antenna temperature, noise, and sensitivity have been explained in detail.

## 2.2   RADIATION AND RECEPTION

Astronomy is based on the signals from astronomical objects received by humans. A relationship between the power radiated from a location in the sky and the corresponding power received on the Earth must be established. A basic unit of radiation must be defined to describe the dependence of all the variables involved in radiation. This is known as *brightness* is astronomy. From the brightness and distance, one may find the received spectral power on Earth.

### 2.2.1 Sky Brightness

If an object emits electromagnetic waves, and the flux density is measured at a distance, then the *brightness B* of that object is defined as the *received flux density per unit frequency per unit solid angle* from the object. Because the power can vary over a wide range of frequencies, it is defined for a unit bandwidth. Increasing the angular size of the object as seen from the receiving end also increases the flux received. Hence brightness is given per unit solid angle. The units of $B$ are usually Watt m$^{-2}$ Hz$^{-1}$ sr$^{-1}$.

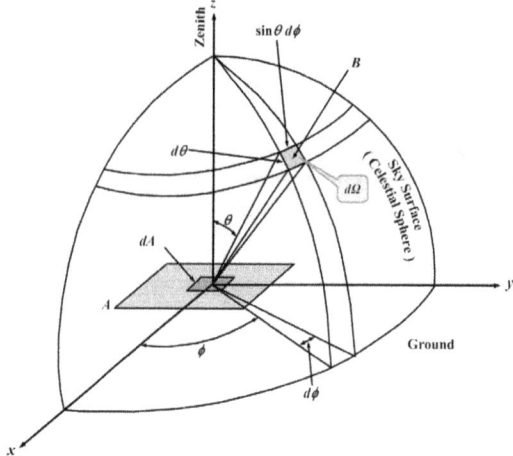

**FIGURE 2.1:** Radiation of brightness $B$ falls on the Earth's surface from a portion of the sky at incident angles of $\theta$ and $\phi$. It forms a solid angle $d\Omega$ = sin $d\theta$ $d\phi$.

Consider the spherical coordinate system shown in Fig. 2.1. Let the sky be a sphere of infinite radius, and the ground be a flat surface in the $x$-$y$ plane. Radiation from distant sources reaches the Earth based on an inverse square law[1]. Assume that the brightness $B$ varies with angular position in the sky $(\theta, \phi)$ and the frequency $v$. Hence, $B$ may be represented as $B(\theta, \phi)$. Let the radiation be falling on a flat area $A$ on the surface of the Earth, as shown. Let the measurement be done over an infinitesimal bandwidth $dv$ Hz. Then an infinitesimal area $dA$ on the aperture $A$ receives an infinitesimal amount of power $dW$, which is obtained from the small region of the sky subtending an infinitesimal solid angle $d\Omega = $ sin $d\theta$ $d\phi$. This is given in Eq. (2.1).

$$dW = B(\theta, \phi)\cos\theta \, d\Omega \, dA \, dv \qquad (2.1)$$

If $\theta$ and $\phi$ are in radians, $d\Omega$ is in steradians (sr), $dv$ is in Hz, and $dA$ in m$^2$, then $dW$ is in Watts. Note that the power will be a maximum when $\theta$ is zero. This is because the wave fronts will now be parallel to the receiving aperture area $A$.

---

1. The flux density emitted by a source is attenuated proportional to the reciprocal of the square of the distance traveled, which is known as the *inverse square law*.

## 2.2.2 Received Power

If the sky is assumed to have infinite radius, $dw$ becomes independent of the position of $dA$. Hence the total power $dW_A$ received by the entire surface $A$ can be written as in Eq. (2.2).

$$dW_A = AB(\theta, \phi)\cos\theta \, d\Omega \, dv \tag{2.2}$$

By integrating Eq. (2.2) over a bandwidth $\Delta v$ and solid angle $\Omega$, we can obtain the received power $W$ as given in Eq. (2.3).

$$W = A\int_v^{v+\Delta v} \iint_\Omega B(\theta,\phi)\cos\theta \, d\Omega \, dv \tag{2.3}$$

## 2.2.3 Total Brightness and Total Radio Brightness

The variation of brightness $B$ as a function of frequency is called the *brightness spectrum*. Integrating $B$ over a bandwidth $\Delta v$ Hz gives the total brightness $B'$ of this frequency range. This is expressed in Eq. (2.4). If integrated over the entire radio-frequency range, one obtains the *total radio brightness*

$$B' = \int_v^{v+\Delta v} B(\theta,\phi)dv. \tag{2.4}$$

Using the total brightness $B'$, one can also express the received power $W$ on the aperture area A over a bandwidth $\Delta v$ and solid angle $\Omega$ as expressed in Eq. (2.5). It must be noted that $B'$ is also a function of the angles of incidence $(\theta, \phi)$

$$W = A\iint_\Omega B'(\theta,\phi)\cos\theta \, d\Omega \tag{2.5}$$

## 2.2.4 Spectral Power

The *spectral power dw* is defined as the *power available per unit bandwidth*. It is measured in Watt $Hz^{-1}$. It is expressed in Eq. (2.6)

$$dW = B(\theta, \phi)\cos\theta \, d\Omega \, dA. \tag{2.6}$$

Because for a distant source $dw$ is independent of the position of $dA$ on the surface $A$, the total spectral power received on the surface $A$ can be expressed as given in Eq. (2.7)

$$dW = AB(\theta, \phi)\cos\theta \, d\Omega \tag{2.7}$$

Integrating Eq. (2.7), one gets the spectral power from a solid angle $\Omega$. This is expressed in Eq. (2.8), where $w$ is the spectral power in Watt $Hz^{-1}$.

$$w = A \iint_{\Omega} B(\theta, \phi) \cos \theta \, d\Omega \qquad (2.8)$$

## 2.3   RECEPTION BY TELESCOPE ANTENNA

The antenna of a *radio telescope* collects radiation from astronomical objects. Thus the radiation pattern modifies the equations derived so far. If the effective aperture area of the telescope antenna is $A_e$ and its normalized power pattern is $P_n(\theta, \phi)$, then the spectral power $w$ received is given by Eq. (2.9). The limits of $\Omega$ must be chosen such that the integration does not miss the nonzero points of the antenna response $P_n(\theta, \phi)$. The factor of 1/2 indicates that the antenna responds to only one polarization of the incoming radiation, which contains only half of the power. The total power received across the bandwidth $\Delta v$ is expressed as in Eq. (2.10).

$$w = \frac{1}{2} \left( A_e \iint_{\Omega} B(\theta, \phi) P_n(\theta, \phi) d\Omega \right) \qquad (2.9)$$

$$W = \frac{1}{2} \left( A_e \int_{v}^{v+\Delta v} \iint_{\Omega} B(\theta, \phi) P_n(\theta, \phi) d\Omega \, dv \right) \qquad (2.10)$$

## 2.4   OBSERVING DISCRETE RADIO SOURCES

So far we have talked about radiation that is continuously distributed across the entire celestial sphere (sky). Let us now consider tiny radio sources in the sky. These sources may be broadly categorized into three types: (*i*) *point* sources, (*ii*) *localized* sources, and (*iii*) *extended* sources. A point source is conceptual. It may be defined as a source that has no extent (zero solid angle). Practically, all radio sources subtend a nonzero solid angle. It may often be convenient to categorize sources of very small extent as point sources if these are much smaller than the antenna beam solid angle. A small source having a finite dimension is called a localized source. Sometimes a localized point source is referred to as a *radio star*. They may or may not be optically luminous like stars. Extended sources are also discrete sources but have larger extents. Sometimes astronomers regard sources subtending an angle less than 1° as localized and those exceeding 1° as extended. A large source possessing an angular extent of many degrees may be regarded as an extended source. An extended source may also be regarded as a *discrete* source if its boundary is well defined. Hence, a source can be both extended and discrete. For example, the Sun may be taken as a discrete source. The integration of the brightness performed over the surface area of a discrete source gives its total flux density $S$. This is expressed in Eq. (2.11), where $\Omega_S$ is the solid angle subtended by the source

$$S = \iint_{\Omega_S} B(\theta,\phi)d\Omega \qquad (2.11)$$

When observed using an antenna possessing a normalized power pattern $P_n(\theta, \phi)$, the measured flux density $S_0$ will be as shown in Eq. (2.12). Note that $S_0$ is always less than $S$ due to the directional property of the antenna used. However, if the source has an extremely small extent we may approximate $S_0 \simeq S$.

$$S_0 = \iint_{\Omega_S} B(\theta,\phi)P_n(\theta,\phi)d\Omega \qquad (2.12)$$

Let the source be of very large extent having uniform brightness. Provided the beam is within the source extent, the brightness will be constant over the main lobe. This is shown in Eq. (2.13), where $\Omega_M$ is the solid angle subtended by the major lobe.

$$S_0 = B(\theta,\phi)\iint P_n(\theta,\phi)d\Omega \simeq B(\theta,\phi)\Omega_M \qquad (2.13)$$

This is, however, only an assumption. In reality, the telescope does not give a true picture of the source but a modified result. However narrow the major lobe of the antenna pattern is, it is never of zero width like a delta function. When the beam is focused slightly away from the source edges, it still receives some power from the source. Thus the flux density does not fall off like a step function at the source edges. Practical antennas produce a gradual and continuous decrease in flux which depends on the sharpness of the antenna beam pattern.

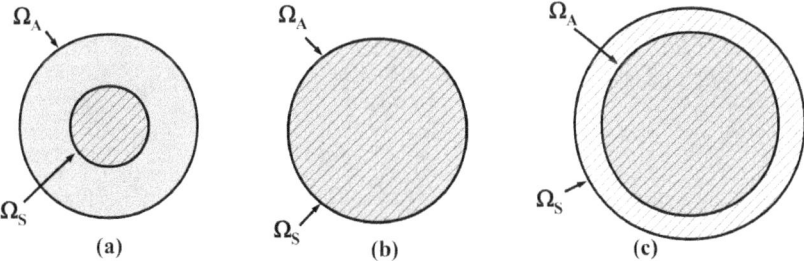

**FIGURE 2.2:** Extents of solid angles with the antenna's major lobe directed on the radio source. (a) Source extent is less than the major lobe of the antenna beam. (b) Source extent and major lobe extents are equal. (c) Source extent exceeds the major lobe.

Consider a radio source in the sky with the center of the antenna beam pointed to the source's center. This is shown in Fig. 2.2. Three possibilities are seen: (*i*) the source solid angle is less than the antenna beam, (*ii*) the source solid angle equals the antenna beam, and (*iii*) the source solid angle is greater than the antenna beam. In any case, the flux density observed is always less than the true value. In general, the observed or *apparent brightness* $B_e$ can be expressed as in Eq. (2.14), where $\Omega_A$ is the antenna beam solid angle. Let us treat the three cases separately.

$$B_e = \frac{\iint B(\theta,\phi)P_n(\theta,\phi)d\Omega}{\iint P_n(\theta,\phi)d\Omega} = \frac{S_0}{\Omega_A} \tag{2.14}$$

**Case (*i*):** When the radio source is sufficiently smaller than the beam solid angle (Fig. 2.2a), the observed flux density will be equal to the mean of the actual flux density of the source. Hence, the average brightness $B_{avg}$ can be written as in Eq. (2.15). If the source possesses a uniform brightness $B = B_{avg}$, then using Eq. (2.15) the apparent brightness $B_e$ can be rewritten as in Eq. (2.16)

$$B_{avg} = \frac{S}{\Omega_S} = \frac{1}{\Omega_S}\iint B(\theta,\phi)d\Omega \tag{2.15}$$

$$B_e = \frac{S}{\Omega_A} = \frac{\Omega_S}{\Omega_A}B . \tag{2.16}$$

**Case (*ii*):** When the extent of the source coincides with the first null of the antenna beam (Fig. 2.2b), then $B_e$ can be expressed as in Eq. (2.17), where $\Omega_M$ is the main beam solid angle

$$B_e = \frac{S}{\Omega_A} = \frac{\Omega_M}{\Omega_A}B = \iint_{\Omega_M} P_n(\theta,\phi)d\Omega . \tag{2.17}$$

**Case (*iii*):** When the source solid angle is larger than that of the main lobe of the antenna beam but occupies a finite portion of the sky, the apparent brightness $B_e$ can be expressed as in Eq. (2.18). Here $\Omega'_M$ is the beam solid angle of the major and side lobes of the antenna that fall within the angular extent of the source. The antenna beam solid angles are related as shown in Eq. (2.19). These are called *beam efficiencies*.

$$B_e = \frac{S}{\Omega_A} = \frac{\Omega_M}{\Omega_A}B = \iint P_n(\theta,\phi)d\Omega \tag{2.18}$$

$$\frac{\Omega_M}{\Omega_A} < \frac{\Omega'_M}{\Omega_A} < 1 \tag{2.19}$$

Many engineering applications require the flux density in Watt m$^{-2}$. This requires integration of the flux density $S$ over the associated bandwidth. This is termed *total flux density S* and is given in Eq. (2.20), where $\Delta\nu$ is the bandwidth of the measuring instrument. The total power $W$ observed over a bandwidth of $\Delta\nu$

is given by Eq. (2.21). Generally, over a small bandwidth $\Delta\nu$, the flux density is nearly uniform. In such cases, Eq. (2.21) can be approximated as Eq. (2.22), where $S_0$ is the observed total flux density in Watt m$^{-2}$ and $S_0$ is the observed flux density in Watt m$^{-2}$ Hz$^{-1}$.

$$S' = \int_{\nu}^{\nu+\Delta\nu} S\, d\nu \tag{2.20}$$

$$\left. \begin{aligned} W &= \frac{1}{2} A_e S_0' = \frac{1}{2} A_e \int_{\nu}^{\nu+\Delta\nu} S_0\, d\nu \\ &= \frac{1}{2} A_e \int_{\nu}^{\nu+\Delta\nu} \iint_{\Omega_S} B(\theta,\phi) P_n(\theta,\phi) d\Omega\, d\nu \end{aligned} \right\} \tag{2.21}$$

$$W = \frac{1}{2} A_e S_0 \Delta\nu \tag{2.22}$$

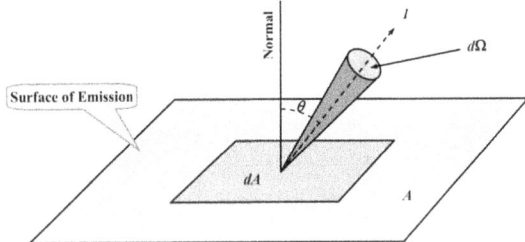

**FIGURE 2.3:** Emission from a flat surface showing radiance or intensity $I$.

## 2.5   RADIANCE (INTENSITY)

The terms *radiance* and *intensity* carry identical meanings. Consider a flat surface emitting electromagnetic radiation as shown in Fig. 2.3. The spectral power $dw$ escaping the surface element $dA$ and appearing within the solid angle $d\Omega$ is given by Eq. (2.23). Here, $dw$ is measured as power per unit bandwidth (Watt m$^{-2}$ Hz$^{-1}$), $I$ is the intensity of radiation from the surface (Watt m$^{-2}$ Hz$^{-1}$ rad$^{-2}$), and $\theta$ is the angle it makes with the normal to the surface (rad). The term $I$ is the radiance. It has a dimension of power per unit area per unit solid angle per unit bandwidth. When radiation from the entire area $A$ is uniform, then Eq. (2.23) may be rewritten as Eq. (2.24). Note that brightness and intensity have the same units. However, intensity is defined for radiation and brightness for reception. Hence, $I = -B$.

$$dw = I \cos\theta d\Omega dA \tag{2.23}$$

$$dw = AI \cos\theta d\Omega \tag{2.24}$$

## 2.6  GOVERNING LAWS OF A BLACK BODY

Any object at a temperature larger than 0 K radiates. The radiation intensity and wavelength are temperature dependent. A black body is a conceptual object whose properties are: (*i*) It completely absorbs radiation of any wavelength; (*ii*) It does not reflect any radiation and is also opaque; (*iii*) If its temperature is greater than 0K, it radiates. Several laws governing black-body characteristics are presented below.

### 2.6.1  Planck's Law (Spectral Radiance)

Planck's law of spectral radiance governs electromagnetic radiation at all wavelengths from a black body at temperature $T$(K) as a function of frequency $\nu$(Hz) or wavelength $\lambda$(m). It is given in Eq. (2.25), where $c$ is the speed of light $(2.998 \times 10^8$ m/s$)$, $h$ is Planck's constant $(6.626069 \times 10^{-34}$ J s$)$, $k_B$ is the Boltzmann constant $(1.38065 \times 10^{-23}$ J/K$)$, and $B$ is the brightness of the object. The radiation spectra of a black body is shown in Fig. 2.4 for various temperatures governed by Planck's law of spectral radiance.

$$B(\nu) = \frac{2h\nu^3}{c^2}\left(\frac{1}{e^{\frac{h\nu}{k_B T}} - 1}\right)$$

also

$$B(\lambda) = \frac{2hc^2}{\lambda^5}\left(\frac{1}{e^{\frac{hc}{\lambda k_B T}} - 1}\right)$$

(2.25)

### 2.6.2  Kirchhoff's Law (Thermal Radiation)

Kirchhoff's law of thermal radiation states that *if a black body is in thermal equilibrium with its surrounding, then its emissivity is equal to its absorptivity.* In other words, the amount of radiation it absorbs is equal to the amount it radiates.

### 2.6.3  Stefan-Boltzmann Law

The Stefan-Boltzmann law states that *the total energy radiated per unit surface area of a black body per unit time per unit solid angle is directly proportional to the fourth power of the black-body's absolute temperature T.* It is expressed in Eq. (2.26), where, $\sigma_{B'}$ $(1.8047 \times 10^{-8}$ Watt m$^{-2}$ rad$^{-2}$ K$^{-4})$ is the Stefan-Boltzmann constant for brightness at all wavelengths. Note that the energy radiated per unit surface area per unit time per unit solid angle is total brightness $B'$.

$$B' = \sigma_{B'} T^4 \quad \text{Watt m}^{-2} \text{ rad}^{-2}$$

(2.26)

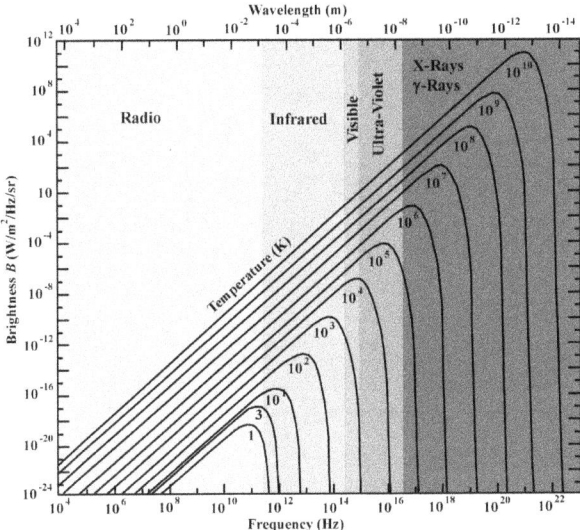

**FIGURE 2.4:** Radiation spectra at different temperatures obtained by Plank's radiation law.

The Stefan-Boltzmann law for calculating the luminosity $L$ of a star is expressed in Eq. (2.27), where $\sigma = 5.67 \times 10^{-8}$ Watt m$^{-2}$ K$^{-4}$, and $R$ is the radius of the star. Note that the proportionality constant and its units are different here.

$$\left.\begin{aligned} L = 4\pi R^2 \sigma T^4 \text{ Watts} \\ \text{where, } \sigma = \pi \sigma_{B'} \end{aligned}\right\} \qquad (2.27)$$

### 2.6.4 Wien's Displacement Law

Wien's displacement law states that *the wavelength $\lambda_{peak}$ at which emission from a black body is maximum is inversely proportional to its temperature T*. This is expressed in Eq. (2.28), where $\lambda_{peak}$ is in meters and $T$ is in K.

$$\lambda_{peak} = \frac{2.88 \times 10^{-3}}{T} \text{ m} \qquad (2.28)$$

### 2.6.5 Approximate Laws

These laws can also predict the radiation from a black body at a given temperature, but are limited to a certain range of frequencies. The two laws are (*i*) Wien's radiation law, and (*ii*) the Rayleigh-Jeans radiation law. The Wien's and Rayleigh-Jeans laws, respectively, approximate the higher- and lower-frequency parts of the spectrum provided by Planck's law. This is shown in Fig. 2.5.

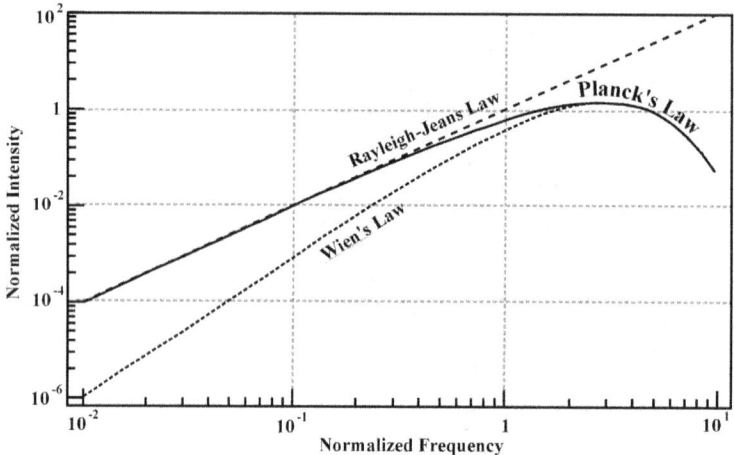

**FIGURE 2.5:** Comparison of Rayleigh-Jeans and Wien's radiation laws with Planck's law.

(*i*) Wien's Radiation Law

Wien's radiation law approximates Planck's law of spectral radiance at shorter wavelengths (higher frequencies). It is given in Eq. (2.29), where $B$ is the object's brightness, $h$ is Planck's constant, $k_B$ is Boltzmann constant, $\lambda$ is the emitted wavelength, $\nu$ is the frequency, and $T$ is the object's absolute temperature.

$$B = \frac{2h\nu}{\lambda^2} e^{-\frac{h\nu}{k_B T}}$$
(2.29)

(*ii*) Rayleigh-Jeans Radiation Law

The Rayleigh-Jeans radiation law approximates Planck's law of spectral radiance at longer wavelengths (lower frequencies). It is given in Eq. (2.30), where $B$ is the object's brightness, $h$ is Planck's constant, $k_B$ is the Boltzmann constant, $T$ is the object's absolute temperature, and $\lambda$ is the emitted wavelength. This expression is of great use in radio astronomy because it covers the radio wavelengths.

$$B = \frac{2k_B T}{\lambda^2}$$
(2.30)

## 2.7  ABSORPTION AND EMISSION OF EM WAVES IN A MEDIUM

Electromagnetic waves travel extremely large distances through space before reaching the Earth. The medium need not be purely empty space. For example, it may be an interstellar medium containing gas and dust clouds. The elements of these gases affect the electromagnetic waves passing through by producing (*i*) absorption, (*ii*) emission, and (*iii*) both. These effects are explained below.

## 2.7.1 Absorption of EM Waves

In free space, the flux density of an electromagnetic wave emitted by a localized source changes as the inverse square of the distance. In a small region distant from the source, the flux-density change is negligible because the wave fronts are almost flat (parallel rays). If the space contains a gas with absorbing elements, the flux density will attenuate as the waves propagate through it.

Consider a small region of length $x$ shown in Fig. 2.6. If $S_1$ is a flux density entering this region, then after traveling a distance $x$ (in meters) the flux density $S$ is given by Eq. (2.31), where $\alpha$ is the attenuation constant (in Nepers per meter) and $\tau = \alpha x$ is known as the *optical depth* (dimensionless). The value of $\tau$ is found from Eq. (2.32). The distance $x$ at which $S/S_1 = 1/e = 0.368$ is termed the *depth of penetration*. These relations hold true even if $S$ and $S_1$ are replaced by brightness terms, as shown in Eq. (2.33), where $B_{src}$ is the actual brightness of the source and $B$ is the observed brightness after the wave has traveled a distance $x$ through the medium.

$$S = S_1\, e^{-\alpha x} = S_1\, e^{-\tau} \tag{2.31}$$

$$\tau = \ln\left(\frac{S_1}{S}\right) = 2.3 \log\left(\frac{S_1}{S}\right) \tag{2.32}$$

$$B = B_{src}\, e^{-\alpha x} = B_{src}\, e^{-\tau} \tag{2.33}$$

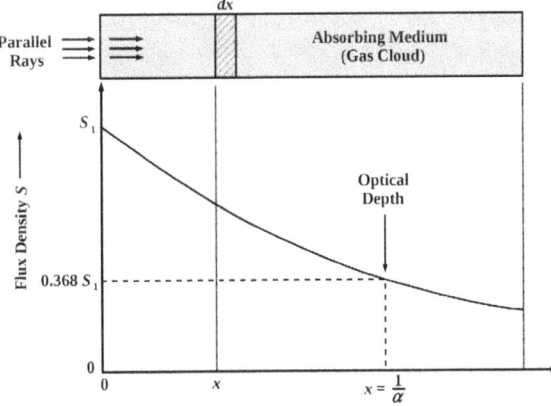

**FIGURE 2.6:** Attenuation of electromagnetic waves in an absorbing medium.

The attenuation constant $\alpha$ can be expressed as a product of the gas density $\rho$ and its absorption coefficient $K$ as given in Eq. (2.34). Usually $K$ is in $m^2kg^{-1}$ and $\rho$ is in kg $m^{-3}$. If $\rho$ is a nonuniform function of $x$, the optical depth is expressed as in Eq. (2.35), where the length of the medium is $x_1$ along the direction of wave propagation.

$$\alpha = Kr \tag{2.34}$$

$$\tau = \int_0^{x_1} K\rho(x)\,dx \tag{2.35}$$

### 2.7.2 Emission of EM Waves

Electromagnetic radiation is also emitted from regions of space filled with hot gases. Consider a gas of infinitesimal volume $dv$ as shown in Fig. 2.7. It subtends an infinitesimal solid angle $d\Omega$ from the observation point $P$ at a distance $r$. Let $j$ be the rate of energy emission per unit volume per unit mass per unit bandwidth (emission coefficient). It has units of Watt $kg^{-1}$ $Hz^{-1}$. The infinitesimal flux density $dS$ and brightness $dB$ observed at $P$ are given, respectively, by Eqs. (2.36) and (2.37), where $dw = j\rho dv$ (Watt $Hz^{-1}$) is the power emitted per unit bandwidth. For the emitting matter enclosed between radii $r_1$ and $r_2$, the brightness $B$ is given by Eq. (2.38).

$$dS = \frac{dw}{4\pi r^2} = \frac{j\rho\,dv}{4\pi r^2} \tag{2.36}$$

$$\left.\begin{aligned} dB &= \frac{dS}{d\Omega} = \frac{j\rho\,dv}{4\pi r^2 d\Omega} = \frac{j\rho\,dr}{4\pi} \\ \text{since, } dv &= r^2 dr\, d\Omega \end{aligned}\right\} \tag{2.37}$$

$$B = \frac{1}{4\pi}\int_{r_1}^{r_2} j\rho\,dr \tag{2.38}$$

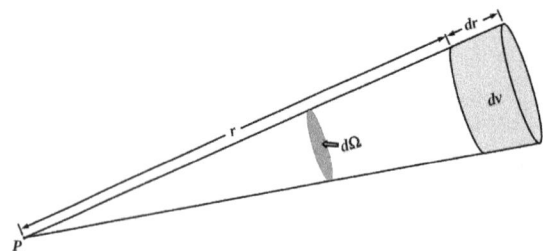

**FIGURE 2.7:** Radiating volume $dv$ of gas subtending a solid angle $d\Omega$ from the observation point $P$ at a distance $r$

### 2.7.3 Absorption and Emission of EM Waves

Consider a medium containing matter that emits as well as absorbs electromagnetic waves. There are two cases: (*i*) internal emission and absorption by a cloud, and (*ii*) an external radio source observed through an emitting and absorbing cloud.

(*i*) Internal Emission and Absorption by a Cloud

Consider the geometry shown in Fig. 2.8. Let no radiation enter from any source outside this region. If there is no absorption, the infinitesimal brightness $dB$ from the emitting region of volume $dv$ will be identical to Eq. (2.37). We considered the region as absorptive. Thus, if the infinitesimal volume $dv$ emits as well as absorbs, then the corresponding brightness $dB$ observed at the point $P$ is given by Eq. (2.39). The brightness as a result of the total thickness (0 to $r_1$) of the cloud is given by Eq. (2.40). Here $B$ is the apparent (observed) brightness of the cloud (Watt m$^{-2}$ Hz$^{-1}$ str$^{-1}$), $B_i = j/4\pi k$ is its intrinsic brightness (Watt m$^{-2}$ Hz$^{-1}$ str$^{-1}$), and $\tau_c$ is its optical depth (dimensionless). From the Rayleigh-Jeans law it is known that the brightness of hot objects at radio wavelengths is proportional to the temperature $T$. Hence Eq. (2.41) follows from Eq. (2.40), where, $T_b$ and $T_c$ are, respectively, the observed and actual temperatures of the cloud in K.

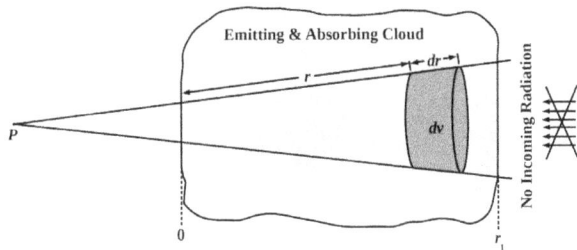

**FIGURE 2.8:** Internal absorption and emission inside a cloud.

$$dB = d\text{Emission} - d\text{Attenuation} = \left(\frac{1}{4\pi}j\rho\right) \times \left(e^{-\tau}\right)dr \qquad (2.39)$$

$$\text{where, } \tau = \int_0^r K\rho\, dr$$

$$B = \int dB = \frac{j}{4\pi K}\left(1 - e^{-\tau_c}\right), \quad \text{where, } \tau_c = \int_0^{r_1} K\rho\, dr \qquad (2.40)$$

$$\text{i.e., } B = B_i\left(1 - e^{-\tau_c}\right)$$

$$T_b = T_c\left(1 - e^{-\tau_c}\right) \qquad (2.41)$$

*(ii)* External Radio Source and an Emitting and Absorbing Cloud

A radio source having brightness $B_S$ is observed through a cloud that emits and also absorbs. The geometry is shown in Fig. 2.9. In the absence of the cloud, the source brightness is independent of optical depth. With the cloud present, the brightness deterioration for the distance $a$ may be neglected. However, the emission and absorption by the cloud ($r = 0$ to $r_1$) affects the apparent brightness of the radio source. An infinitesimal brightness change $dB$ by a volume of length $dr$ is given in Eq. (2.42), whose solution is given by Eq. (2.43), where $B$ is the apparent brightness (Watt $m^{-2}$ $Hz^{-1}$ $str^{-1}$), $B_S$ is the actual brightness of the source (Watt $m^{-2}$ $Hz^{-1}$ $str^{-1}$), $j$ is the emission coefficient (Watt $kg^{-1}$ $Hz^{-1}$), $K$ is the absorption coefficient ($m^2$ $kg^{-1}$), $\rho$ is the density ($kgm^{-3}$), $dr$ is the elemental distance (m), and $\tau_c$ is the optical depth (dimensionless). Using the Rayleigh-Jeans law, the relationship with temperature is given by Eq. (2.44), where, $T_b$ is the observed brightness temperature of the cloud (K), $T_S$ is the source temperature, and $T_c$ is the temperature of the cloud (K).

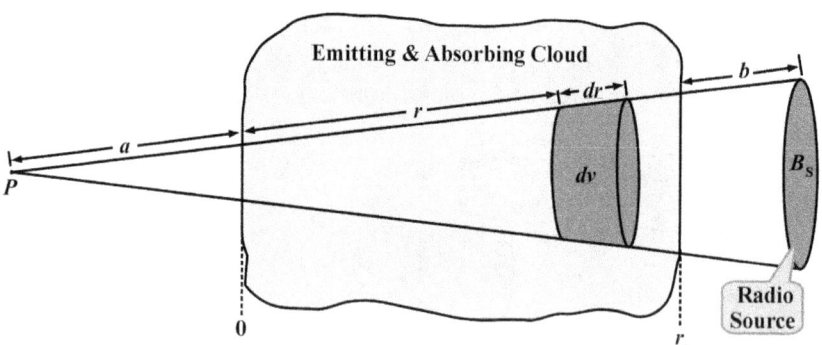

**FIGURE 2.9:** A radio source of brightness $B_S$ is observed through an absorbing and emitting medium.

$$\frac{dB}{dr} + K\rho B = \frac{j\rho}{4\pi} \tag{2.42}$$

$$B = B_S\, e^{-\tau_c} + \frac{j}{4\pi K}\left(1 - e^{-\tau_c}\right) = B_S\, e^{-\tau_c} + B_i\left(1 - e^{-\tau_c}\right) \tag{2.43}$$

$$T_b = T_S\, e^{-\tau_c} + T_c\left(1 - e^{-\tau_c}\right) \tag{2.44}$$

## 2.8   RELATING BRIGHTNESS TO ANTENNA TEMPERATURE

Antennas are the primary instruments in radio astronomy. A relationship exists between the source brightness, source temperature (if a black body), and antenna output when pointed to the source.

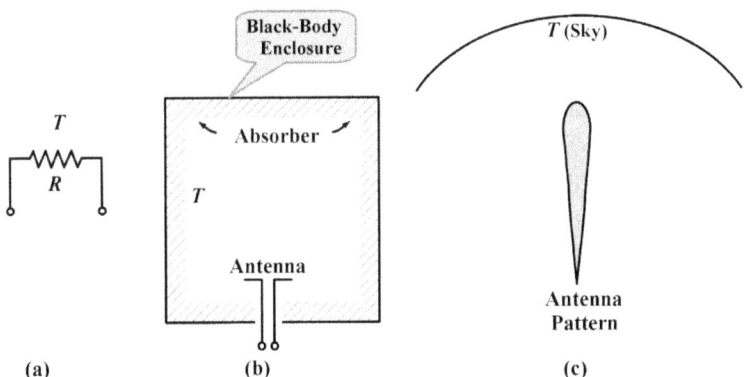

**FIGURE 2.10:** (a) A resistor at temperature $T$ produces thermal noise. (b) An antenna inside a black-body enclosure at temperature $T$. (c) The brightness temperature $T$ of the sky is sensed by an antenna whose beam is pointed to it.

As shown in Fig. 2.10a, a resistor with $R$ Ohms is at temperature $T$. The noise power per unit bandwidth generated across the resistor is given by Eq. (2.45), where $w$ is the spectral power (Watt $Hz^{-1}$), $k_B$ is the Boltzmann constant, and $T$ is the temperature (K).

$$w = k_B T \tag{2.45}$$

Let $R$ be the real part of the impedance (radiation resistance) of a lossless antenna. Let the imaginary part of the impedance be cancelled out using a suitable impedance across the antenna terminals. If the antenna is pointed at a region of a black body having a temperature $T$, the power measured at its terminals remains identical to Eq. (2.45), as in the case of a heated resistor. The output power in terms of the normalized antenna power pattern $P_n(\theta, \phi)$ and source brightness $B(\theta, \phi)$ is given by Eq. (2.9). This is reproduced in Eq. (2.46).

$$w = \frac{1}{2} A_e \iint B(\theta,\phi) P_n(\theta,\phi) d\Omega \tag{2.46}$$

We now enclose an antenna inside a black body maintained at a temperature $T$. This is shown in Fig. 2.10b. In any direction from the black body, its brightness $B_c$ will be a constant and is given by the Rayleigh-Jeans law as expressed in Eq. (2.47). Substituting this expression into Eq. (2.46), the spectral power $w$ (Watt $Hz^{-1}$) can be expressed as in Eq. (2.48), where $A_e$ is the effective aperture area of the antenna ($m^2$), $k_B$ is the Boltzmann constant, $T$ is the absolute temperature (K), $\lambda$ is the wavelength (m), and $\Omega_A$ is the solid angle subtended by the antenna beam (str). Note that $w$ is identical for both the antenna and the resistor.

$$B(\theta,\phi) = B_c = \frac{2k_B T}{\lambda^2} \tag{2.47}$$

$$w = \frac{k_B T}{\lambda^2} A_e \Omega_A = k_B T, \text{ since } A_e \Omega_A = \lambda^2 \tag{2.48}$$

The black-body inner walls produce radiance (brightness) $B$ given by Eq. (2.47). As shown in Fig. 2.10b, the antenna receives and converts this brightness into power, which is related to the temperature $T$ of the black body. The temperature of the radiation resistance is known as the *antenna temperature* $T_A$. Note that $T_A$ is dependent on the temperature of the emitting region that the antenna sees through its beam. It is independent of the temperature of the antenna structure. In this way, the antenna can sense the equivalent black-body temperature of a remote object. From Eq. (2.48), $w$ can be written as $w = k_B T_A$ Combining this result with Eq. (2.46) we obtain Eq. (2.49), which is the received power per unit bandwidth. Having substituted from Eq. (2.49), the observed flux density $S$ (Watt m$^{-2}$ Hz$^{-1}$) is given by Eq. (2.50), where $k_B$ is the Boltzmann constant, $T_A$ is the antenna temperature, and $A_e$ is the effective antenna aperture area (m$^2$).

$$w = \frac{1}{2} A_e \iint B(\theta,\phi) P_n(\theta,\phi) d\Omega = k_B T_A \tag{2.49}$$

$$S = \iint B(\theta,\phi) P_n(\theta,\phi) d\Omega = \frac{2k_B T_A}{A_e} \tag{2.50}$$

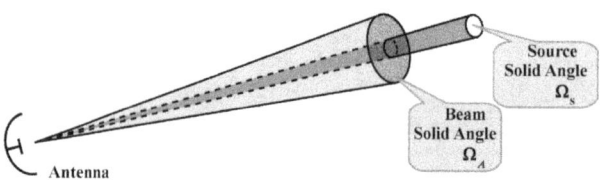

**FIGURE 2.11:** A common situation where the source solid angle is less than the beam solid angle.

Until now, we have assumed a celestial radio source whose extent was larger than the antenna beam. In practice, astronomical radio source extents are much smaller than antenna beams. This situation is shown in Fig. 2.11. Here, only a small portion of the antenna beam is occupied by the source solid angle $\Omega_S$. The remaining part of the beam area is occupied by the background sky. Assuming that the background sky has a uniform brightness distribution, the antenna temperature obtained is denoted as $T_{\text{Sky}+\text{Src}}$. Say the antenna is now focussed slightly away from the radio source such that the source goes out of the antenna beam. The corresponding antenna temperature measured is $T_{\text{Sky}}$. The difference between these two temperatures $\Delta T_A$ is known as the *incremental temperature* and is given by Eq. (2.51). Using Eq. (2.50), the apparent source flux density $S$ may be obtained as in Eq. (2.52).

$$\Delta T_A = T_{Sky+Src} - T_{Sky} \tag{2.51}$$

$$S = \frac{2 k_B \Delta T_A}{A_e} \tag{2.52}$$

If the antenna beam solid angle $\Omega_A$ and the source solid angle $\Omega_S$ are known, the temperature of the source can be expressed as in Eq. (2.53). The source temperature $T_S$ may represent the physical temperature of the object if the emission mechanism is like a black body. However, the emission could be generated by other mechanisms, such as a celestial plasma cloud with oscillating electrons with an equivalent temperature of thousands of K, although the physical temperature of the cloud may not be the same. Thus in these types of cases, it may be said the antenna shows an equivalent black-body noise temperature of $T_S$ K.

$$T_S = \frac{\Omega_A}{\Omega_S} \Delta T_A \tag{2.53}$$

Until now we have studied two extreme cases where the source extent is: (*i*) much larger than the antenna beam solid angle; and (*ii*) much smaller than the antenna beam solid angle. In general, for any size of source and antenna beam solid angle, the antenna temperature $T_A$ is given by Eq. (2.54), where $T_S(\theta, \phi)$ is the sky temperature distribution, $P_n(\theta, \phi)$ is the normalized antenna beam pattern, and $d\Omega = \sin\theta \, d\theta d\phi$ is an infinitesimal solid angle (str).

$$T_A = \frac{1}{\Omega_A} \int_0^\pi \int_0^{2\pi} T_S(\theta, \phi) P_n(\theta, \phi) \, d\Omega \tag{2.54}$$

## 2.9 SENSITIVITY AND NOISE IN RADIO TELESCOPES

We have so far considered only lossless antennas. However, all practical antennas produce losses of their own, which effectively contributes some noise at the antenna output. For practical purposes, the received signal must be greater than this noise. Moreover, antennas are connected to a receiving chain consisting of amplifiers, filters, mixers, detectors, etc., for further signal processing. The first stage in the receiver chain is a low-noise amplifier (LNA), which is the major noise contributor from the receiving end. Noise contribution also comes from the RF-cable between the antenna and the LNA. Thus the *minimum detectable signal* of a radio telescope primarily depends on the antenna noise, noise from the LNA and the RF-cable. However, the integrator at the receiving end can reduce this to some extent with longer integration times. The entire radio telescope may be viewed as a system formed with (*i*) an antenna, (*i,i*) a transmission line, and (*iii*) a receiving system whose noise is slightly more than the LNA. Thus the noise that appears at the antenna and receiver junction is known as *system noise*. The corresponding temperature is called the *system temperature*. If the antenna temperature is lower than the system temperature, the radio telescope will fail to operate. We now describe the calculations of (*i*) system temperature, and (*ii*) the minimum detectable temperature.

### 2.9.1 System Temperature

A general model of a simple radio telescope is shown in Fig. 2.12. If $T_A$ is the antenna noise temperature (K), $T_{AP}$ is its physical temperature , and $\varepsilon_A$ is its dimensionless thermal efficiency ($0 \leq \varepsilon_A \leq 1$), then the system temperature $T_{Sys}$ (K) is given by Eq. (2.55), where $T_R$ is the receiver-noise temperature (K), $T_{LP}$ is the physical temperature of the transmission line (K), and $\varepsilon_{TxLi}$ is the dimensionless efficiency of the transmission line ($0 \leq \varepsilon_{TxLi} \leq 1$). The transmission line efficiency $\varepsilon_{TxLi}$ depends on the physical length of the line and is given by Eq. (2.56), where $\alpha$ is the *attenuation constant* (Npm$^{-1}$) and *l* is the line length (m). The receiver temperature $T_R$ is expressed in Eq. (2.57), where $G_{LNA} \gg 1$ is the dimensionless power gain of LNA, $T_{LNA}$ is the noise temperature of the LNA (K), and $T_{others}$ is the noise temperature of rest of the receiving chain (K). Generally, $G_{LNA} \gg (T_{others}/T_{LNA})$. This implies $T_R \simeq T_{LNA}$. The LNA noise temperature is given in Eq. (2.58), where $T_{LNA\_Phy}$ is the physical temperature of the LNA and *F* is the *noise factor* (dimensionless) of the LNA. To increase the sensitivity, the transmission line length and the LNA noise temperature should be minimized. It is also important to keep the physical temperature of the LNA and transmission line as small as possible.

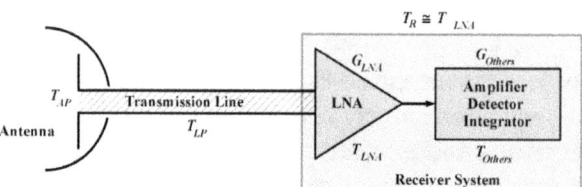

**FIGURE 2.12:** Model of a single-antenna radio telescope. The antenna is connected to the receiving system through a transmission line. The receiving system has a LNA followed by other amplifier-detector and integrator circuits.

$$T_{Sys} = T_A + T_{AP}\left(\frac{1}{\varepsilon_A} - 1\right) + T_{LP}\left(\frac{1}{\varepsilon_{TxLi}} - 1\right) + \frac{1}{\varepsilon_{TxLi}}T_R \qquad (2.55)$$

$$\varepsilon_{TxLi} = e^{-\alpha l} \tag{2.56}$$

$$\left. \begin{aligned} T_R &= T_{LNA} + \frac{T_{Others}}{G_{LNA}} \\[2mm] &\square\ T_{LNA}\ \ if\ G_{LNA}\ \square\ \left(\frac{T_{Others}}{T_{LNA}}\right) \end{aligned} \right\} \tag{2.57}$$

$$T_{LNA} = (F-1)T_{LNA\_Phy} \tag{2.58}$$

## 2.9.2 Minimum Detectable Temperature

The sensitivity of radio telescopes is limited by the minimum detectable temperature $\Delta T_{min}$. The detection is restricted by the fluctuations at the receiver output, whose value is proportional to the system temperature $T_{Sys}$. As shown in Eq. (2.55), the system temperature $T_{Sys}$ is mainly dependent on the antenna temperature $T_A$ and the receiver temperature $T_R$. Theoretically, the system noise may be reduced to any extent by (*i*) increasing the postdetection integration time, (*ii*) increasing the predetection bandwidth, (*iii*) taking the mean value of two or more observations. Practically, it is impossible to increase the integration time beyond a certain limit, for it distorts the true profile of the observed radio source. Larger bandwidths can (*i*) cause loss of spectral information, and (*ii*) increase the risk of local radio interference. The minimum detectable temperature $\Delta T_{min}$(K) is expressed in Eq. (2.59), where $\Delta v$ is the system bandwidth in Hz, $K_{sys}$ is known as the *system sensitivity constant* (dimensionless), $\tau$ is the postdetection integration time (seconds), $\Delta T_{rms}$ is the rms noise temperature of the system (K), and *n* is the number of records averaged. The corresponding *minimum detectable brightness* can be obtained by applying the Rayleigh-Jeans law. This is shown in Eq. (2.61), where $\lambda$ is the wavelength in meters and $k_B$ is Boltzmann constant. Similarly, the *minimum detectable flux density* $\Delta S_{min}$ is given by Eq. (2.61).

$$\Delta T_{min} = K_{Sys} \frac{T_{Sys}}{\sqrt{\Delta v n \tau}} = \Delta T_{rms} \tag{2.59}$$

$$\Delta B_{min} = \frac{2k_B}{\lambda^2} \Delta T_{min} = \frac{2k_B}{\lambda^2}\left(K_{Sys} \frac{T_{Sys}}{\sqrt{\Delta v n \tau}}\right) \tag{2.60}$$

$$\Delta S_{min} = \frac{2k_B}{A_e} \Delta T_{min} = \frac{2k_B}{A_e}\left(K_{Sys} \frac{T_{Sys}}{\sqrt{\Delta v n \tau}}\right) \tag{2.61}$$

## 2.10   LOSSES FROM THE ANTENNA BEAM SHAPE

An ideal antenna gives correct information about the radiative properties of a source. However, practical antennas lose some information due to their beam shapes, which act as low-pass filters. We study these first in one dimension and then extend the analysis to two dimensions.

### 2.10.1 Convolution of Beam Pattern with Extended Sources

Consider the geometry shown in Fig. 2.13a, where for a given frequency $\nu$, an antenna possess a finite beam width having a normalized radiation pattern $P_n(\varphi)$ as a function of the one-dimensional angle $\varphi$. Let $T(\varphi)$ be the sky temperature distribution. The antenna absorbs the power falling within its beam from the sky. The antenna power output consists of the integrated power from the sky within the beam extent, modified by the beam shape.

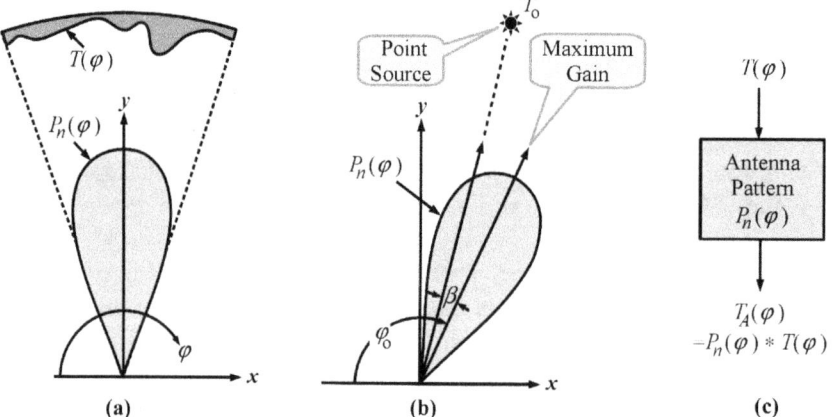

**FIGURE 2.13:** (a) An antenna absorbs the power falling within its beam from a nonuniformly bright sky. (b) An antenna beam points at an angle $\varphi_0$, but may receive radiation from a point source at a different angle $\varphi_0 - \beta$. (c) The block shows the relationship between the antenna temperature, the radiation pattern, and the source brightness temperature distribution.

In Fig. 2.13b, a point radio source is positioned at an angle $\beta$ from the center of the antenna beam (aligned at an angle $\varphi_0$ with respect to the x-axis). If the point source is at temperature $T_0$ (at an angle $\varphi_0 - \beta$ with the x-axis), then the temperature response of the antenna $T_A(\varphi_0)$ towards the source can be expressed as a function of $P_n(\varphi_0 - \beta)$ and $T_0$. This is given in Eq. (2.62).

$$T_A(\varphi_0) = P_n(\varphi_0 - \beta)T_0 \tag{2.62}$$

Across the beam, the temperature distribution of the sky is a function of $\beta$. Instead of a point source, if the temperature distribution of the sky $T(\beta)$ is considered, we may generalize the antenna temperature $T_A(\varphi)$ as given in Eq. (2.63), which is a convolution (*) of the normalized antenna pattern with the temperature distribution across the sky.

$$T_A(\varphi) = \int P_n(\varphi - \beta) T(\beta) d\beta = P_n(\varphi) * T(\varphi) \qquad (2.63)$$

We may thus state that *the antenna temperature is the convolution of the normalized radiation pattern with the angular distribution of the source's temperature.* Fig. 2.13c is a block diagram of this relationship.

Practically, the source temperature, antenna temperature, and the antenna pattern are two dimensional. Let the antenna aperture coincide with the *x-y* plane and be centered at $x = 0$, $y = 0$. Using the dummy variables $\xi$ and $\eta$, we may represent the convolution as shown below in Eq. (2.64).

$$T_A(x,y) = \int_{-\infty}^{\infty} P_n(x - \xi, y - \eta) T(\xi, \eta) d\xi d\eta = P_n(x,y) * T(x,y) \qquad (2.64)$$

### 2.10.2 Product of Spatial Antenna Spectra with Spatial Source Spectra

The convolution relationship discussed above can be expressed as a product relationship in the spectral domain. If $T_A(S)$, $P_n(S)$, and $T(S)$ are, respectively, the Fourier transforms of $T_A(\varphi)$, $P_n(\varphi)$, and $T(\varphi)$, we may express them in the spectral domain using Eq. (2.65), where $s$ is the spatial frequency.

$$\bar{T}_A(s) = \bar{P}_n(s)\bar{T}(s) \qquad (2.65)$$

The individual Fourier transforms of $T_A(\varphi)$, $P_n(\varphi)$, and $T(\varphi)$ are expressed in Eqs. (2.66) through (2.68), as shown below.

$$\bar{T}_A(s) = \int_{-\infty}^{\infty} T_A(\varphi) e^{-j2\pi s\varphi} d\varphi \qquad (2.66)$$

$$\bar{P}_n(s) = \int_{-\infty}^{\infty} P_n(\varphi) e^{-j2\pi s\varphi} d\varphi \qquad (2.67)$$

$$\bar{T}(s) = \int_{-\infty}^{\infty} T(\varphi) e^{-j2\pi s\varphi} d\varphi \qquad (2.68)$$

Here, $s$ is called the *spatial frequency* because it gives the power distribution as a function of angular spacing. The highest frequency component present within $P_n(\varphi)$ is known as the *cut-off frequency* $S_c$. This is given in Eq. (2.69), where $w$ is the width of the aperture along $\varphi$ and $\lambda$ is the wavelength. The inverse of $S_c$ is called the *cut-off period*. It is shown in Eq. (2.70).

$$s_c = \frac{w}{\lambda} \qquad (2.69)$$

$$\varphi_c = \frac{1}{s_c} = \frac{\lambda}{w} \tag{2.70}$$

For an antenna aperture placed in the $x$-$y$ plane, the two-dimensional Fourier transform relationships are shown in Eqs. (2.71) through (2.74). The spatial-frequency coordinates $(u,v)$ are parallel to the aperture plane.

$$\bar{T}_A(u,v) = \bar{P}_n(u,v)\bar{T}(u,v) \tag{2.71}$$

$$\bar{T}_A(u,v) = \int_{-\infty}^{\infty}\int_{-\infty}^{\infty} T_A(x,y)e^{-j\,2\pi(ux+vy)}\,dx\,dy \tag{2.72}$$

$$\bar{P}_n(u,v) = \int_{-\infty}^{\infty}\int_{-\infty}^{\infty} P_n(x,y)e^{-j\,2\pi(ux+vy)}\,dx\,dy \tag{2.73}$$

$$\bar{T}(u,v) = \int_{-\infty}^{\infty}\int_{-\infty}^{\infty} T(x,y)e^{-j\,2\pi(ux+vy)}\,dx\,dy \tag{2.74}$$

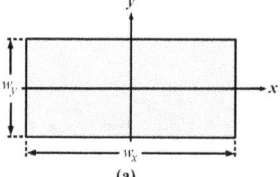 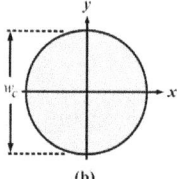

(a)           (b)

**FIGURE 2.14:** (a) A rectangular aperture. (b) A circular aperture.

There exist two cut-off frequencies determined from the shape and size of the aperture. Fig. 2.14 shows two simple apertures, namely, rectangular and circular. We express their cut-off frequencies as follows. For the rectangular aperture (Fig. 2.14a), the two cut-off frequencies are $S_{cx}$ and $S_{cy}$, respectively, along the $x$- and $y$-axes. These are given in Eqs. (2.75) and (2.76). The circular aperture has a cut-off frequency $S_{cc}$, which is identical in the $x$ and $y$ directions. This is shown in Eq. (2.77).

$$S_{cx} = \frac{w_x}{\lambda} \tag{2.75}$$

$$S_{cy} = \frac{w_y}{\lambda} \tag{2.76}$$

$$S_{cc} = \frac{w_c}{\lambda} \tag{2.77}$$

### 2.10.3 Aerial Smoothing and Loss of Spectral Information

From Eq. (2.65) we find that $\bar{T}_A$ is the product of $\bar{P}_n(s)$ and $\bar{T}(s)$. It is understood that for those spatial frequencies $S_k$ at which $\bar{T}_A(s_k) = 0$, the value of $\bar{T}(s_k)$ is lost.

Hence, depending on the shape of its radiation pattern, the antenna may loose certain spatial-frequency components of the source.

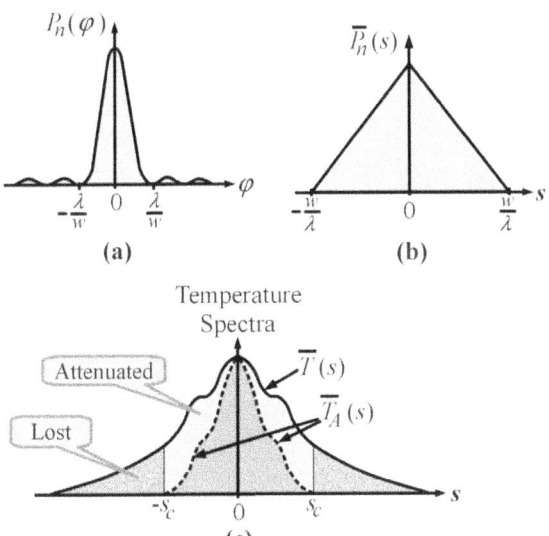

FIGURE 2.15: (a) Radiation pattern of an antenna. (b) Fourier transform of the radiation pattern. (c) Filtration of spectral components.

Consider a normalized pattern $P_n(\varphi)$ in single angular dimension $\varphi$ as shown in Fig. 2.15a, which is given by Eq. (2.78). Here $w$ is the width of the aperture along $\varphi$, $\lambda$ is the wavelength, and $k_1$ is a constant. Integration of the radiation pattern over $(\theta, \varphi)$ gives the effective antenna beam-width $\Omega_A$, as shown in Eq. (2.79).

$$P_n(\varphi) = k_1 \frac{\lambda}{w} \left[ \frac{\sin(\pi \varphi w / \lambda)}{\pi \varphi} \right]^2 \tag{2.78}$$

$$\int_0^{2\pi} \int_0^{\pi} P_n(\varphi) \sin \theta \, d\theta d\varphi = \Omega_A \tag{2.79}$$

The main-lobe beam width between the first nulls is $2\lambda/w$, and its half-power beam width is $0.89\lambda/w$. Fig. 2.15b shows $\overline{P}_n(s)$, which is the Fourier transform of the antenna pattern. Note that for $S > w/\lambda$, $\overline{P}_n(s)$ is zero. This is the cut-off frequency $S_c$, defined as $S_c = w/\lambda$. Beyond $S_c$, the antenna does not receive, which indicates that the antenna has a low-pass response. Fig. 2.15c shows a spectral brightness temperature distribution. The dotted lines shows the spectral antenna temperature distribution $\overline{T}_A(s)$. Note that the spectral antenna temperature does not contain any spectral information of the temperature distribution for $S > S_c$.

Also note that the spectral information within the passband $S = \pm S_c$ is attenuated. Thus we see that the antenna pattern acts as a low-pass filter with cut-off frequency $\pm S_c$, and depending on its band shape, it also attenuates some components within its passband.

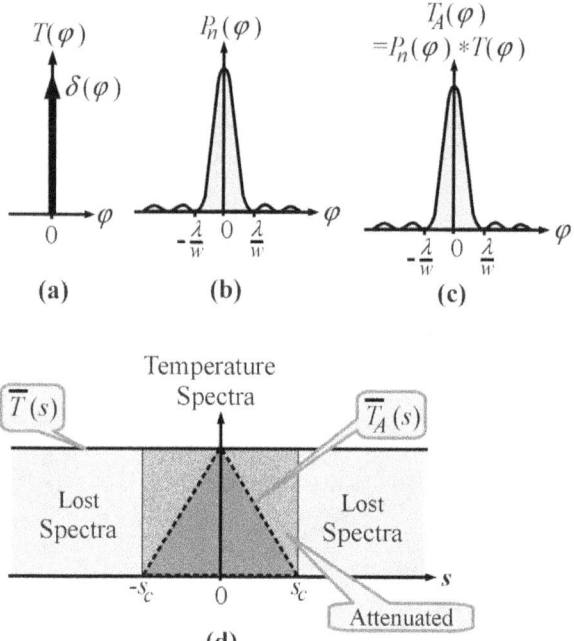

**FIGURE 2.16:** (a) A point source is represented by a delta function $\delta(\varphi)$. (b) Normalized radiation pattern of an antenna. (c) Convolution of $\delta(\varphi)$ with the radiation pattern retrieves the pattern. (d) Temperature spectra of the point source and the antenna.

For a better understanding, consider a point source $\delta(\varphi)$ shown in Fig. 2.16a. Fig. 2.16b shows the normalized radiation pattern $P_n(\varphi)$ of the receiving antenna, which is given by Eq. (2.68). Fig. 2.16c shows the convolution of $\delta(\varphi)$ (delta function) with the radiation pattern $P_n(\varphi)$, which is the same as $P_n(\varphi)$. Fig. 2.16d shows the temperature spectra of the point source and the antenna. Note that the Fourier transform of the point source exists at all spatial frequencies, but the transform for the convolved function ($T_A(\varphi) = P_n(\varphi) * T(\varphi)$) is limited within the cutoff frequencies $\pm S_c = \pm(w/\lambda)$. Hence a major portion of the spectral components are absent in the received signal and many of these components are attenuated by the spectral shape of the antenna pattern.

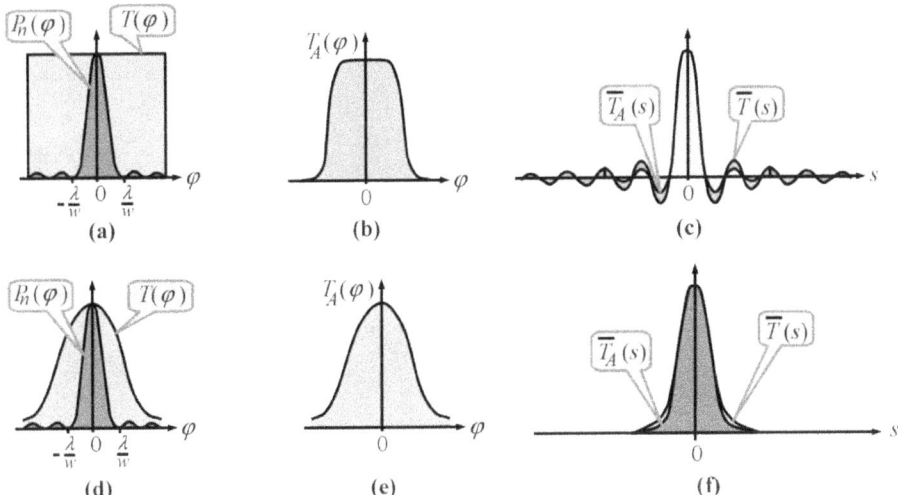

**FIGURE 2.17:** (a) Patterns of a rectangular source and the antenna. (b) Convolution of the rectangular source with the antenna pattern. (c) Spatial spectra of the rectangular source and the antenna temperature. (d) Patterns of a Gaussian source and the antenna. (e) Convolution of the Gaussian source with the antenna pattern. (f) Spatial spectra of the Gaussian source and antenna temperature.

We shall now graphically illustrate the effects of using an antenna to scan a source pattern that has sharp edges. Fig. 2.17 illustrates the effect on the antenna pattern of scanning a rectangular and a Gaussian source. Fig. 2.17a shows the pattern of a discrete rectangular source and the antenna. Fig. 2.17b shows the antenna temperature response obtained by convoluting the rectangular source with the antenna pattern. Fig. 2.17c shows the corresponding spatial spectra. Note that the shape of the antenna temperature pattern is not same as the source pattern. It appears to have smoothened the edges of the source pattern. The pattern has also been widened. This phenomena in which the edges of a radio source are smoothed by the receiving antenna is termed *aerial smoothing* by Bracewell and Roberts.

Let us now investigate what happens when the edges of the source are smooth. The patterns of a Gaussian source and an antenna are shown in Fig. 2.17d. The antenna temperature response is shown in Fig. 2.17e. Fig. 2.17f illustrates the spatial spectra of the Gaussian source and the antenna temperature. Note that the shape of the Gaussian source is more or less preserved in the antenna temperature pattern. The spectral loss is also much less. The above illustrations are equally applicable in three dimensions.

## 2.11   SAMPLING THEOREMS OF OBSERVING ANGLE AND SYNTHESIZED APERTURE

An extended source may be observed by scanning the antenna beam, and the corresponding power that is obtained is plotted as a function of the scanning angle. Instead of a continuous scan, the power may be measured at discrete angular intervals, and image construction may be attempted. If the angular points are widely separated, the information between any two adjacent points is lost. The minimum angular separation intervals are also related to the width of the antenna pattern. This principle is primarily described by the Nyquist-Shannon sampling theorem used in analog-to-digital conversion. For the present purpose, we will refer to this as the *sampling theorem of the observing angle*.

### 2.11.1 One-Dimensional Sampling Theorem of the Observing Angle

According to Bracewell the sampling theorem for one dimension may be stated as *an observed distribution is completely determined by measurements spaced at equal discrete intervals that are at least as narrow as $1/2S_c$, where $S_c$ is the spatial cut-off frequency of the antenna aperture*.

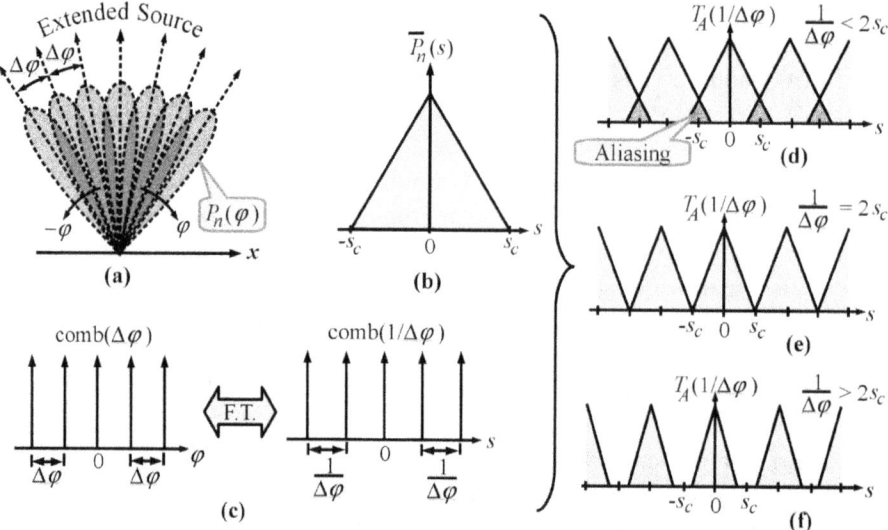

**FIGURE 2.18:** (a) An antenna scans an extended source at angular intervals of $\Delta\varphi$. (b) Spatial spectrum of the antenna in one dimension. (c) Discrete angular observation points with spatial Fourier transforms in one dimension. (d) The scanning interval is larger than the minimum required interval, resulting in aliasing. (e) The optimum angular scanning interval results in the true spectrum. (f) The angular scanning interval is less than the minimum required interval.

Fig. 2.18 graphically illustrates the one-dimensional sampling theorem of the observing angle. Let an antenna scan an extended source at angular intervals of $\Delta\varphi$ (Fig. 2.18a). Along the scanning direction, let the spatial spectrum be a low-pass function with a cut-off frequency $S_c$ (Fig. 2.18b). Data are collected from the antenna at discrete intervals. These points may be approximated by a train of impulse functions $\delta$ separated by $\Delta\varphi$, as shown in Fig. 2.18c. The impulse train (known as a *comb* function) is mathematically expressed in Eq. (2.80), where $m$ is an integer.

$$\text{comb}\left(\Delta\varphi\right) = \sum_{m=-\infty}^{\infty} \delta\left(\varphi - m\,\Delta\varphi\right) \qquad (2.80)$$

The Fourier transform of a comb function is another comb function in the spectral domain, as shown. This is expressed in Eq. (2.81). Note that the impulses are separated by the reciprocal of the angular scanning intervals.

$$\text{comb}\left(\frac{1}{\Delta\varphi}\right) = \sum_{m=-\infty}^{\infty} \delta\left(s - m\frac{1}{\Delta\varphi}\right) \qquad (2.81)$$

Because the antenna beam is convoluted with the source power in the angular domain, they appear as a product in the spatial-frequency domain. Due to the antenna pattern's low-pass characteristics, the spatial-frequency spectrum at the antenna output is restricted within the cut-off frequency $S_c$. The antenna temperature spectrum $\overline{T}_A(s)$ is the product of the comb function given in Eq. (2.81) with the band-limited temperature spectrum $k_1 \overline{P}_n(s)$. It is expressed in Eq. (2.82), where $k_1$ is a proportionality constant.

$$\overline{T}_A(s) = k_1 \overline{P}_n(s) \sum_{m=-\infty}^{\infty} \delta\left(s - m\frac{1}{\Delta\varphi}\right) = k_1 \sum_{m=-\infty}^{\infty} \overline{P}_n\left(s - m\frac{1}{\Delta\varphi}\right) \qquad (2.82)$$

Figs. 2.18d through 2.18f show three distinct cases of $\overline{T}_A(s)$: (*i*) Undersampling, where $(\Delta\varphi)^{-1} < 2S_c$, (*ii*) Optimal sampling, where $(\Delta\varphi)^{-1} < 2S_c$, and (*iii*) Oversampling, where $(\Delta\varphi)^{-1} < 2S_c$. Undersampling results in aliasing, which causes data corruption. Oversampling may be unnecessary and expensive. It is best to sample at a rate that is slightly higher than the optimum rate. To recover the data correctly, a rectangular low-pass filter function having a cut-off frequency $S_c$ is used.

## 2.11.2 Two-Dimensional Sampling Theorem for a Synthesized Aperture

In small apertures, the field is continuous across the aperture. The construction of single large-sized apertures are limited mainly by cost and mechanical support. However, a huge aperture may be synthesized using a large number of small antennas. The sampling theorem can be extended to two dimensions and applied here.

Suppose individual small antennas appear at discrete locations across a synthe-sized aperture plane. The values of the fields are now obtained from these antennas at discrete positions over the synthesized aperture. Let the synthesized antenna aperture be aligned with the $x$-$y$ plane. The location of the discrete points (small antennas) may be represented by a two-dimensional comb function[2] constructed from a two-dimensional delta function $\delta_{\mathrm{II}}$, as expressed in Eq. (2.83), where $\Delta x$ and $\Delta y$ are the units of spacings between the impulses along the $x$- and $y$-axes, respectively, and $m_x$ and $m_y$ are integers. This is illustrated in Fig. 2.19a.

$$\mathrm{comb}(\Delta x, \Delta y) = \sum_{m_x=-\infty}^{\infty} \sum_{m_y=-\infty}^{\infty} \delta_{\mathrm{II}}\left(x - m_x\,\Delta x,\ y - m_y\,\Delta y\right) \tag{2.83}$$

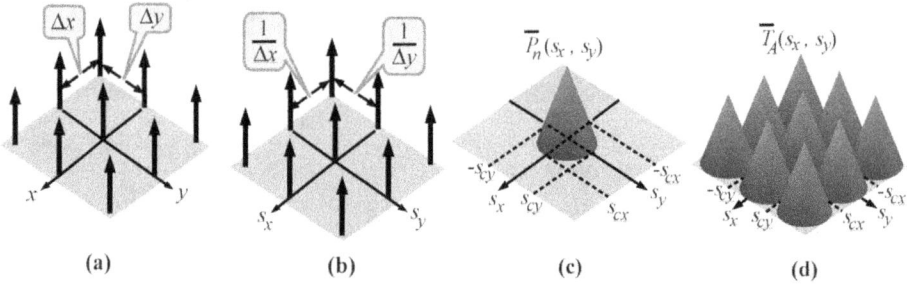

(a)  (b)  (c)  (d)

**FIGURE 2.19:** (a) A two-dimensional comb function. (b) Spatial Fourier transform of the comb function shown in part a. (c) Two-dimensional spatial spectral pattern of the antenna. (d) Two-dimensional spatial spectral pattern of the synthesized antenna.

Eq. (2.84) gives the Fourier transform of Eq. (2.83), where $S_x$ and $S_y$ are, re-spectively, the spatial frequencies corresponding to the $x$- and $y$-axes. Fig. 2.19b shows this function, which is another two-dimensional comb function in the spa-tial-frequency domain. The Fourier transform of the two-dimensional radiation patten $P_n(s_x, s_y)$ obtained from the aperture field has cut-off frequencies $S_{cx}$ and $S_{cy}$, respectively, along the $S_x$ and $S_y$ spatial-frequency axes. Note that the response is a low-pass along both the frequency axes. This is illustrated in Fig. 2.19c.

$$\mathrm{comb}\left(\frac{1}{\Delta x}, \frac{1}{\Delta y}\right) = \sum_{m_x=-\infty}^{\infty} \sum_{m_y=-\infty}^{\infty} \delta_{\mathrm{II}}\left(s_x - m_x\,\frac{1}{\Delta x},\ s_y - m_y\,\frac{1}{\Delta y}\right) \tag{2.84}$$

The antenna temperature spectrum $\overline{T}_A(s_x, s_y)$ of the sampled aperture fields is the product of the comb function given in Eq. (2.84) and the spatial spectrum of the

2. This function is also described as a shah function and symbolized as III in many texts.

antenna pattern $\overline{P}_n(s_x, s_y)$, as shown in Fig. 2.19d. This is expressed in Eq. (2.85), where $k_2$ is the constant of proportionality.

$$\overline{T}_A\left(s_x, s_y\right) = k_2 \sum_{m_x=-\infty}^{\infty} \sum_{m_y=-\infty}^{\infty} \overline{P}_n\left(s_x - m_x \frac{1}{\Delta x}, s_y - m_y \frac{1}{\Delta y}\right) \quad (2.85)$$

Similar to the one-dimensional case, there are three distinct cases of $\overline{T}_A(s_x, s_y)$: (*i*) Undersampling, (*ii*) Optimal sampling, and (*iii*) Oversampling. Undersampling results in aliasing in one or both dimensions, depending on the chosen values of $\Delta x$ and $\Delta y$. If $(\Delta x)^{-1} < 2S_{cx}$ aliasing is along $S_x$, and if $(\Delta y)^{-1} < 2S_{cy}$ aliasing is along $S_y$. Optimal sampling occurs along both *x*- and *y*-axes if $(\Delta x)^{-1} = 2S_{cx}$ and $(\Delta y)^{-1} = 2S_{cy}$, respectively. Oversampling results along the *x*- and *y*-axes if $\left(\Delta x\right)^{-1} > 2s_{cx}$ $(\Delta x)^{-1} > 2S_{cx}$ and $(\Delta y)^{-1} > 2S_{cy}$, respectively.

## 2.12 UNIFORM-GAIN POWER-SPECTRUM ANTENNA-PATTERN THEOREM

Instruments for power-spectrum monitoring use wide-band antennas. Certain applications require wide-angle coverage with uniform antenna illumination. Examples are terrestrial radio-spectrum monitors, radio-burst detectors (Sun and Jupiter), magnitude spectrographs, and the spectrum monitoring of satellites. The magnitude of the spectrum is of prime interest in these instruments. Some of these applications prefer omnidirectional coverage (0° to 360°) in the azimuth. They require a uniform power pattern across the angular coverage. Individual antennas fail to meet these requirements. However, using two or more antennas, it may be possible to create a uniform power-spectrum antenna pattern, as stated by Joardar and Bhattacharya. This is described in the next subsection.

### 2.12.1 Two-Antenna Uniform-Gain Power-Spectrum Antenna-Pattern Theorem

*This theorem states that if two electrically identical antennas possessing maximum individual gains $G_0$, positioned in free space in a plane, such that they subtend an angle $\alpha_v$ (equal to their half-power beam-widths) with respect to each other, then the antenna system effectively produces an amplitude power spectrum with uniform antenna gain $G_0$ and a uniform signal-to-noise ratio across the angle $\alpha_v$ if the amplitudes of the power spectra of the individual antennas are added.*

*If the exponents of the imaginary phase values of every frequency channel obtained from one of the antennas are multiplied with the corresponding channel's sum of the* *amplitude spectrum and an inverse Fourier transform is applied, the time-domain signal thus produced is effectively the signal that would be obtained from a single antenna possessing a uniform gain $G_0$ across an angle $\alpha_v$.*

The above is valid when the individual antenna patterns are identical and the antennas are placed in free space (Fig. 2.20a). In practical conditions, the presence of the ground alters the radiation patterns. Hence, antennas are positioned in a plane parallel to the ground so that the patterns are symmetrical in the horizontal plane. This is shown in Fig. 2.20b. This type of setup is suitable for monitoring the terrestrial spectrum. If the requirement is to point the antennas towards the sky, the effect of the ground can be reduced by positioning the antenna over a suitably designed metallic hemisphere kept above the ground. This is shown in Fig. 2.20c. Care should be taken that the patterns of individual antennas within the plane of combination should be symmetrical and identical. Also, the maximum side-lobe level should be much below the main-beam half-power points, preferably by 10 dB or more, which is typical of a fairly designed antenna.

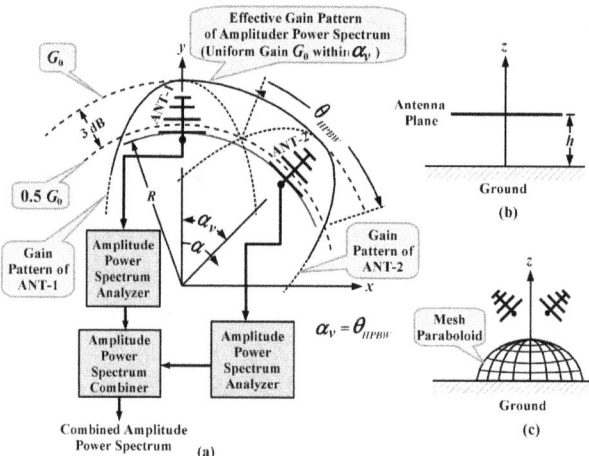

**FIGURE 2.20:** (a) Antennas in free space along with the additional electronic hardware required so that they might be treated as a single antenna-pair cell. (b) Antennas for monitoring the surrounding power spectrum are kept in this plane. (c) Antennas for monitoring the power spectra of sky signals.

### *2.12.1.1 Effective-Gain Equation*
The antenna gain $G$ as a function of angle $(\theta, \phi)$ within the major lobes of a directional antenna is mathematically modeled in Eq. (2.86), where $G_0$ is the maximum gain, $\kappa$ is the angle scale factor[3], and $(\theta, \phi)$ are spherical coordinates. Consider the

---

[3] $\kappa$ is introduced for modeling low-gain (wide major-lobes) log-periodic antennas whose first nulls might appear beyond $90°$ $\kappa = 1$ for higher gain antennas.

geometry of the two-antenna system shown in Fig. 2.20a. With reference to the $y$-axis, the power-gain pattern of Ant-1 ($x$-$y$ plane) is given in Eq. (2.87). On the $y$-axis $\alpha = 0$, and it increases clockwise. The power gain pattern for Ant-2 is similarly expressed in Eq. (2.88). Effectively, the amplitude power spectrum at the spectrum combiner's output can be thought of as being produced from a single antenna. This single antenna will have a gain pattern equal to the summation of the gain patterns of $G_1$ (Ant-1) and $G_2$ (Ant-2) displaced by an angle $\alpha_v$. The *effective-gain equation* is given as Eq. (2.89).

$$G(\theta,\phi) = G_0 \cos^n(\kappa\phi), \; \theta = \frac{\pi}{2}, 0 \le \phi \le 2\pi, n \ge 1, 0 \le \kappa \le 1 \qquad (2.86)$$

$$G_1(\alpha) = G_0 \cos^n(\kappa\alpha), 0 \le \alpha \le \alpha_v, n \ge 1, 0 \le \kappa \le 1 \qquad (2.87)$$

$$G_2(\alpha) = G_0 \cos^n[\kappa(\alpha_v - \alpha)], 0 \le \alpha \le \alpha_v, n \ge 1, 0 \le \kappa \le 1 \qquad (2.88)$$

$$\left.\begin{aligned}
&G(\alpha) = G_1(\alpha) + G_2(\alpha) \\
&\text{or, } G(\alpha) = G_0[\cos^n(\kappa\alpha) + \cos^n\{\kappa(\alpha_v - \alpha)\}] \\
&\quad 0 \le \alpha \le \alpha_v, n \ge 1, 0 \le \kappa \le 1
\end{aligned}\right\} \qquad (2.89)$$

### 2.12.1.2 Half-Power Equation

To model an antenna pattern using Eq. (2.86), one needs to find the value of $n$ from the measured pattern. Substitute the left-hand side of Eq. (2.87) with $0.5G_0$, replace $\alpha$ with $0.5_{\text{HPBW}}$ on the right-hand side, and then solve to find find the value of $n$. Here, $\theta_{\text{HPBW}}$ is the angular half-power beam width. The *half-power equation* is given as Eq. (2.90). The value of $n$ thus obtained is given in Eq. (2.91). The bases of the logarithms must be identical.

$$\cos^n\left(0.5\,\kappa\,\theta_{\text{HPBW}}\right) = 0.5 \qquad (2.90)$$

$$n = \frac{\log(0.5)}{\log\left[\cos\left(0.5\kappa\theta_{\text{HPBW}}\right)\right]} \qquad (2.91)$$

After having substituted $n$ from Eq. (2.91), the effective combined gain is evaluated from Eq. (2.89) using a computer. Eq. (2.89) shows a nearly uniform gain (close to $G_0$) within the angle $\alpha_v$. The percentage deviations of the computed gain from $G_0$ across the angle $\alpha_v$ can be expressed as in Eq. (2.92). It is plotted for antenna pairs having various HPBW in Fig. 2.21, and the values are listed in Table 2.1. The error is less for wider beams. It is minimized for $\theta_{\text{HPBW}} = 90°$ ($n = 2$). For narrower beams, $n$ increases, but the error tends to stabilize at a value close to 8%.

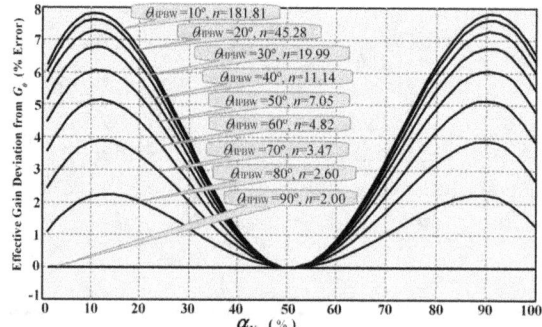

**FIGURE 2.21:** A simulated plot showing the percentage effective antenna gain deviations from $G_0$ as a function of percentage of operating angle $\alpha_v$. As a special case, if the elemental antennas posses $n = 2$, the errors vanish.

$$\text{Percentage Error} = 100 \times \frac{G(\alpha) - G_0}{G_0} \qquad (2.92)$$

**TABLE 2.1:** Maximum and minimum percentage errors of the effective antenna gain pattern (for the amplitude power spectrum) constructed with antennas having different half-power beam widths.

| $\alpha_v = \theta_{\text{HPBW}}$ (Degrees) | $n$ ($\kappa = 1$) | Min. Error (percentage) | Max. Error (percentage) |
|---|---|---|---|
| 10 | 181.81 | $2.22 \times 10^{-14}$ | 7.79 |
| 20 | 45.28 | $2.22 \times 10^{-14}$ | 7.59 |
| 30 | 19.99 | $2.22 \times 10^{-14}$ | 7.25 |
| 40 | 11.14 | $2.22 \times 10^{-14}$ | 6.75 |
| 50 | 7.05 | $2.22 \times 10^{-14}$ | 6.05 |
| 60 | 4.82 | $2.22 \times 10^{-14}$ | 5.12 |
| 70 | 3.47 | $2.22 \times 10^{-14}$ | 3.88 |
| 80 | 2.60 | 0.00 | 2.24 |
| 90 | 2.00 | $-1.11 \times 10^{-14}$ | 0.00 |

### 2.12.1.3 Directional Signal-to-Noise Ratio

The time-averaged noise power output from the antennas at a given physical temperature may be assumed constant. The total mean noise at the spectrum combiner's output is $\sqrt{2}\,\overline{N}_0$, because noise from different antennas is incoherent. Here, $\overline{N}_0$ (Watt) is the mean thermal noise power over a significant period of time obtained from the individual antennas. The signal power is proportional to the antenna gain, which is a function of $\alpha$. For any antenna, the SNR (signal-to-noise ratio) varies with the direction ($\alpha$) of the incoming electromagnetic waves. It is therefore termed a *Directional SNR* or DSNR. If $S$ is the flux density (Watt m$^{-2}$), $\lambda$

is the wavelength (m), $\eta$ is the antenna efficiency (dimensionless), then the DSNRs of Ant-1 and Ant-2 are given, respectively, by Eqs. (2.93) and (2.94). Because the effective gain of the combined pattern is more or less constant over $\alpha_v$, the SNR is independent of $\alpha$ and is given by Eq. (2.95). The SNR for a single antenna averaged across $\alpha_v$ is given in Eq. (2.96). Comparing Eq. (2.95) with (2.96), an improvement in SNR by a factor of $\sqrt{2}$ is seen in the combined pattern.

$$SNR_1(\alpha) = \left(\frac{S\,\eta\lambda^2}{4\pi}\right)\left(\frac{G_1(\alpha)}{N_0}\right) = \left(\frac{S\,\eta\,\lambda^2 G_0}{4\pi N_0}\right)\cos^n(\kappa\alpha) \qquad (2.93)$$

$$SNR_2(\alpha) = \left(\frac{S\,\eta\lambda^2}{4\pi}\right)\left(\frac{G_2(\alpha)}{N_0}\right) = \left(\frac{S\,\eta\,\lambda^2 G_0}{4\pi N_0}\right)\cos^n[\kappa(\alpha_v - \alpha)] \qquad (2.94)$$

$$\left.\begin{array}{l} SNR(\alpha) = \left(\frac{S\,\eta\,\lambda^2}{4\pi}\right)\left(\frac{G_1(\alpha)+G_2(\alpha)}{\sqrt{2}\,N_0}\right) \approx \left(\frac{S\,\eta\,\lambda^2}{4\pi}\right)\left(\frac{G_0}{\sqrt{2}\,N_0}\right), \\[4mm] 0 \le \alpha \le \alpha_v, n \ge 1, 0 \le \kappa \le 1 \end{array}\right\} \qquad (2.95)$$

$$\overline{SNR_1} = \overline{SNR_2} \approx \left(\frac{S\,\eta\,\lambda^2}{4\pi}\right)\left(\frac{G_0}{2N_0}\right) \qquad (2.96)$$

### 2.12.1.4 Solution to Practical Problems
The properties of ultra-wideband antennas vary with frequency. The ratio of maximum to minimum gain lies within 1.5 dB for a well-designed antenna. The antenna-pair cell (Fig. 2.20a) might be constructed considering the minimum gain with $\alpha_v$ equal to HPBW. Calibration for compensating the extra gain is required. Fig. 2.22 shows some some theoretical effective gain deviations of antenna cells using frequency-independent antennas. Computed results from antennas having minimum gains between 5 and 12 $dB_i$ and a maximum gain deviation of 1.5 dB are shown. The error increases with antenna gain.

### 2.12.1.5 Recovery of the Time-Domain Signal
Here we describe a possible method for adding phase information to the combined spectrum in an attempt to recover the time-domain signal. Consider a sinusoidal signal $s(t)$ with amplitude $A_0$ and frequency $v_0$ being radiated from a distant location. From the geometry shown in Fig. 2.20a, let $S_1(t)$ and $S_2(t)$ be the signals obtained from antennas Ant-1 and Ant-2, respectively. These are represented as in Eq. (2.97), where $A_0$ represents their transmitted amplitude, $\phi_1$ and $\phi_2$ represent the introduced phase delays in the received signals due to propagation in space and phase shifts occurring within the receiver antennas, and $\sigma_1$

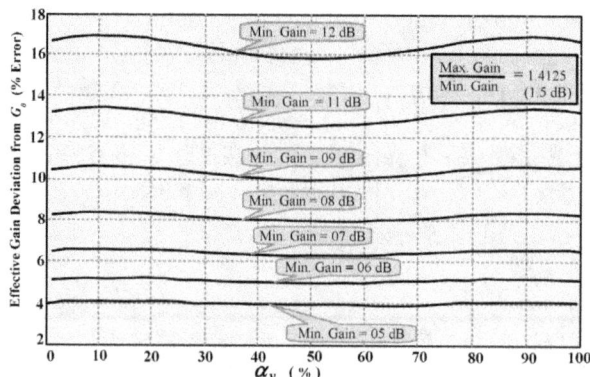

**FIGURE 2.22:** A simulation showing percentage deviation of the effective gain as a function of operating angle $\alpha_\nu$ using wideband antennas whose individual gain varies within a limited range as a function of frequency.

and $\sigma_2$ are the proportionality constants. The Fourier transforms of these are given in Eq. (2.98). Unless $\phi_1 = \phi_2$, the signals $S_1(t)$ and $S_2(t)$ cannot be directly added. The phase difference of the different Fourier components might be calculated as shown in Eq. (2.99). Alternatively, we can combine the amplitudes of the individual power spectra and associate this with the phase values of any one of the antennas (reference antenna). This is expressed in Eq. (2.100), where Ant-1 is taken as the reference antenna. Here $S_{\text{total}}(\nu)$ is the resultant spectrum. The effective time-domain signal is obtained from the inverse Fourier transform, as shown in Eq. (2.101).

$$
\left.
\begin{aligned}
s(t) &= A_0 \exp(j2\pi v_0\, t) \\
s_1(t) &= \sigma_1 A_0 \exp[j(2\pi v_0\, t - \phi_1)] = \sigma_1 s(t)\exp[-j\phi_1] \\
s_2(t) &= \sigma_2 A_0 \exp[j(2\pi v_0\, t - \phi_2)] = \sigma_2 s(t)\exp[-j\phi_2]
\end{aligned}
\right\}
\tag{2.97}
$$

$$
\left.
\begin{aligned}
S(v) &= A_0\, \delta(v - v_0) \\
S_1(v) &= \sigma_1 A_0\, \delta(v - v_0) \exp(-j\phi_1) = \sigma_1 S(v)\exp(-j\phi_1) \\
S_2(v) &= \sigma_2 A_0\, \delta(v - v_0) \exp(-j\phi_2) = \sigma_2 S(v)\exp(-j\phi_2)
\end{aligned}
\right\}
\tag{2.98}
$$

$$
\text{Phase Difference} = \angle S_1(v) - \angle S_2(v) = \phi_2 - \phi_1
\tag{2.99}
$$

$$
\left.
\begin{aligned}
S_{\text{total}}(v) &= S_1(v) + S_2(v)\exp[j\,(\text{Phase Difference})] \\
&\approx [\sigma_1 + \sigma_2] S(v)\exp(-j\phi_1)
\end{aligned}
\right\}
\tag{2.100}
$$

$$
s_{\text{total}}(t) = [\sigma_1 + \sigma_2]A_0 \exp[j(2\pi v_0 t - \phi_1)] = [\sigma_1 + \sigma_2]s(t)\exp[-j\phi_1]
\tag{2.101}
$$

The signals can be sampled and discritized. The above operations can then be performed in the discrete frequency and time domains. Fig. 2.23 shows the possible blocks. Filters are suggested for band-limiting the signals before sampling to satisfy the Nyquist criterion. Fast Fourier transforms (FFT-1 and FFT-2) are applied, and their magnitude parts are added. Next they are multiplied with the exponent of the imaginary phase components from FFT-1. The real part of the inverse Fourier transform of this represents the combined signal in the time domain. If more antennas are present, their magnitude spectra can also be added and associated with the common phase derived from the reference antenna.

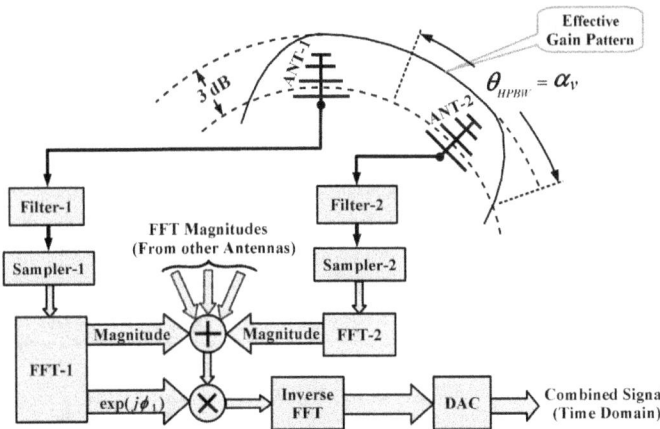

**FIGURE 2.23:** A suggested scheme for combining the magnitude spectra and phase such that the system behaves effectively as a single antenna with flat angular coverage within $\alpha_v$. Ant-1 is the reference antenna. If more antennas are present, their magnitude spectra should be added directly using the adder.

### 2.12.2 Complete Omnidirectional Power-Spectrum Pattern

Based on the uniform-gain power-spectrum antenna-pattern theorem, a complete omnidirectional ultra-wideband antenna pattern can be synthesized using a number of identical frequency-independent antennas. They should be pointed in various directions on a plane parallel to the ground and must have equal angular separations, as shown in Fig. 2.24. To cover the omnidirectional plane completely, the antenna gains must be properly chosen such that the ratio of $360°$ to $\theta_{\text{HPBW}}$ is an integer $N$ given by Eq. (2.102). The number of antennas required is $N$.

$$N = \frac{360°}{\theta_{\text{HPBW}}} \qquad (2.102)$$

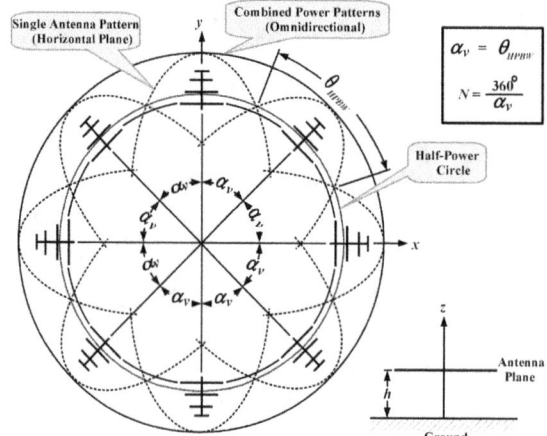

**FIGURE 2.24:** Several antennas with identical characteristics are used to form a complete omnidirectional power-spectrum antenna pattern.

It is speculated that this theorem shall find applications in tracking artificial satellites, radio spectrographs for the Sun and Jupiter, pulsar survey (incoherent beam), etc., apart from spectrum monitoring.

## REVIEW QUESTIONS

1. Describe the terms (*i*) brightness, (*ii*) total brightness, (*iii*) total radio brightness, and (*iv*) spectral power.

2. What is *radiance*? Explain the relationship between radiance and brightness.

3. Describe the terms (*i*) point source, (*ii*) localized source, and (*iii*) extended source.

4. Using Planck's law, write the equations of black-body radiation as functions of (*i*) frequency, and (*ii*) wavelength. Draw a tentative curve illustrating brightness as a function of frequency for a temperature of 5800 K.

5. Write the expression for Wien's displacement law. At which wavelength does the Sun show maximum intensity? Assume the temperature of the Sun to be 5800 K.

6. Write down the expressions of (*i*) the Rayleigh-Jeans law, and (*ii*) Wien's radiation law. Why are these said to be approximate laws?

7. The Rayleigh-Jeans law is preferred over Planck's law in radio astronomy. Why?

8. Write down the expression of the Stefan-Boltzmann law and explain. Assuming the temperature of the Sun is 5800 K, calculate the total brightness.

9. Using a diagram, compare the Rayleigh-Jeans law and Wien's radiation law with Planck's law of spectral radiance.

10. Using equations, explain the meaning of (*i*) optical depth, and (*ii*) depth of penetration.

11. Let a source having brightness $B_S$ be observed through a cloud that absorbs as well as emits, as shown in Fig. 2.9. The change in brightness $dB$ produced over a volume of length $dr$ is given by Eq. (2.103), where $B$ is the apparent brightness, $K$ is the absorption coefficient, $j$ is the emission coefficient, and $\rho$ is the matter density within the cloud. Assume a local thermodynamic equilibrium and apply Kirchhoff's law to show that the apparent brightness $B$ is given by Eq. (2.104).      [*Hint: $dB = 0$*]

12. An object in the sky appears circular with diameter 0.049°. It emits like a black body. An antenna having a HPBW of 0.115° is employed to measure the object's temperature. The incremental temperature measured is 0.239 K. Assuming $\kappa_B = 0.8$, calculate the temperature of the object.

   [*Hint:* Use Eqs. (B.10) and (2.53)]

13. Calculate the source flux density for problem no. 12 if the antenna efficiency is 100% and the wavelength is 3 cm.

   [*Hint:* Use Eqs. (B.23), (B.29) and (2.52)]

14. The model of a single-antenna radio telescope is shown in Fig. 2.12. It is operated at room temperature (300 K). The transmission-line length (between antenna and receiver) is 2 m and has an attenuation constant of 0.05 Nepers per meter. If the antenna noise temperature is 55 K, the antenna efficiency is 98%, and the receiver temperature is 70 K, find the system temperature.      [*Hint:* Use Eqs. (2.55) and (2.56)]

15. A LNA (low-noise amplifier) operates at room temperature (300 K). It has a noise factor of 1.2. Calculate the noise temperature of the LNA.

   [*Hint*: Use Eq. (2.58)]

16. Let the gain of the LNA described in problem 15 be 1000. If 500 K is the temperature of the remaining system (after the LNA), find the receiver temperature. Compare this with the LNA temperature.

   [*Hint*: Use Eq. (2.57)]

17. The system temperature of a radio telescope is 90 K. The operating frequency is 1420 MHz and the bandwidth is 30 MHz. It has a system sensitivity constant of 0.7 and an integration time of 1 second. If the number of averaged records is 20, calculate the (*i*) minimum detectable temperature, (*ii*) minimum detectable brightness, and the (*iii*) minimum detectable flux density. [***Hint:*** Use Eqs. (2.59) through (2.61)]

18. How are the antenna temperature, antenna pattern, and sky temperature distribution related? [***Hint:*** See Fig. 2.13]

19. What is meant by *aerial smoothing*? Using simple equations and some figures to explain how the loss of temperature spectra takes place due to the low-pass characteristics of the antenna pattern.

[***Hint***: See Figs. 2.15 and 2.16]

20. Loss of spectral components are less when an extended Gaussian source is observed as compared to an extended rectangular source. Justify with simple reasons. [***Hint***: See Fig. 2.17]

21. Describe in detail the one-dimensional sampling theorem of the observing angle. Where is it used?

22. Describe in detail the two-dimensional sampling theorem of synthesized aperture. Where is it used?

23. Describe in detail the two-antenna uniform-gain power-spectrum antenna-pattern theorem. Where is it used?

24. Explain how to make an omnidirectional system using linearly polarized antennas in the presence of the ground?

# POLARIZATION ANALYSIS

## 3.1   INTRODUCTION

Different emission mechanisms give rise to different polarization types. For example, black-body emission is randomly polarized, synchrotron radiation is linearly polarized, pulsars emit mainly linearly and partly circularly polarized waves, Zeeman splitting of spectral lines generates two circularly polarized components, Faraday rotation changes the plane of polarization, etc. Thus, astronomical radio sources produce a variety of polarizations. Measurements of polarization are necessary in studies relating to magnetic fields, esoteric scattering geometries, gas pressure in the interstellar medium, etc.

## 3.2   POLARIZATION TYPES

There are three basic polarization types: (*i*) linear, (*ii*) circular, and (*iii*) elliptical. Assume a plane wave is traveling out of the page along the positive $z$ direction. To identify the polarization type, the movement of the electric field (as a function of time) should be observed on the $x$-$y$ plane for any fixed value of $z$.

The wave is said to be *linearly polarized* provided the orientation of the electric field vector with respect to the $x$- or $y$-axis does not change with time. A linearly polarized wave along the $y$-axis in the plane $z = 0$ is shown in Fig. 3.1a. The electric field $E_y$ along the $y$-axis is given by Eq. (3.1), where $\nu$ represents frequency and $\lambda$ represents wavelength. If the electric field vector in the $x$-$y$ plane keeps rotating with time such that the tip of the vector describes a circle in every frequency cycle, the wave is said to be *circularly polarized*. This is shown in Fig. 3.1b. If the tip of the electric field vector describes an ellipse, the wave is said to be *elliptically polarized,* as shown in Fig. 3.1c.

$$E_y = E_2 \sin(\omega t - \beta z) \left.\begin{array}{c} \\ \\ \end{array}\right\}$$

$$\text{where, } \omega = 2\pi v \text{ and } \beta = \frac{2\pi}{\lambda}$$

(3.1)

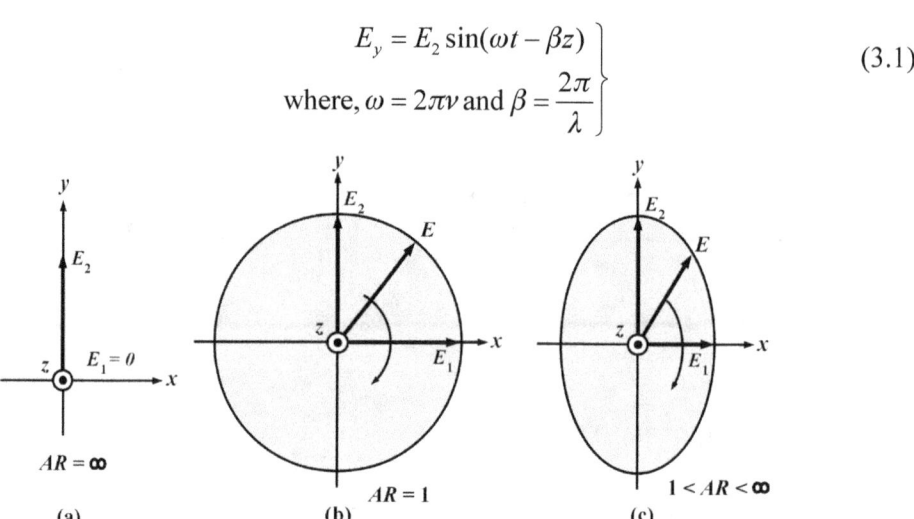

**FIGURE 3.1:** The three basic types of polarization: (a) linear, (b) circular, and (c) elliptical.

Elliptical polarization represents the general case of polarization. The ellipse described by the tip of the vector is called the *polarization ellipse*. The ratio of its semimajor axis to its semiminor axis is called the *axial ratio (AR)* of the polarization ellipse. It is shown in Eq. (3.2). For linear polarizations, $AR = \infty$; for circular polarizations, $AR = 1$; and for elliptical polarizations, $1 < AR < \infty$. Depending on the type of polarization, a phase difference may exist between the x- and y-components of the electric field. The direction of vector rotation is based on this.

$$AR = \frac{E_2}{E_1}$$

(3.2)

## 3.3   THE POLARIZATION ELLIPSE

Fig. 3.2 shows the polarization ellipse formed by the tip of the electric field vector $E$ in the x-y plane for a fixed value of z. The major and minor axes are shown (marked). The ellipse is enclosed within a rectangle whose diagonal makes an angle $\varepsilon$ with the major axis. Any form of polarization resulting from a coherent source (single frequency) can be analyzed using the polarization ellipse.

Let an elliptically polarized wave propagate along the z-direction (out of the page). It may be expressed as a vector sum of two linearly polarized waves (along the x- and y-axes). These are given in Eqs. (3.3) and (3.4), where $E_1$ and $E_2$, respectively, represent the maximum magnitudes of the x- and y-components of $E$, and $\delta$ is the time-phase angle by which $E_y$ leads over $E_x$. Combining Eqs. (3.3) and (3.4), we obtain the instantaneous total electric field vector $\vec{E}$, as shown in Eq. (3.5), where $\hat{x}$ and $\hat{y}$ represent unit vectors along the x- and y-axes, respectively.

$$E_x = E_1 \sin(\omega t - \beta z) \tag{3.3}$$

$$E_y = E_2 \sin(\omega t - \beta z + \delta) \tag{3.4}$$

$$\vec{E} = \hat{x} E_1 \sin(\omega t - \beta z) + \hat{y} E_2 \sin(\omega t - \beta z + \delta) \tag{3.5}$$

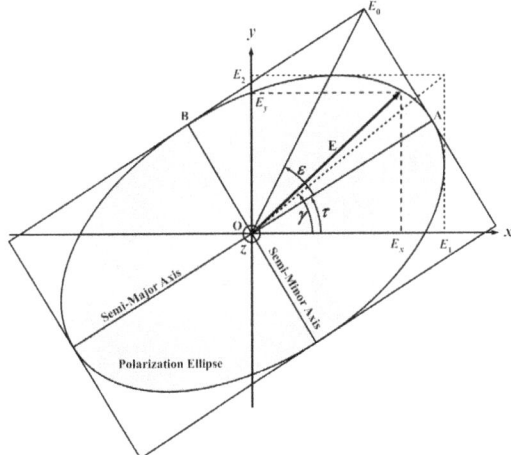

FIGURE 3.2: Polarization ellipse.

Let us choose the x-y plane at $z = 0$. We get $E_x = E_1 \sin \omega t$ and $E_y = E_2 \sin(\omega t + \delta)$. We expand $E_y$ as shown in Eq. (3.6). From Eq. (3.3) at $z = 0$, we have $\sin \omega t = E_x / E_1$ and $\cos \omega t = \sqrt{1 - (E_x / E_1)^2}$. By substituting these into Eq. (3.6) and rearranging, we get Eq. (3.7). This may be re-expressed as Eq. (3.8), which describes the polarization ellipse (Fig. 3.2). The line segments $OA$ and $OB$ represent the semimajor and the semiminor axes, respectively. Eq. (3.9) gives $AR$. The tilt angle is denoted by $\tau$. Eqs. (3.10) and (3.11), respectively, express $\gamma$ and $\varepsilon$. The average pointing vector $S_{av}$ is shown in Eq. (3.12), where $E$ is the amplitude of the total electric field, $\hat{z}$ is a unit vector along the z-axis, and $Z_0$ is the free-space impedance (medium of propagation). Applying these expressions, some possible values of $E_1$, $E_2$, $\tau$, and $\delta$ for some specific cases of linear and circular polarization have been produced in Tables 3.1 and 3.2, respectively.

$$E_y = E_2(\sin \omega t \cos \delta + \cos \omega t \sin \delta) \tag{3.6}$$

$$\left(\frac{E_x}{E_1}\right)^2 - 2\left(\frac{E_x}{E_1}\right)\left(\frac{E_y}{E_2}\right)\cos \delta + \left(\frac{E_y}{E_2}\right)^2 = \sin^2 \delta \tag{3.7}$$

$$aE_x^2 + bE_xE_y + cE_y^2 = 1 \qquad (3.8)$$

$$\text{where, } a = \frac{1}{E_1^2 \sin^2 \delta} \quad b = \frac{2\cos\delta}{E_1E_2 \sin^2 \delta} \quad c = \frac{1}{E_2^2 \sin^2 \delta}$$

$$AR = \frac{OA}{OB}, \quad 1 \le AR \le \infty \qquad (3.9)$$

$$\gamma = \tan^{-1}\left(\frac{E_2}{E_1}\right), \quad 0° \le \gamma \le 90° \qquad (3.10)$$

$$\varepsilon = \cot^{-1}(\pm AR), \quad -45° \le \varepsilon \le +45° \qquad (3.11)$$

$$S_{av} = \hat{z}\frac{1}{2}\left(\frac{E_1^2 + E_2^2}{Z_0}\right) = \hat{z}\frac{1}{2}\left(\frac{E^2}{Z_0}\right) \qquad (3.12)$$

$$\text{where, } E = \sqrt{E_1^2 + E_2^2}$$

**TABLE 3.1:** Linear polarization.

| Axis of Linear polarization | $E_1$ | $E_2$ | $\tau$ | $\delta$ |
|---|---|---|---|---|
| $y$-axis | 0 | > 0 | 90° | — |
| $x$-axis | > 0 | 0 | 0° | — |
| 45° with $x$- and $y$-axes | $= E_2$ | $= E_1$ | 45° | 0° |

**TABLE 3.2:** Circular polarization.

| Type of circular polarization | $E_1$ | $E_2$ | $\tau$ | $\delta$ |
|---|---|---|---|---|
| Left Circular (IEEE standards) | $= E_2$ | $= E_1$ | — | +90° |
| Right Circular (IEEE standards) | $= E_2$ | $= E_1$ | — | −90° |

## 3.4   PARTIAL POLARIZATION

So far, completely polarized waves have been described, where $E_1$, $E_2$, and $\delta$ were assumed constant or slowly varying. In other words, a single frequency (mono-chromatic) source was used. Celestial radio sources, on the other hand, are extremely wideband. These signals consist of a superposition of numerous statistically independent waves possessing a variety of polarizations within any nonzero finite bandwidth $\Delta v$ and are termed *randomly polarized*. The waves are partially polarized in the most general situations. They may be broken into two components: (*i*) completely polarized, and (*ii*) randomly polarized. All states of polarization can be graphically represented using the Poincaré sphere. Partial polarizations are often analyzed using the Stokes parameters.

## 3.5  POINCARÉ SPHERE AND POLARIZATION ELLIPSE

The Poincaré sphere is a graphical representation for visualizing the state of polarization. It was invented by Jules Henry Poincaré (1854–1912). A spherical coordinate system is used. The polarization state is described by a point on the sphere (Fig. 3.3) whose longitude and latitude are related to the polarization ellipse (Fig. 3.2). These mathematical expressions are given in Eqs. (3.13) through (3.18). A Cartesian coordinate system is also shown with axes marked $Q$, $U$, and $V$. This will be used later for deriving the Stokes parameters $I$, $Q$, $U$, and $V$.

$$\text{Longitude} = 2\tau \tag{3.13}$$

$$\text{Latitude} = 2\varepsilon \tag{3.14}$$

$$\text{Great circle angle} = 2\gamma \tag{3.15}$$

$$\text{Equator to great circle angle} = \delta \tag{3.16}$$

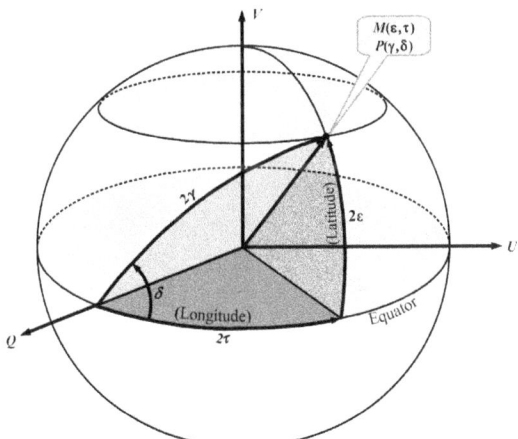

**FIGURE 3.3:** Poincaré sphere showing relationships among the angles $\varepsilon$, $\tau$, $\delta$, and $\gamma$.

The quantities $\varepsilon$, $\tau$, $\delta$, and $\gamma$ known as *polarization parameters*. Eqs. (3.17) through (3.20) use spherical trigonometry to show their relationships. From the geometry, one can easily find $\gamma$ and $\delta$ if $\varepsilon$ and $\tau$ are known. To describe a point on the Poincaré sphere, only one set of angles $((\varepsilon\varepsilon, \tau)$ or $(\gamma, \delta))$ is required. Let the polarization states on the Poincaré sphere for the set of angles $(\varepsilon\varepsilon, \tau)$ and $(\gamma, \delta)$ be represented by $M(\varepsilon, \tau)$ and $P(\gamma, \delta)$, respectively, as shown in Fig. 3.3.

$$\cos 2\gamma = \cos 2\varepsilon \ \cos 2\tau \qquad (3.17)$$

$$\tan \delta = \frac{\tan 2\varepsilon}{\sin 2\tau} \qquad (3.18)$$

$$\tan 2\tau = \tan 2\gamma \ \cos \delta \qquad (3.19)$$

$$\sin 2\varepsilon = \sin 2\gamma \ \sin \delta \qquad (3.20)$$

Fig. 3.4 maps the polarization types on the Poincaré sphere. Let $\varepsilon = 0, \tau = 0$ be the origin point of polarization. This is identical to $\delta = 0, \gamma = 0$ on the polarization ellipse. We now analyze the three types of polarization:

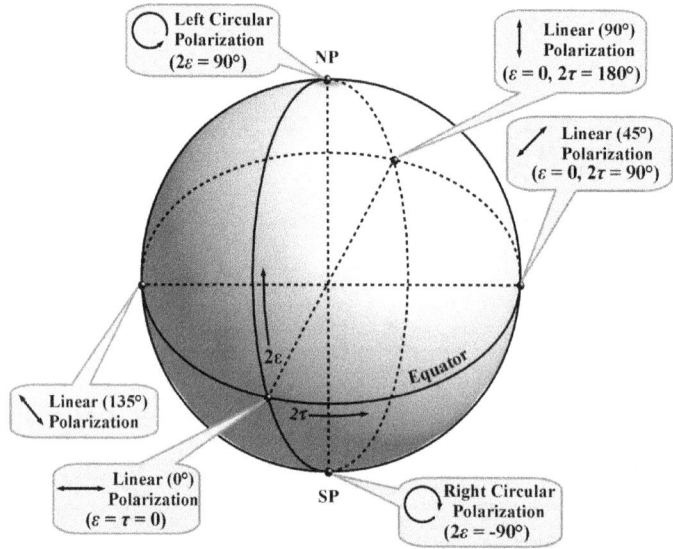

**FIGURE 3.4:** Mapping of basic polarization types on the Poincaré Sphere.

**Case (i):** At $\delta = 0$, or at $\delta = \pm 180°$, $E_x$ and $E_y$ are either in exact phase or completely out of phase. Any point on the equator of the Poincaré sphere thus represents a state of linear polarization. The polarization is linear and horizontal at the origin ($\varepsilon = 0, \tau = 0$ or $\delta = 0, \gamma = 0$). At $\varepsilon = 0, 2\tau = 90°$, the polarization is linear having a 45° tilt. At $\varepsilon = 0, 2\tau = 180°$, it is linear and vertical. Similarly, it is 135° at $\varepsilon = 0, 2\tau = 270°$, and so on.

**Case (*ii*):** At the poles ($2\varepsilon = \pm 90°$), which is same as $\delta = \pm 90°$, the polarizations are circular. The north and south poles, respectively, represent the left and right circular polarizations.

**Case (*iii*):** The limiting conditions are described in the above two cases. In general, a point on the northern hemisphere describes a left elliptically polarized wave. It transforms from a pure left circular polarization at the pole to linear polarization at the equator. Similarly, any point on the southern hemisphere describes a right elliptically polarized wave. It transforms from pure right circular polarization at the pole to linear polarization at the equator.

## 3.6   STOKES PARAMETERS

It is impossible to obtain the instantaneous tilt angle $\tau$ or the axial ratio for unpolarized or partially polarized waves. Thus, the polarization ellipse is not applicable here. To overcome these limitations, in 1852 George Gabriel Stokes introduced a set of time-averaged components obtained from the wave, known as the *Stokes parameters*. They consist of four variables, namely, *I, Q, U,* and *V*. They can be related with the polarization ellipse and the Poincaré sphere. The time average of the waves is given by Eq. (3.21), where *i* and *j* represent *x* or *y,* or both.

$$\langle E_i(t) E_j(t) \rangle = \lim_{T \to \infty} \frac{1}{T} \int_0^T E_i(t) E_j(t) dt, \quad i, j = x, y \tag{3.21}$$

The time averages of completely polarized waves give their effective values. With the aid of time-averaged components, the polarization ellipse (Eq. 3.7) can be expressed as in Eq. 3.22.

$$\left[ \langle E_1^2 \rangle + \langle E_2^2 \rangle \right]^2 = \left[ \langle E_1^2 \rangle - \langle E_2^2 \rangle \right]^2 + \left[ 2\langle E_1 E_2 \cos\delta \rangle \right]^2 + \left[ 2\langle E_1 E_2 \sin\delta \rangle \right]^2 \tag{3.22}$$

Dividing the effective values of the field components ( $E_{1eff} = E_1/\sqrt{2}$ and $E_{2eff} = E_2/\sqrt{2}$ ) by the square of the characteristic impedance $Z_0$ of free space on both sides, we obtain Eq. (3.23).

$$\left. \begin{aligned} &\left[ \frac{\langle E_{1eff}^2 \rangle + \langle E_{2eff}^2 \rangle}{Z_0} \right]^2 = \left[ \frac{\langle E_{1eff}^2 \rangle - \langle E_{2eff}^2 \rangle}{Z_0} \right]^2 \\ &+ \left[ \frac{2}{Z_0} \langle E_{1eff} E_{2eff} \cos\delta \rangle \right]^2 + \left[ \frac{2}{Z_0} \langle E_{1eff} E_{2eff} \sin\delta \rangle \right]^2 \end{aligned} \right\} \tag{3.23}$$

We may express the above equation symbolically, as given below in Eq. (3.24), for which the individual terms are given in Eqs. (3.25) through (3.36).

$$I^2 = Q^2 + U^2 + V^2 \tag{3.24}$$

$$I = \frac{\left\langle E_{1eff}^2 \right\rangle + \left\langle E_{2eff}^2 \right\rangle}{Z_0} = S_0 \tag{3.25}$$

$$Q = \frac{\left\langle E_{1eff}^2 \right\rangle - \left\langle E_{2eff}^2 \right\rangle}{Z_0} = S_1 \tag{3.26}$$

$$U = \frac{2}{Z_0}\left\langle E_{1eff} E_{2eff} \cos\delta \right\rangle = S_2 \tag{3.27}$$

$$V = \frac{2}{Z_0}\left\langle E_{1eff} E_{2eff} \sin\delta \right\rangle = S_3 \tag{3.28}$$

The last three quantities are marked on the Poincaré sphere (Fig. 3.3). Note that the radius of the Poincaré sphere is equal to $S_0$, or $I$. We will now find the Stokes parameters for (*i*) completely polarized waves, (*ii*) completely unpolarized waves, and (*iii*) partially polarized waves.

### 3.6.1 Completely Polarized Waves

From the geometry of the polarization ellipse (Fig. 3.5), we can express the electric field equations as given in Eqs. (3.29) and (3.30), where $\delta_1$ and $\delta_2$ represent the individual phases, and $\delta_1 - \delta_2$ is the phase difference. To relate the Stokes parameters with linear components along the *x*- and *y*-axes using $E_1$ and $E_2$, we place two dipole antennas (dipole-cross) oriented along the *x*- and *y*-axes.

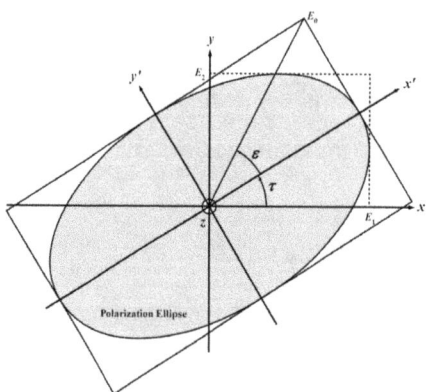

**FIGURE 3.5:** Relationships among the axes of the polarization ellipse (*x′, y′*) and the axes of the reference (*x, y*).

$$E_x = E_1 \sin(\omega t - \delta_1) \tag{3.29}$$

$$E_y = E_2 \sin(\omega t - \delta_2) \tag{3.30}$$

The $x'$- and $y'$-axes are coincident with the ellipse. The electric fields along these axes are given in Eqs. (3.31) and (3.32).

$$E_{x'} = E_0 \cos \epsilon \sin \omega t \tag{3.31}$$

$$E_{y'} = E_0 \sin \epsilon \cos \omega t \tag{3.32}$$

Expressing the electric fields $E_x$ and $E_y$ in terms of $E_x'$ and $E_y'$, we obtain Eqs. (3.33) and (3.34), as shown below.

$$E_x = E_{x'} \cos \tau - E_{y'} \sin \tau \tag{3.33}$$

$$E_y = E_{x'} \sin \tau + E_{y'} \cos \tau \tag{3.34}$$

Relating Eqs. (3.33) and (3.34) with Eqs. (3.31) and (3.32), and after some algebraic manipulation, we get Eqs. (3.35) and (3.36), as shown below.

$$E_x = E_0[\cos \epsilon \cos \tau \sin \omega t - \sin \epsilon \sin \tau \cos \omega t] \tag{3.35}$$

$$E_y = E_0[\cos \epsilon \sin \tau \sin \omega t - \sin \epsilon \cos \tau \cos \omega t] \tag{3.36}$$

Expanding Eq. (3.29), we obtain $E_x = (E_1 \cos \delta_1) \sin \omega t - (E_1 \sin \delta_1) \cos \omega t$. Again, $E_x = (E_0 \cos \epsilon \cos \tau) \sin \omega t - (E_0 \sin \epsilon \sin \tau) \cos \omega t$ from Eq. (3.35). Equating the terms containing $\sin \omega t$ we obtain $E_1 \cos \delta_1 = E_0 \cos \epsilon \cos \tau$. Equating the terms containing $\cos \omega t$, we obtain $E_1 \sin \delta_1 = E_0 \sin \epsilon \sin \tau$. By squaring and adding these two terms, we eliminate $\sin \delta_1$ and $\cos \delta_1$. After simplification, we obtain Eq. (3.37). Similarly, using Eqs. (3.30) and (3.36), we get Eq. (3.38), as given below.

$$E_1 = E_0 \sqrt{\cos^2 \epsilon \cos^2 \tau + \sin^2 \epsilon \sin^2 \tau} \tag{3.37}$$

$$E_2 = E_0 \sqrt{\cos^2 \epsilon \sin^2 \tau + \sin^2 \epsilon \cos^2 \tau} \tag{3.38}$$

By squaring and adding Eqs. (3.37) and (3.38), we get $E_1^2 + E_2^2 = E_0^2$. Let $E_{1eff}$, $E_{2eff}$, and $E_{0eff}$ be, respectively, the effective values of $E_x$, $E_y$, and $E_0$. That is, $E_{1eff} = E_1/\sqrt{2}$, $E_{2eff} = E_2/\sqrt{2}$, and $E_{0eff} = E_0/\sqrt{2}$. The magnitude of the total Poynting vector $S$ can be expressed in terms of these effective values, as shown in Eq. (3.39), where $Z_0$ is the impedance of the medium. The polarized wave components of the Poynting vector along the $x$ and $y$ directions are, respectively, $S_x$ and $S_y$.

$$S = S_x + S_y = \frac{E_{1eff}^2}{Z_0} + \frac{E_{2eff}^2}{Z_0} = \frac{E_{0eff}^2}{Z_0} \tag{3.39}$$

Substituting Eqs. (3.37) and (3.38) into Eq. (3.39), we obtain the values of $S_x$ and $S_y$, as shown in Eqs. (3.40) and (3.41).

$$S_x = \frac{E_{1eff}^2}{Z_0} = S(\cos^2 \epsilon \cos^2 \tau + \sin^2 \epsilon \sin^2 \tau) \tag{3.40}$$

$$S_y = \frac{E_{2eff}^2}{Z_0} = S(\cos^2 \epsilon \sin^2 \tau + \sin^2 \epsilon \cos^2 \tau) \tag{3.41}$$

For completely polarized waves, the Stokes parameters $I$, $Q$, $U$, and $V$ are now given in Eqs. (3.42) through (3.45).

$$I = S = S_x + S_y = \frac{E_{1eff}^2}{Z_0} + \frac{E_{2eff}^2}{Z_0} \tag{3.42}$$

$$Q = S_x - S_y = \frac{E_{1eff}^2}{Z_0} - \frac{E_{2eff}^2}{Z_0} = S \cos 2\epsilon \cos 2\tau \tag{3.43}$$

$$U = \left(S_x - S_y\right) \tan 2\tau = S \cos 2\epsilon \sin 2\tau = 2\frac{E_{1eff} E_{2eff}}{Z_0} \cos(\delta_1 - \delta_2) \tag{3.44}$$

$$V = \left(S_x - S_y\right) \tan 2\epsilon \sec 2\tau = S \sin 2\epsilon = 2\frac{E_{1eff} E_{2eff}}{Z_0} \sin(\delta_1 - \delta_2) \tag{3.45}$$

All four of the Stokes parameters are related in Eq. (3.46). This may be verified using Eqs. (3.42) through (3.45). The angular relations (for the polarization ellipse) are shown in Eqs. (3.47) and (3.48). We express these equations (the Stokes parameters for completely polarized waves) in matrix form, as shown in Eq. (3.49).

$$I = \sqrt{Q^2 + U^2 + V^2} \tag{3.46}$$

$$\frac{U}{Q} = \tan 2\tau \tag{3.47}$$

$$\frac{V}{S} = \sin 2\epsilon = \frac{V}{\sqrt{Q^2 + U^2 + V^2}} \tag{3.48}$$

$$\begin{bmatrix} I \\ Q \\ U \\ V \end{bmatrix} = \begin{bmatrix} S_x + S_y \\ \left(S_x - S_y\right)\tan 2\tau \\ \left(S_x - S_y\right)\tan 2\tau \sec 2\tau \\ S_x + S_y \end{bmatrix} = \begin{bmatrix} S \\ S \cos 2\epsilon \cos 2\tau \\ 2\dfrac{E_{1eff} E_{2eff}}{Z_0} \cos(\delta_1 - \delta_2) \\ 2\dfrac{E_{1eff} E_{2eff}}{Z_0} \sin(\delta_1 - \delta_2) \end{bmatrix} \tag{3.49}$$

The Stokes parameters of completely polarized waves for some different states of polarization are listed in Table 3.3.

**TABLE 3.3:** Stokes parameters for completely polarized waves.

| Stokes parameters | Left Circular | Right Circular | Linear Horizontal | Linear Vertical | Linear | Linear |
|---|---|---|---|---|---|---|
| $I$ | $S$ | $S$ | $S$ | $S$ | $S$ | $S$ |
| $Q$ | 0 | 0 | $S$ | $-S$ | 0 | 0 |
| $U$ | 0 | 0 | 0 | 0 | $S$ | $-S$ |
| $V$ | $S$ | $-S$ | 0 | 0 | 0 | 0 |

### 3.6.2 Completely Unpolarized Waves

The phases of both $E_x(t)$ and $E_y(t)$ vary randomly with time. We first express the values of $E_x(t)$ and $E_y(t)$ as in Eqs. (3.50) and (3.51). We then attempt to relate the Stokes parameters with the linear $x$- and $y$-components using $E_1$ and $E_2$ (Fig. 3.5). This is done by placing two dipole antennas (dipole-cross) oriented along the $x$- and $y$-axes.

$$E_x = E_1(t)\sin(\omega t - \delta_1(t)) \tag{3.50}$$

$$E_y = E_2(t)\sin(\omega t - \delta_2(t)) \tag{3.51}$$

It is necessary to consider the mean variations of the phases $\delta_1$ and $\delta_2$, because they vary randomly with time. The Stokes parameters are given by Eqs. (3.52) through (3.55), where $\delta$ is defined in Eq. (3.56) and <. . .> represents the time average. The time-averaged effective values of the electric fields are given in Eq. (3.57), as shown below.

$$I = \frac{\left\langle E_{1eff}^2 \right\rangle}{Z_0} + \frac{\left\langle E_{2eff}^2 \right\rangle}{Z_0} = S_x + S_y = S E \tag{3.52}$$

$$Q = \frac{\left\langle E_{1eff}^2 \right\rangle}{Z_0} - \frac{\left\langle E_{2eff}^2 \right\rangle}{Z_0} = S_x - S_y = S\left\langle \cos 2\varepsilon \cos 2\tau \right\rangle \tag{3.53}$$

$$U = 2\frac{\left\langle E_{1eff} E_{2eff} \cos\delta \right\rangle}{Z_0} = S\left\langle \cos 2\varepsilon \sin 2\tau \right\rangle \tag{3.54}$$

$$V = 2\frac{\left\langle E_{1eff} E_{2eff} \sin\delta \right\rangle}{Z_0} = S\left\langle \sin 2\varepsilon \right\rangle \tag{3.55}$$

$$\delta = \delta_1 - \delta_2 \tag{3.56}$$

$$\left\langle E_{1eff}^{2} \right\rangle = \frac{1}{T} \int_{0}^{T} \left[ E_{1eff}(t) \right]^{2} dt$$ 

$$\left\langle E_{2eff}^{2} \right\rangle = \frac{1}{T} \int_{0}^{T} \left[ E_{2eff}(t) \right]^{2} dt$$ (3.57)

Unlike the completely polarized waves, here the sum-equality may or may not be true. This is shown in Eq. (3.58). Eq. (3.59) shows the Stokes parameters in matrix form.

$$I \geq \sqrt{Q^2 + U^2 + V^2}$$ (3.58)

$$\begin{bmatrix} I \\ Q \\ U \\ V \end{bmatrix} = \frac{1}{Z_0} \begin{bmatrix} \left\langle E_{1eff}^2 \right\rangle + \left\langle E_{2eff}^2 \right\rangle \\ \left\langle E_{1eff}^2 \right\rangle - \left\langle E_{2eff}^2 \right\rangle \\ 2 \left\langle E_{1eff} E_{2eff} \cos \delta \right\rangle \\ 2 \left\langle E_{1eff} E_{2eff} \sin \delta \right\rangle \end{bmatrix} = \begin{bmatrix} S \\ S \left\langle \cos 2\epsilon \cos 2\tau \right\rangle \\ S \left\langle \cos 2\varepsilon \sin 2\tau \right\rangle \\ S \left\langle \sin 2\varepsilon \right\rangle \end{bmatrix}$$ (3.59)

The Stokes parameters from the above equations may be interpreted as: (*i*) the total power (sum of the autocorrelation of the *x*- and *y*-components) is represented by *I*; (*ii*) the difference between the autocorrelations of the *x*- and *y*-components is represented by *Q*; (*iii*) *U* is the crosscorrelation between $E_1$ and $E_2$ cos δ; (*iv*) *V* is the crosscorrelation between $E_1$ and $E_2$ sin δ. Fig. 3.6 shows these interpretations.

$$I \sim R(E_1) + R(E_2) \qquad U \sim r(E_1, E_2 \cos \delta)$$

$$Q \sim R(E_1) - R(E_2) \qquad V \sim r(E_1, E_2 \sin \delta)$$

Functions:   *R* - Auto-correlation,   *r* - Cross-correlation

FIGURE 3.6: Summarized expressions of the Stokes parameters.

Here, correlation does not exist between $E_{1eff}$ and $E_{2eff}$, although the magnitudes are equal ($S_x = S_y$). Hence, $\left\langle E_{1eff} E_{2eff} \cos \delta \right\rangle = 0$. Similarly, we have $\left\langle E_{1eff} E_{2eff} \sin \delta \right\rangle = 0$. Table 3.4 lists the Stokes parameters for completely polarized waves. Note that $Q = U = V = 0$, because there are no polarized components (linear or circular). However, $I > 0$ because when averaged over a period of time, both the *x*- and *y*-components are nonzero and real. Because imaginary components do not exist, $U = 0$. The *degree of polarization d* is defined as *the ratio of the completely polarized power to the total power,* as shown in Eq. (3.60).

$$d = \frac{\text{polarized power}}{\text{total power}} = \frac{\sqrt{Q^2 + U^2 + V^2}}{I}, \quad 0 \le d \le 1 \qquad (3.60)$$

**TABLE 3.4:** Stokes parameters for completely unpolarized waves.

| Stokes parameters | Values |
|:---:|:---:|
| $I$ | $S$ |
| $Q$ | 0 |
| $U$ | 0 |
| $V$ | 0 |

### 3.6.3 Partially Polarized Waves

Partially polarized waves may be treated as the sum of (*i*) completely polarized waves, and (*ii*) completely unpolarized waves. Unlike completely polarized waves, where $Q = U = V = 0$, for partially polarized waves at least one ($Q$ or $U$ or $V$) is nonzero. The flux density $S$ (Watt m$^{-2}$) obtained from the Poynting vector may thus be treated as being composed of the three flux densities: (*i*) $S_u$ (for unpolarized components), (*ii*) $S_{xp}$ (for components polarized along the $x$-axis), and (*iii*) $S_{yp}$ (components polarized along the $y$-axis). This is expressed in Eq. (3.61).

$$S = S_u + S_{xp} + S_{yp} \qquad (3.61)$$

Note that this is a generic equation capable of representing any type of polarization. The Stokes parameters are expressed as in Eqs. (3.62) through (3.65) as shown below.

$$I = S = S_u + S_{xp} + S_{yp} \qquad (3.62)$$

$$Q = S_{xp} - S_{yp} \qquad (3.63)$$

$$U = \left(S_{xp} - S_{yp}\right) \tan 2\tau \qquad (3.64)$$

$$V = \left(S_{xp} - S_{yp}\right) \tan 2\varepsilon \, \sec 2\tau \qquad (3.65)$$

## 3.7  NORMALIZED STOKES PARAMETERS

Eqs. (3.66) through (3.69) express the generic Stokes parameters (for partially polarized waves) in normalized form. These are obtained by dividing Eqs. (3.62) through (3.65) by $S$.

$$s_0 = \frac{I}{S} = 1 \tag{3.66}$$

$$s_1 = \frac{Q}{S} \tag{3.67}$$

$$s_2 = \frac{U}{S} \tag{3.68}$$

$$s_3 = \frac{V}{S} \tag{3.69}$$

These can be expressed as a matrix, as shown in Eq. (3.70), where the degree of polarization $d$ is given by Eq. (3.71).

$$\begin{bmatrix} I \\ Q \\ U \\ V \end{bmatrix} = S \begin{bmatrix} s_0 \\ s_1 \\ s_2 \\ s_3 \end{bmatrix} = S \begin{bmatrix} 1-d \\ 0 \\ 0 \\ 0 \end{bmatrix} + S \begin{bmatrix} d \\ d\cos 2\varepsilon \cos 2\tau \\ d\cos 2\varepsilon \sin 2\tau \\ d\sin 2\varepsilon \end{bmatrix} \tag{3.70}$$

$$d = \frac{\sqrt{Q^2 + U^2 + V^2}}{I} = \sqrt{s_1^2 + s_2^2 + s_3^2} \tag{3.71}$$

## 3.8   AN ALTERNATIVE APPROACH

So far, the Stokes parameters have been described in terms of correlations of linearly polarized components. From Eq. (3.59), $I$ may be also be visualized as the total power obtained along the $x$- and $y$-axes. We may also visualize $Q$ as the difference in the powers obtained along the $x$- and $y$-axes. Similarly, $U$ may be visualized as the difference between the components obtained at 45° and 135°. This is because $Q$ and $U$ contain orthogonal components of $2\tau$ (cos $2\tau$ and sin $2\tau$). Here, we express the Stokes parameters in terms of circularly polarized waves.

Let a wave be composed of two oppositely rotating field components: (*i*) $E_r$ (right circular), and (*ii*) $E_l$ (left circular). These are given by Eqs. (3.72) and (3.73).

$$E_r = E_R\, e^{j\omega t} \tag{3.72}$$

$$E_l = E_L\, e^{-j(\omega t + \delta')} \tag{3.73}$$

Eq. (3.74) expresses $AR$ in terms of $E_{Leff}$ and $E_{Reff}$, which are, respectively, the effective magnitudes of the left and right circular polarized components.

$$AR = \frac{E_{Leff} + E_{Reff}}{E_{Leff} - E_{Reff}} = \cot \epsilon \qquad (3.74)$$

When $E_L > E_R$, $AR$ is positive, which indicates that the polarization is left circular. Similarly, if $E_L < E_R$, $AR$ is negative, the polarization is right circular. Relating Eq. (3.74) with the Poincaré sphere, we get Eqs. (3.75) and (3.76), as shown below. We also obtain $\delta = 2\tau$.

$$\cos 2\epsilon = \frac{2 E_{Leff} E_{Reff}}{E_{Leff}^2 + E_{Reff}^2} = \frac{(AR)^2 - 1}{(AR)^2 + 1} \qquad (3.75)$$

$$\sin 2\epsilon = \frac{E_{Leff}^2 - E_{Reff}^2}{E_{Leff}^2 + E_{Reff}^2} = \frac{2(AR)}{(AR)^2 + 1} \qquad (3.76)$$

The Stokes parameters are now be expressed in terms of the circular polarized components. These are given in Eqs. (3.77) through (3.80).

$$I = \frac{\langle E_{Leff}^2 \rangle}{Z_0} + \frac{\langle E_{Reff}^2 \rangle}{Z_0} = S_L + S_R = S \qquad (3.77)$$

$$Q = 2 \frac{\langle E_{Leff} E_{Reff} \cos \delta' \rangle}{Z_0} \qquad (3.78)$$

$$U = 2 \frac{\langle E_{Leff} E_{Reff} \sin \delta' \rangle}{Z_0} \qquad (3.79)$$

$$V = \frac{\langle E_{Leff}^2 \rangle}{Z_0} - \frac{\langle E_{Reff}^2 \rangle}{Z_0} = S_L - S_R \qquad (3.80)$$

The above equations can be expressed in matrix form, as shown in Eq. (3.81).

$$\begin{bmatrix} I \\ Q \\ U \\ V \end{bmatrix} = \begin{bmatrix} \dfrac{\langle E_{Leff}^2 \rangle}{Z_0} + \dfrac{\langle E_{Reff}^2 \rangle}{Z_0} \\[2mm] 2\dfrac{\langle E_{Leff} E_{Reff} \cos \delta' \rangle}{Z_0} \\[2mm] 2\dfrac{\langle E_{Leff} E_{Reff} \sin \delta' \rangle}{Z_0} \\[2mm] \dfrac{\langle E_{Leff}^2 \rangle}{Z_0} - \dfrac{\langle E_{Reff}^2 \rangle}{Z_0} \end{bmatrix} = \begin{bmatrix} S_L + S_R \\[2mm] 2\dfrac{\langle E_{Leff} E_{Reff} \cos 2\tau \rangle}{Z_0} \\[2mm] 2\dfrac{\langle E_{Leff} E_{Reff} \sin 2\tau \rangle}{Z_0} \\[2mm] S_L - S_R \end{bmatrix} \qquad (3.81)$$

Hence, $I$ may be seen as the power sum of two opposite circular polarized waves, $V$ is seen as the power difference of two opposite circular polarized components, and $Q$ and $V$ can be seen as the cross-correlation of two opposite circular polarized components.

Recall the values of the Stokes parameters for a completely polarized wave, which were listed in Table 3.3. They match the current equations. The Stokes parameters for a completely unpolarized wave are listed in Table 3.4. The recently derived values of $I$ and $V$ also match these results. Here, $Q$ and $U$ are zero because they form like cross-correlations between the right and left circular components.

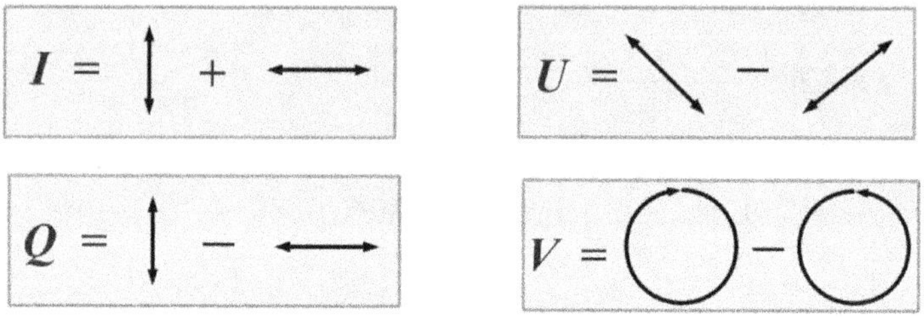

**FIGURE 3.7:** Symbolic representation of the Stokes parameters.

Comparing Eqs. (3.59) and (3.81), we interpret the Stokes parameters as: (*i*) $I$ is the total power (sum of the $x$- and $y$-components), (*ii*) $Q$ is the difference of the $x$- and $y$-components, (*iii*) $U$ is the difference of time-averaged components obtained at 45° and 135°, and (*iv*) $V$ is the difference between the two opposite circular polarized components. These are shown in Fig. 3.7.

## 3.9 STOKES PARAMETERS AND ANTENNA APERTURE

We relate the Stokes parameters in matrix form with the aperture of an antenna pair receiving arbitrary polarized waves. Assume that both antennas in the pair have the same effective aperture area $A_e$. If they are linearly polarized, assume their polarization directions are perpendicular to each other. If they are circularly polarized, assume the two to be oppositely polarized (right and left circular). The effective aperture area $A_e$ of the pair have four components, which are expressed in a matrix form in Eq. (3.82), where the $a_i$ represent the Stokes parameters.

$$A_e\left[a_i\right] = A_e \begin{bmatrix} a_0 \\ a_1 \\ a_2 \\ a_3 \end{bmatrix} \qquad (3.82)$$

### 3.9.1 Transmitting Mode

When the antennas are transmitting, $a_i$ is given by Eqs. (3.83) through (3.86), where $S_t$ is the transmitted power, $E_{1t}$ and $E_{2t}$ are the effective values of the electric fields, and $\delta$ is the phase difference between them.

$$a_0 = 1 \tag{3.83}$$

$$a_1 = \frac{1}{Z\,S_t}\left(E_{1t}^2 - E_{2t}^2\right) \tag{3.84}$$

$$a_2 = \frac{2}{Z\,S_t}\left(E_{1t}\,E_{2t}\cos\delta\right) \tag{3.85}$$

$$a_3 = \frac{2}{Z\,S_t}\left(E_{1t}\,E_{2t}\sin\delta\right) \tag{3.86}$$

### 3.9.2 Receiving Mode

When the antennas are receiving, and a wave of polarization $S[s_i]$ is incident, the power $W$ available from the antennas is given by Eq. (3.87).

$$W = \frac{1}{2}S\,A_e\,[a_i]^T\,[s_i] = \frac{1}{2}S\,A_e\,[a_0 \quad a_1 \quad a_2 \quad a_3]\begin{bmatrix} s_0 \\ s_1 \\ s_2 \\ s_3 \end{bmatrix} = \frac{1}{2}S\,A_e\sum_{i=0}^{3} a_i\,s_i \tag{3.87}$$

## 3.10   STOKES PARAMETERS FOR INTERFEROMETERS

Modern radio telescopes use multiple interferometers consisting of many antennas. Using the correlations between the signals received by the antennas, a technique called *aperture synthesis* is used for imaging the radio source. An interferometer may be described as two antennas separated by a distance $d$ and tracking a radio source whose output signals are multiplied in phase, followed by integration over time, and then saved as *visibility* data (see Chapter 5). Let $v_1(t)$ and $v_2(t)$ be the antenna signal voltages after phase correction. The output of the interferometer will be $\langle v_1(t)v_2(t)\rangle$. In order to obtain complete information about the source, the radio telescope receives both polarizations using dual polarized antenna feeds at the foci of the dishes. Thus four correlated outputs (two self and two cross) are produced instead of just one. The polarization of the antenna feeds can be linear or circular. Let both antenna gains be identical. If the antenna feeds are linearly polarized, the correlations in relation to the Stokes parameters will be as shown in Eqs. (3.88) through (3.91), where $X$ and $Y$ are the two polarized output voltages from each antenna in the equatorial reference frame of the source, $Z_{TL}$ is the characteristic impedance of the transmission line matched to the antenna, and $A_e$ is the effective aperture area of each antenna. Both antennas are assumed to be identical in all their characteristics. Note that the self products $XX$ and $YY$ represent, respectively, the cross-correlations between the $x$-component of antenna 1 with the $x$-component of antenna 2 and that of the $y$-component of antenna 1 with the $y$-component of antenna 2. The cross products $XY$ and $YX$ represent, respectively, the correlations between the $x$-component of antenna 1 with the $y$-component of antenna 2 and the $y$-component of antenna 1 with the $x$-component of antenna 2.

$$XX\left[\frac{1}{Z_{TL}\,A_e}\right] = I + Q \tag{3.88}$$

$$YY\left[\frac{1}{Z_{TL}\,A_e}\right] = I - Q \tag{3.89}$$

$$XY\left[\frac{1}{Z_{TL}\,A_e}\right] = U + jV \tag{3.90}$$

$$YX\left[\frac{1}{Z_{TL}\,A_e}\right] = U - jV \tag{3.91}$$

We shall now consider interferometers using dual circularly polarized antenna feeds. If $R$ and $L$ represent, respectively, the right and left circular polarized voltages obtained from the antennas, their relationship with the Stokes parameters can be as shown in Eqs. (3.92) through (3.95). The self- and cross-correlation conventions are the same as before.

$$RR\left[\frac{1}{Z_{TL}\,A_e}\right] = I + V \tag{3.92}$$

$$LL\left[\frac{1}{Z_{TL}\,A_e}\right] = I - V \tag{3.93}$$

$$RL\left[\frac{1}{Z_{TL}\,A_e}\right] = Q + jU \tag{3.94}$$

$$LR\left[\frac{1}{Z_{TL}\,A_e}\right] = Q - jU \tag{3.95}$$

In the above equations, we have assumed the antennas are ideal. In practice, antenna feeds are not perfectly polarized. Hence, an unpolarized source appears as polarized when observed using an antenna feed. Furthermore, while tracking the source, the antenna feeds rotate relative to the equatorial frame of the source. Thus, the above equations have to be modified accordingly before usage. A polarization calibration is generally done to remove both of these effects. After calibration, only the source polarization information remains in the data.

**The South Pole Telescope**

**FIGURE 3.8** The South Pole Telescope. Image Credit: Keith Vanderlinde, National Science Foundation

The South Pole Telescope (SPT) shown in Figure 3.8 is a 10-meter diameter offset Gregorian telescope located at the Amundsen-Scott South Pole station in Antarctica. The telescope has been operational since 2007. Its position on the Earth's rotational axis enables observations over extended time durations and with exceptional atmospheric transparency and stability. Research collaborations at the SPT involve over ten institutions, with primary funding provided by the U.S. National Science Foundation (NSF).

Current high-resolution images of the cosmic microwave background (CMB) are obtained at millimeter wavelengths using the SPT 966-pixel bolometer camera. Large-scale surveys have been conducted at 1.3, 2, and 3 mm, searching for galaxy clusters by measuring the Sunyaev–Zel'dovich (SZ) effect. The SZ effect, also known as the inverse Compton effect, results from the scattering of low-energy CMB photons with high-energy electrons in galaxy clusters. This effect was first measured from clusters of galaxies in 1983. The first discovery of galaxy clusters by the SZ effect was conducted at the SPT in 2008.

One goal of the SPT is to conduct measurements of small-scale anisotropies in the CMB as well as the B-mode polarization signature. B-mode polarization has curl similar to that of a magnetic field (hence the name B-mode). E-mode polarization diverges like an electric field. Proposed measurements of the B-mode CMB anisotropy will serve as a probe of the gravitational wave background that was produced during inflation of the early universe. Because the B-mode polarization is also generated by gravitational lensing of the E-mode, measurements over different angular scales will be necessary to separate the contributions of each polarization component.

## REVIEW QUESTIONS

1. What is meant by polarization of an electromagnetic wave? Describe the terms: (*i*) linear polarization, (*ii*) circular polarization, and (*iii*) elliptical polarization.

2. Describe an unpolarized wave.

3. Explain the polarization ellipse, using a suitable diagram.

[***Hint***: Use Fig. 3.2]

4. If $E_1$ and $E_2$, respectively, are the peak magnitudes of the *x*- and *y*-components of the electric field, the angular frequency is $\omega$, $\delta$ is the time-phase angle by which $E_y$ leads over $E_x$, and $\hat{x}$ and $\hat{y}$ are, respectively, the unit vectors along the *x* and *y* directions, find the expression for the total electric field $\vec{E}$.

[***Hint***: see Eq. (3.5)]

5. Using the same symbolic representation described in question 4, prove that

$$aE_x^2 + bE_xE_y + cE_y^2 = 1 \text{ where } a = \frac{1}{E_1^2 \sin^2 \delta}, \quad b = \frac{1}{E_1E_2 \sin^2 \delta} \quad c = \frac{1}{E_2^2 \sin^2 \delta}.$$

[***Hint***: see Eq. (3.8)]

6. What is meant by the axial ratio of a polarization ellipse?

7. Explain the Poincaré Sphere. How is it related to the polarization ellipse?

8. Describe in detail the four Stokes parameters. Tabulate the values of the Stokes parameters for (*i*) linear, and (*ii*) circular polarization.

9. How many cross-correlations are formed between two antennas having two orthogonally polarized outputs?

10. Relate the cross-correlations of problem 9 with the Stokes parameters when the polarization is (*i*) linear, and (*ii*) circular.

[***Hint***: See Eqs. (3.88) through (3.95)]

# DESIGNING SINGLE DISHES AND PHASED ARRAYS

## 4.1  INTRODUCTION

We now look at the aspects of designing single-dish radio telescopes and phased arrays. We first learn about filled-aperture radio telescopes, beginning with the angular resolution of a single-dish radio telescope and its operation and design, and then move on to aperture array radio telescopes.

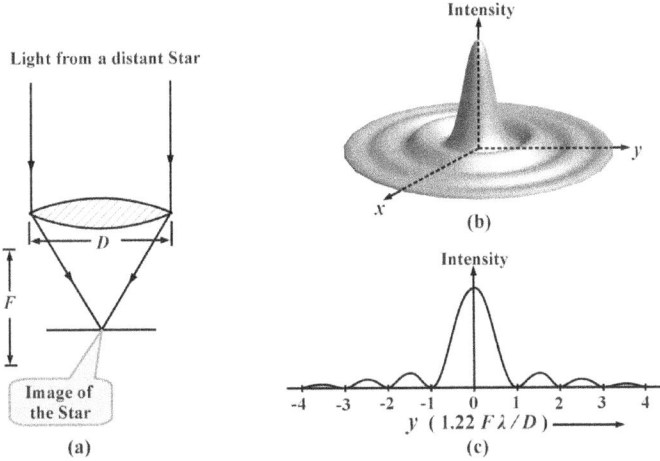

**FIGURE 4.1:** (a) Monochromatic light from a distant star is focused using a convex lens. (b) Intensity distribution of the image on the screen. (c) Intensity distribution of the image along one axis.

## 4.2 RAYLEIGH CRITERION: RESOLVING POWER

The Rayleigh criterion describes the angular resolving power of aperture tele-scopes. Fig. 4.1a shows a simple optical convex lens used to form an image of a distant star. It has a diameter $D$ and a focal length $F$. Let the star emit at a single wavelength $\lambda$ (monochromatic). The diffraction pattern of the lens results in an im-age of the star that has a variation in intensity (Fig. 4.1b and 4.1c). Because a star is like a point source, this intensity distribution of the image is known as the *point spread function*. If another star exists at an angle $\alpha'$, as shown in Fig. 4.2a, the im-age will be formed from the superimposition of two such patterns. This is shown in Fig. 4.2b. The dotted lines show the superposed images. The two images can be distinctly resolved if the maxima of one image lies over the first minima of the other. This condition occurs if $\alpha' = 1.22\lambda/D$ (radians) and is known as the *Rayleigh criterion*. If $\alpha'$ is reduced, the two images tend to overlap more, and the two point sources are difficult to identify. On the other hand, if $\alpha'$ is increased, the point sources appear more and more distinct on the image. If the star emits at more than one frequency, more images will be formed corresponding to each wavelength, and the composite image will become blurred. Hence, the Rayleigh criterion is valid for only a single wavelength. The resolution of any circular aperture tele-scope is limited by the Rayleigh criterion.

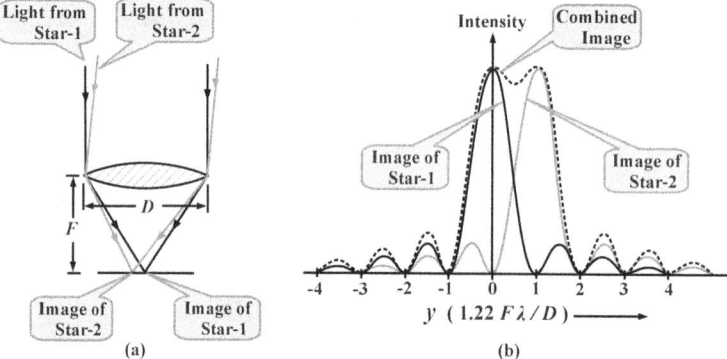

**FIGURE 4.2:** Monochromatic light from two distant point stars forming an image. (a) Instrumental arrangement. (b) Intensity distribution due to overlapping of the two images.

### 4.2.1 Application to Single-Dish Radio Telescopes

The Rayleigh criterion can also be applied to dish antennas. For this situation, it may be stated as *two point sources can be resolved using a single-dish telescope if the angular separation between the two objects is at least equal to or greater than*

1.22 $\lambda/D$. This is expressed in Eq. (4.1), where $D$ is the aperture diameter of the telescope, $\lambda$ is the wavelength, and $\alpha'$ is the minimum angle of resolution between the two point sources (radians). Note that $D$ and $\lambda$ should have the same units.

$$\alpha' = 1.22 \frac{\lambda}{D} \approx \frac{\lambda}{D} \qquad (4.1)$$

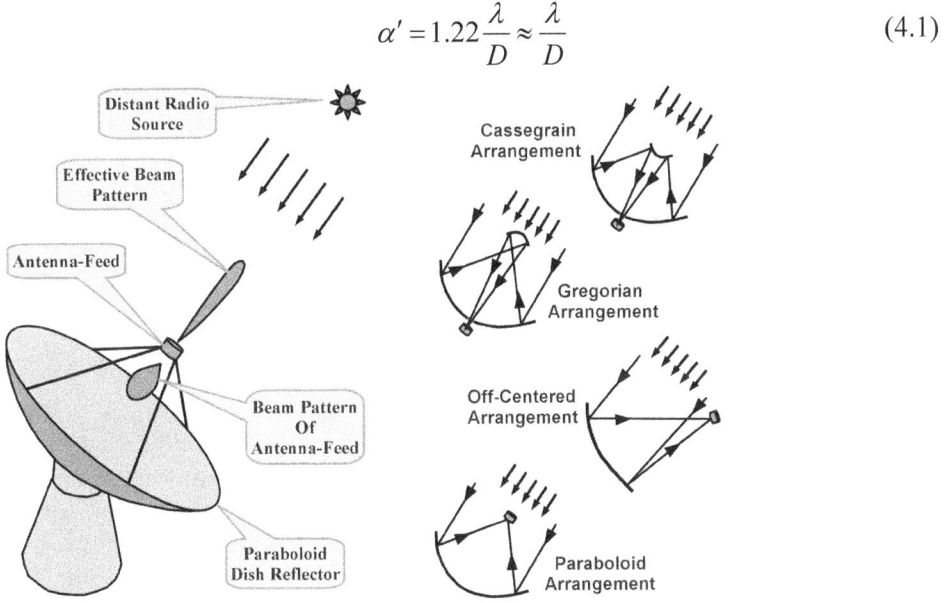

**FIGURE 4.3:** Common radio telescope reflector and antenna configurations.

## 4.3   SINGLE-DISH TELESCOPE ANTENNAS

Single-dish antennas fall in the category of reflector antennas. Broadly speaking, these antennas consist of two parts: (*i*) a reflector to focus the beam to a point, and (*ii*) an antenna feed placed at the focus that collects the signal. Fig. 4.3 shows several type of arrangements for collecting the signal. Examples include the Cassegrain, Gregorian, off-centered, or simple parabolic arrangements. The antenna beam pattern appears effectively like a pencil beam (high gain and reduced side lobes). A parabolic dish antenna uses two coordinates to point to a source, namely (*i*) altitude, and (*ii*) azimuth. These are determined with the help of a computer, using the instantaneous equatorial coordinates of any radio source under observation (see Appendix A.11.2). This information is fed to the azimuth and elevation controls of the dish. Hence the dish keeps tracking the moving source across the sky. Each antenna feed gives two outputs for two cross polarizations. These outputs are amplified using two LNA and fed to the receiving chain, which further processes the signals for final use.

### 4.3.1 Aperture Illumination of Dish Antennas

The cross-sectional area bounded by the edge of a dish is known as its *aperture area*. Theoretically, this should be equal to the effective aperture area of the antenna $A_e$. However, $A_e$ is always less than the dish cross-sectional area. This is because of diffraction, obscuring of the aperture by the shadow of the antenna feed and its mechanical supports, and also the nonuniform pattern of the antenna feed. Due to the large collecting area of the dish, the overall antenna gain (dish plus antenna feed) is enhanced. The antenna directivity depends on how large the dish aperture area can be for a given wavelength $\lambda$. Because of this reason, the directivity of a dish increases at higher operating frequencies. Fig. 4.4a shows the ideal dish illumination. Here, the pattern of the antenna feed uniformly illuminates the dish area, and no power is lost in any other direction. Unfortunately, due to discontinuity of the field at the edges of the dish, significant side lobes are generated in the effective beam pattern (dish plus antenna feed). Fig. 4.4b shows the dish illumination by a practical antenna feed.

Practical antenna-feed patterns always contain minor lobes and a back lobe. These are not useful for illuminating the dish. They also pick up unwanted signals from other directions. The antenna feed does not illuminate the dish uniformly, resulting in inefficient use of the dish aperture. As shown in Fig. 4.4b, the central zone (dish) is illuminated with the largest power, and power decreases gradually towards the edges. This may be termed *inefficient dish illumination*. What little power falls out of the dish is called as *spillover*. With the dish pointed to the zenith, the spillover and some side lobes gather ground temperature and contribute to the antenna temperature. The rest of the side lobes, along with the back lobe, pick up sky signals. They also contribute to the antenna temperature. Practical dishes are not perfect reflectors, for they may be built using a metallic mesh to reduce cost and weight. Some signals from the ground penetrate the mesh and contribute to antenna temperature. To uniformly illuminate the dish, if the beam width of antenna feed is broadened, then spillover increases, and the dish is said to be *over illuminated*. On the other hand, if the beam width of the antenna feed is reduced, the effective usage of the dish aperture is reduced, and dish is said to be *under illuminated*. The phase of waves reaching the focal point is affected by the irregularity of the dish surface. Some power is lost depending on the surface variations. A certain amount of power is diffracted in unwanted directions when the surface is rough.

To summarize, there are six important factors that affect the performance of a dish antenna system. These are: (*i*) efficiency $\eta_a$ of the antenna feed (see Appendix B.8), (*ii*) efficiency of the spillover $\eta_{sp}$, (*iii*) efficiency of the dish leakage $\eta_{\mathrm{msh}}$,

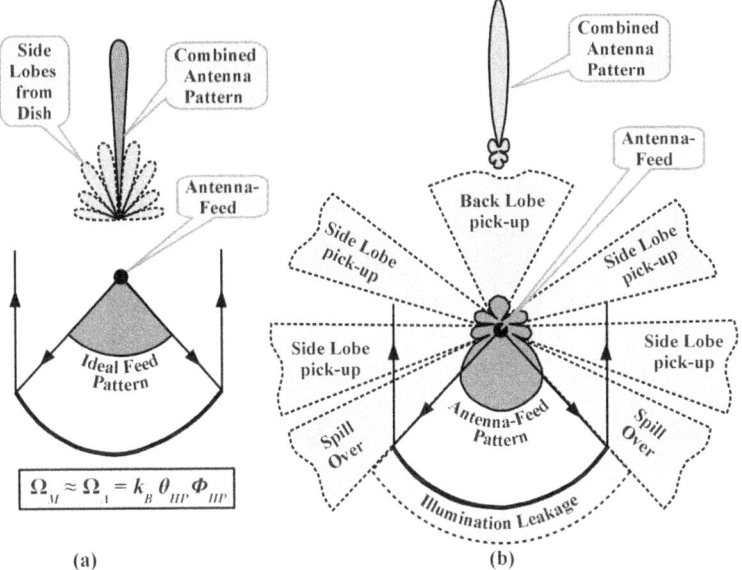

(a)

(b)

**FIGURE 4.4:** Comparison between ideal and practical dish illumination. (a) An ideal antenna feed uniformly illuminates the dish. Due to discontinuity at the dish edges, the side lobes are high. (b) A practical antenna feed produces nonuniform illumination. Side lobes are lower, but spillover may not be negligible.

which depends on the quality of the mesh, (*iv*) efficiency of the surface smoothness $\eta_{rms}$, which depends on the root-mean-square variations on the surface of the dish, (*v*) efficiency of the illumination $\eta_{ill}$, which depends on the disribution of the illumination intensity across the dish (produced by the antenna feed), and (*vi*) efficiency of the polarization $\eta_{pol}$. Considering the antenna in radiating mode, the ratio of the power falling on the dish surface to the total power radiated by the antenna feed is the spillover efficiency ($\eta_{sp}$). In the radiating mode, the ratio of the power reflected by the dish (mesh) to the power illuminating the dish (mesh) is the dish leakage efficiency ($\eta_{msh}$). For a leak-proof dish, the ratio of actual power available at the focus to the theoretical power that would be available at the focus if a plane wave front illuminated the dish surface is the smoothness efficiency ($\eta_{rms}$). In the radiating mode, the ratio of the total power illuminating the dish to the total power required for illuminating the dish uniformly at a peak intensity (of the antenna-feed pattern) is considered here as the illumination efficiency ($\eta_{ill}$). For completely polarized waves incident on the dish, the polarization efficiency ($\eta_{pol}$) is defined as the ratio of power actually received to the maximum power that could be received if the antenna had optimum polarization for the received wave. The maximum value of any efficiency is unity.

### 4.3.2 Gain and Aperture-Area Analysis

Consider the dish geometry shown in Fig. 4.5. We define $K_{FD}$ as the ratio of the focal length $F$ to the dish diameter $D$. Based on the dish illumination or from the beam pattern of the antenna feed, the value of $K_{FD}$ is selected. Generally, for most practical designs, $K_{FD}$ is kept near to 0.4, because antenna feeds are designed accordingly. The dish diameter $D$ is given by Eq. (4.2). The depth $d$ of the dish is obtained from $D$ using Eq. (4.3). The angle of the extent of illumination $\beta_d$ of the dish is obtained from Eq. (4.4). As said before, in practice the effective aperture area $A_e$ is always less than the actual cross-section area of the dish by a factor $\eta_A$, known as the *aperture efficiency*. It is expressed in Eq. (4.5). Eq. (4.6) expresses $\eta_A$ in terms of the several other efficiencies described earlier. The effective gain $G_{eff}$ of the antenna system (dish plus antenna feed) is dependent on $\eta_A$, as shown in Eq. (4.7). From $G_{eff}$ we obtain the effective half-power beam-width $\theta_{effHP}$, as in Eq. (4.8).

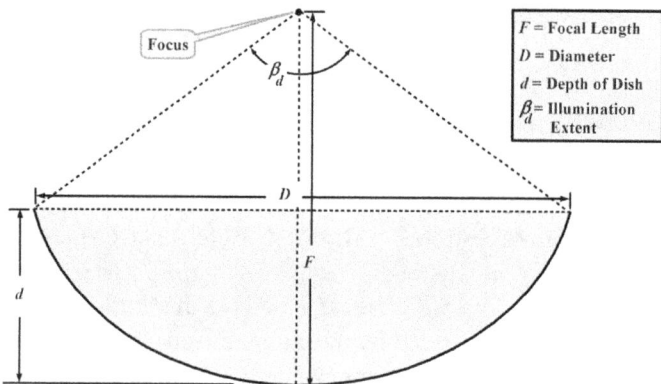

**FIGURE 4.5:** The basic geometry of a dish antenna.

$$D = \frac{F}{K_{FD}} \tag{4.2}$$

$$d = \frac{D^2}{16F} \tag{4.3}$$

$$\beta_d = 2tan^{-1}\left(\frac{D/2}{F-d}\right) \tag{4.4}$$

$$A_e = \eta_A\left(\frac{\pi D^2}{4}\right) \quad \text{where, } \eta_A < 1 \tag{4.5}$$

$$\eta_A = \eta_a\, \eta_{sp}\, \eta_{msh}\, \eta_{rms}\, \eta_{ill}\, \eta_{pol} \tag{4.6}$$

$$G_{\text{eff}} = \eta_A \left( \frac{\pi D}{\lambda} \right)^2 \tag{4.7}$$

$$\theta_{\text{effHP}} = \sqrt{\frac{40000}{G_{\text{eff}}}} \tag{4.8}$$

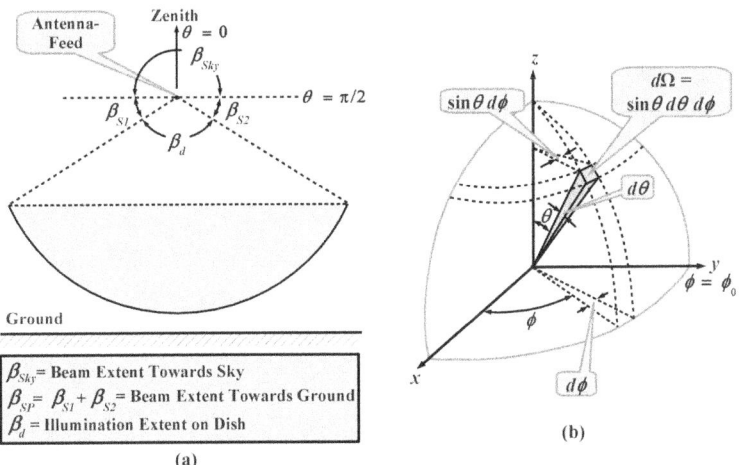

$\beta_{Sky}$ = Beam Extent Towards Sky
$\beta_{SP}$ = $\beta_{S1}$ + $\beta_{S2}$ = Beam Extent Towards Ground
$\beta_d$ = Illumination Extent on Dish

(a)

**FIGURE 4.6:** (a) A dish pointed to the zenith having an antenna feed at its focus. The angular extents indicate the beam pattern distribution from the antenna feed. (b) The three-dimensional coordinate system used for analysis.

### 4.3.3 Antenna Temperature Analysis

Fig. 4.6a shows a paraboloid dish pointed to a radio source at the zenith. Let the source be randomly polarized. Assume the polarization efficiency $\eta_{\text{pol}} = 1$. The antenna temperature receives contribution from: (*i*) source, (*ii*) spillover, (*iii*) sky, and (*iv*) dish leakage. The angles $\beta_d$, $\beta_{SP} = \beta_{S1} + \beta_{S2}$, and $\beta_{Sky}$ represent, respectively, the feed pattern distribution angles towards the dish, ground spillover, and the sky. Assuming the beam pattern is identical for both polarizations, the beam solid angle $\Omega_{\text{Dish}}$ of the antenna system is expressed in Eq. (4.9) as shown below.

$$\Omega_{\text{Dish}} \approx \left( \theta_{\text{effHP}} \right)^2 \tag{4.9}$$

Consider the coordinate system shown in Fig. 4.6b, where $\phi_0$ represents the plane of this page. Let $P_n(\theta, \phi_0)$ represent the normalized power pattern of the antenna feed on the plane $\phi = \phi_0$. Note that $P_n$ is a maximum at $\theta = \pi/2$. Between any two arbitrary values of $\theta$ (that is, $\theta_1$ and $\theta_2$), the effective solid angle $\Omega_{12}$ subtended by the beam is given by Eq. (4.10).

$$\Omega_{12} = \int_0^{2\pi} \int_{\theta_1}^{\theta_2} P_n(\theta, \phi) \sin \theta \, d\theta \, d\phi \tag{4.10}$$

Because $\theta$ lies between 0 and $\pi/2$, we have, $0 \le \theta_1 \le \pi/2$, $0 \le \theta_2 \le \pi/2$. We also have $\theta_1 \le \theta_2$. From our assumption of cylindrical symmetry around the $z$-axis, Eq. (4.10) can be simplified, as shown in Eq. (4.11).

$$\Omega_{12} = 2\pi \int_{\theta_1}^{\theta_2} P_n(\theta, \phi) \sin\theta \, d\theta \tag{4.11}$$

Using Eq. 11, the effective portion of the beam solid angle $\Omega_{\text{Sky}}$ of the antenna feed looking towards the sky (side and back lobes) can be written as in Eq. (4.12). In a similar manner, the effective beam solid angle $\Omega_d$ illuminating the dish is written as in Eq. (4.13). The effective beam solid angle $\Omega_{SP}$ responsible for spillover is given in Eq. (4.14).

$$\Omega_{Sky} \approx 2\pi \left[ \int_0^{\frac{\pi}{2}} P_n(\theta, \phi_0) \sin\theta \, d\theta \right] \tag{4.12}$$

$$\Omega_d \approx 2\pi \left[ \int_{\frac{\pi}{2} + \frac{\beta_{SP}}{2}}^{\pi} P_n(\theta, \phi_0) \sin\theta \, d\theta \right], \quad \text{where } \beta_{S1} = \beta_{S2} = \frac{\beta_{SP}}{2} \tag{4.13}$$

$$\Omega_{SP} \approx 2\pi \left[ \int_{\frac{\pi}{2}}^{\frac{\pi}{2} + \frac{\beta_{SP}}{2}} P_n(\theta, \phi_0) \sin\theta \, d\theta \right] \tag{4.14}$$

The temperature contribution from the dish alone is given in Eq. (4.15), where $\Omega_s$ is the beam solid angle occupied by the source, and $T_{\text{Src}}$ is mean source temperature across $\Omega_s$.

$$T_{\text{Dish}} = \eta_{\text{msh}} \eta_{\text{rms}} \Omega_s \overline{T}_{\text{Src}} + \left( \Omega_{\text{Dish}} - \Omega_s \right) T_{\text{Sky}} \tag{4.15}$$

The complete antenna temperature $T_A$ is given by Eq. (4.16), where $\eta_a$ is the antenna efficiency of the antenna feed, $T_{\text{SkyPickup}}$ is the temperature contribution from sky, $T_{\text{GndPickup}}$ is the temperature contribution from the ground due to spillover, and $T_{\text{GndLeakage}}$ is the temperature contribution from the ground due to mesh leakage.

$$\left. \begin{array}{l} T_A = T_{\text{Dish}} + \eta_a \left[ T_{\text{SkyPickup}} + T_{\text{GndPickup}} + T_{\text{GndLeakage}} \right] \\[4pt] \text{where,} \quad T_{\text{SkyPickup}} = \Omega_{\text{Sky}} T_{\text{Sky}} \\[4pt] T_{\text{GndPickup}} = \Omega_{SP} T_{\text{Gnd}} \\[4pt] T_{\text{GndLeakage}} = \Omega_d T_{\text{Gnd}} (1 - \eta_{\text{msh}}) \end{array} \right\} \tag{4.16}$$

We have now analyzed the antenna temperature with the dish pointing towards the zenith. In actuality, the antenna position keeps changing because it tracks the radio source. Thus, it is difficult to analytically estimate the antenna temperature. Hence these are estimated from measurements.

### 4.3.4 Reduced-Aperture Illumination

The coordinates of the dish keep changing with time as it tracks the source. Effectively, changes occur in $\Omega_{Sky}$ and $\Omega_{SP}$. As illustrated in Fig. 4.4b, although the beam intensity of the antenna feed is less across spillover regions, it is reasonably larger than the side lobes, especially at the dish edges. The spillover can be reduced by increasing the antenna-feed gain (reduced aperture illumination). Under these conditions, the unwanted temperature contributions will be from the side and back lobes. If the side and back lobes are uniformly distributed, the unwanted temperature contributions will remain more or less the same. This is because, as the antenna moves away from the zenith, the ground contribution reduces on one side and increases on the other side. However, reduced dish illumination reduces aperture efficiency $\eta_A$. For any angle of inclination, the above equations hold reasonably well provided the dish leakage efficiency $\eta_{msh}$ (also known as *mesh efficiency*) is close to unity.

## 4.4 ANTENNA FEEDS

Several types of antenna feed are used in radio telescopes. The properties like gain, bandwidth, efficiency, side lobes, etc., vary depending on the design. We start here with a simple dipole feed for grasping the concepts, then we will study some ultrawideband feeds, because they look more promising for current and future technology.

### 4.4.1 Dipole-Cross Antenna Feed

As discussed in Chapter 3, the polarization of both components is essential in radio astronomy. An antenna feed receiving only one polarized component can provide only partial information about the source. Moreover, it will convert only half of the source brightness into power (Section 3.8). In general, dual polarized antenna feeds are used in order to sense both the polarized components of the electric field.

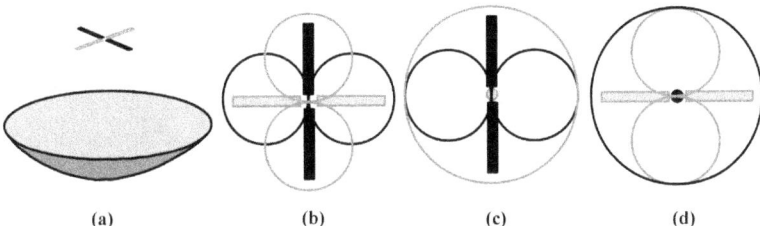

| (a) | (b) | (c) | (d) |

**FIGURE 4.7:** (a) A dipole-cross is mounted at the focus of a dish. (b) Radiation pattern of the dipole-cross as observed from the top. (c) Pattern of the cross as observed along the gray dipole. (d) Pattern of the cross as observed along the black dipole.

Fig. 4.7 shows a dipole-cross, which is the simplest dual polarized antenna feed. It has large back lobes and doesn't utilize the dish efficiently. For understanding a simplified case, we consider two half-wave crossed dipoles are mounted at the focus of a dish reflector. This is shown in Fig. 4.7a. To increase the bandwidth, the dipole thickness may be increased within a limited extent. Increasing it beyond a quarter wavelength will dramatically change the radiation pattern. The maximum bandwidth achieved from a half-wave dipole is roughly 5% of the center frequency. The typical directivity of a dipole antenna is about 2.15 dBi, and its radiation resistance is around 73 ohms. Fig. 4.7 shows the radiation pattern of a dipole-cross from various angles.

As discussed previously, dipole feeds are inefficient in several ways: (*i*) narrow bandwidth, (*ii*) large back lobes, and (*iii*) poor directivity. The back lobe can be reduced using a reflector behind the cross. Antennas with much higher directivities can replace the dipole feed. Examples are log-periodic antennas, reflector-based bow-tie antennas, horn antennas, helical antennas, etc. For astronomical purposes, all of these should be dual polarized.

### 4.4.2 Log-Periodic Antenna Feeds

Log-periodic antennas can be broadly divided into two groups: (*i*) log-periodic dipole arrays, and (*ii*) log-periodic antennas. These can be further subclassified into (*i*) planar, and (*ii*) nonplanar, depending on their structures. Some of them are described below.

#### 4.4.2.1 Nonplanar Log-Periodic Antenna Feeds

Fig. 4.8 shows the design and assembly of a log-periodic antenna feed built on a PCB (printed circuit board). The isometric view of the antenna is shown in Fig. 4.8a. Actually, there are two linearly polarized antennas. Four triangular PCBs are joined together. The zigzag metallic surface forms the antenna. Fig. 4.8 shows the top view. The two metallic pairs 1-1' and 2-2' form two cross-polarized linear antennas. The angle $\gamma$ between any two structures of an antenna is shown in Fig. 4.8c. The geometry of this antenna design is shown in Fig. 4.8d. The directivity depends on $\alpha$ and $\gamma$. The major lobe is directed upwards. While mounting, the apex of the antenna feed is pointed towards the dish for maximum illumination.

A log-periodic antenna essentially consists of a number of half-wave dipoles, each covering a different range of frequencies, thereby increasing the overall bandwidth. All dipoles are connected in phase reversed order to a common transmission line whose impedance also depends on $\beta$ and $\gamma$. The center frequency of any dipole is related to its length, as shown in Eq. (4.17), where $v_n$

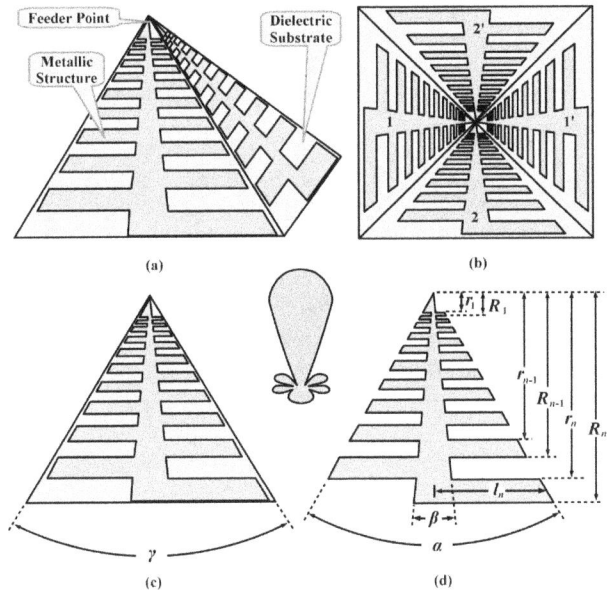

**FIGURE 4.8:** (a) Isometric view of the antenna feed. (b) Top view. (c) Angle between two structures. (d) Design dimensions.

is center frequency of the $n^{\text{th}}$ dipole, $l_n$ is mean length of a half dipole, *and c* is the velocity of light. The smallest and largest dipoles operate, respectively, at the highest and lowest frequencies of the antenna. The ratio of edge distances $\tau$ between any two adjacent dipoles is given by Eq. (4.18), where $r_n$, $r_{n-1}$... represents the top-edge distances from the feed point and $R_n$, $R_{n-1}$,... are the bottom-edge distances. A dipole length is directly proportional to the design wavelength and also inversely proportional to the frequency. Hence, Eq. (4.18) can be rewritten as Eq. (4.19). The number of dipoles $n$ is decided by $v_1$ and $v_n$, which are, respectively, the maximum and minimum frequencies. Eq. (4.20) shows this. Note that any of the dipole's dimensions can be found by proportionally scaling the dimension of a standard dipole with its distance from the apex. The distance between the dipoles increases logarithmically. Also, the electrical and radiation characteristics replicate themselves as a logarithm function of frequency. Hence it is known as a *log-periodic antenna*.

$$v_n = \frac{c}{4l_n} \tag{4.17}$$

$$\tau = \frac{R_{n-1}}{R_n} = \frac{r_{n-1}}{r_n} = \frac{l_{n-1}}{l_n} < 1 \tag{4.18}$$

$$\tau = \frac{v_2}{v_1} = \frac{v_3}{v_2} = \cdots < 1 \quad \text{where,} \quad v_1 > v_2 > v_3 \ldots \tag{4.19}$$

$$\tau^{n-1} = \frac{v_2}{v_1} \times \frac{v_3}{v_2} \times \ldots \times \frac{v_4}{v_3} \times \frac{v_n}{v_{n-1}} = \frac{v_n}{v_1} \tag{4.20}$$

Note that $\alpha \approx \gamma$. The antenna patterns are almost frequency independent for $30°$ $< \gamma < 60°$. The antennas are polarized along the dipoles. The radial currents along the transmission lines generate a small cross-polarized component, typically 18 dB below the main polarized component. The antenna impedance varies from 70 Ohms for $\gamma = 30°$ to nearly 180 Ohms for $\gamma = 180°$.

Fig. 4.9 shows photographs of two nonplanar cross-polarized log-periodic antenna feeds designed by the radio astronomy lab at the University of California, Berkeley. They are to be integrated with cryogenically cooled microwave amplifiers and used in radio astronomy.

### 4.4.2.2 Planar Log-Periodic Antenna Feeds

Fig. 4.10 shows the structure of dual-polarized planar log-periodic antenna feeds. Two designs are possible based on the choice of the reflector: (*i*) *step-lane* reflector (thick lines), and (*ii*) *pyramidal* reflector (dotted lines). The secondary reflector is optional, and it improves the gain near the lowest frequencies. The antenna may be fabricated on a PCB, or a zigzag metallic sheet may be used. Mechanical support to the sheet may be provided by a thin lossless dielectric sheet. To reduce air

**FIGURE 4.9:** Antenna-feed prototypes for integration with cryogenic amplifier. Courtesy: University of California, Radio Astronomy Lab, Berkeley.

resistance, slots can be made in the dielectric (between the dipoles). If a step-lane reflector is chosen, its step widths should be same as the dipole widths. For any particular frequency, the $z$-axis distance between a dipole and the reflector is a quarter of the dipole wavelength. For any plane wave front approaching perpendicular to the antenna surface, there exists a dipole and a reflecting surface at a quarter wavelength to enhance the gain at that frequency. Hence these reflectors are termed the *frequency independent reflectors*. Effectively, the reflector converts the bidirectional radiation pattern of the planar antenna into a unidirectional pattern. The average gain is improved over the designed frequency range by nearly a factor of two.

Consider a single polarized antenna associated with one of the simplified frequency-independent reflectors, as shown in Fig. 4.11 (step-reflector is in dark lines, and plane-reflector is marked dotted). Here, $\lambda_{min}$ and $\lambda_{max}$, respectively, represent the wavelengths corresponding to the smallest and largest dipoles. Let the required gain of the antenna with the reflector be $G_{LPA+Refl}$. The gain $G_{LPA}$ of the planar log-periodic antenna (without reflector) is obtained using Eq. (4.21), where ideally the constant $p \approx 0.5$.

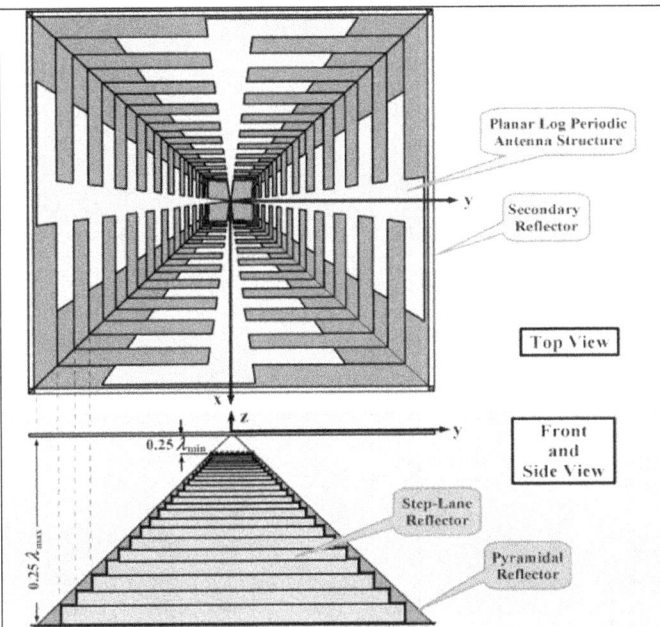

**FIGURE 4.10:** Elevation and plan of the antenna structures associated with a step-lane reflector (dark lines) or a pyramidal reflector (dotted lines). An optional secondary reflector is provided for gain improvement near the lowest frequencies.

$$G_{\text{LPA}} = p\, G_{\text{LPA+Refl}} \tag{4.21}$$

The reflector dimensions are now derived from the planar antenna. Each half-dipole element possesses two long edges. Let $k$ represent the edge numbers. The smallest edge resides at $k = 0$, and the largest edge reside at $k = m$, where $m$ is the total number of steps (including the top and bottom) if a step-reflector is used. Let $V_0$, $V_1$, ..., $V_{m-1}$, $V_m$ be the vertical distances between the dipole edges and the reflector, as shown. Let $\psi$ be the angle formed between the surfaces of the antenna and the reflector, as shown. Eqs. (4.22) through (4.34) are derived from this geometry.

$$\tan\Psi = \left(\frac{\delta V_k}{\delta W_k}\right) \qquad k = 0, 1, ..., m-1, m \tag{4.22}$$

$$\tan\Psi = \left(\frac{\delta V_k}{\delta W_k}\right) \qquad k = 0, 1, ..., m-1, m \tag{4.22}$$

$$\tan\Psi = \left(\frac{V_k}{W_k}\right) \qquad k = 0, 1, ..., m-1, m \tag{4.23}$$

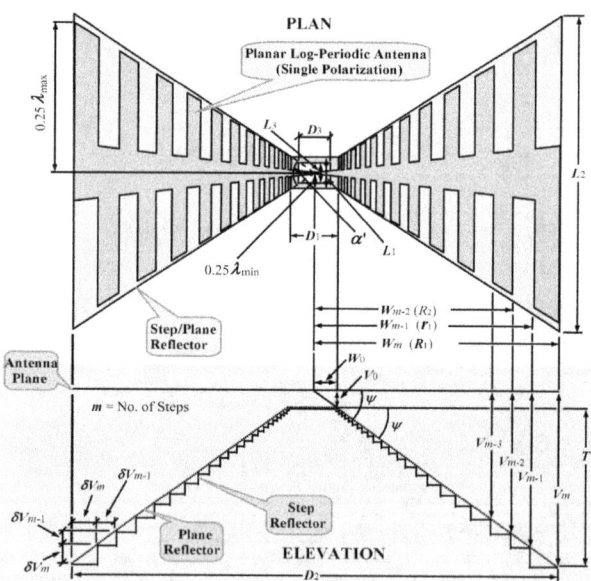

**FIGURE 4.11:** Details of a step reflector (dark lines) and a plane reflector (dotted lines) associated with a single-polarized antenna.

$$V_0 = 0.25\lambda_{min} \tag{4.24}$$

$$V_m = 0.25\lambda_{max} \tag{4.25}$$

The gap $D_1$ between the smallest dipole is related to $\lambda_{min}$ and the apex-angle $\alpha$ as is shown in Eq. (4.26). The total length $D_2$ of the antenna is related to $\lambda_{max}$ and the apex-angle $\alpha$ as is shown in Eq. (4.27). It is also related to the half-length $W_m$ of the antenna, as shown in Eq. (4.28). To reduce the back lobe at lower frequencies, $D_2$ is chosen slightly larger than its minimum value.

$$D_1 = \left( \frac{\lambda_{min}}{2\tan(0.5\alpha)} \right) \tag{4.26}$$

$$D_2 \geq \left( \frac{\lambda_{max}}{2\tan(0.5\alpha)} \right) \tag{4.27}$$

$$D_2 \geq 2W_m \tag{4.28}$$

$$0 \leq D_3 \leq 2W_0 \tag{4.29}$$

$$T = 0.25\left(\lambda_{max} - \lambda_{min}\right) \tag{4.30}$$

$$L_1 \geq 0.5\lambda_{min} \tag{4.31}$$

$$L_2 \geq 0.5\lambda_{max} \tag{4.32}$$

$$0 \leq L_3 \leq 2V_0 \tag{4.33}$$

$$\alpha' \geq \alpha \tag{4.34}$$

The projected reflector angle $\alpha'$ is expressed in Eq. (4.34). For reasons stated above, it is also chosen slightly greater than $\alpha$. The same reasoning is made for the determination of $L_1$, $L_2$, $L_3$, and $D_3$, which are expressed in Eqs. (4.31) through (4.33) and Eq. (4.29), respectively. To reduce the back lobes at the highest frequencies, $L_3$ and $D_3$ should have minimized dimensions. The geometric ratio $\tau$ of the planar antenna structure is given in Eqs. (4.35) through (4.38). They clearly show the log-periodic nature of the steps in a step reflector.

$$\tau = \left( \frac{W_{k-1}}{W_k} \right)^2 \qquad k = 0,1,...,m-1,m \tag{4.35}$$

$$\tau = \left(\frac{V_{k-1}}{V_k}\right)^2 \qquad k = 0,1,...,m-1,m \qquad (4.36)$$

$$\tau^m = \left(\frac{V_0}{V_m}\right)^2 \qquad (4.37)$$

$$\tau^m = \left(\frac{W_0}{W_m}\right)^2 \qquad (4.38)$$

Within the frequency range, the extent to which $G_{\mathrm{LPA+Refl}}$ may vary with respect to $G_{\mathrm{LPA}}$ is given in Eq. (4.39). It is based on the results shown in Fig. 4.11.

$$G_{\mathrm{LPA}} \leq G_{\mathrm{LPA+Refl}} \leq 2.65 G_{\mathrm{LPA}} \qquad (4.39)$$

The plane reflector design is much simpler. It requires only $V_k$ and $W_k$ for $k = 0$, $m$. There are two reflectors (stepped or plane) for a dual-polarized antenna. Extending and joining the two reflectors (for two polarizations) by increasing $L_2$ and $\alpha'$, a single reflector of the shape shown in Fig. 4.10 is formed. The two stepped reflectors join to form a step-lane reflector. Similarly, two plane reflectors join to form a pyramidal reflector.

For 1:10 bandwidth designs, Fig. 4.12 shows the simulated (MATLAB) and measured antenna gains for three models: (*i*) planar LPA with secondary reflector, (*ii*) planar LPA with secondary and step-lane reflectors, and (*iii*) planar LPA with secondary and pyramidal reflectors. Here, $F = 200$ MHz is the minimum frequency. Table 4.1 lists some of the average gains over the frequency range. The polarization isolation varies between 20 dB and 30 dB over the frequency range.

**TABLE 4.1:** Mean of the simulated gains and gain-ratios across the ultra-wideband.

| $G_{\mathrm{LPA}}$ | $G_{\mathrm{LPA+StepLane}}$ | $G_{\mathrm{LPA+Pyramidal}}$ | $\left(\dfrac{G_{\mathrm{LPA+StepLane}}}{G_{\mathrm{LPA}}}\right)$ | $\left(\dfrac{G_{\mathrm{LPA+Pyramidal}}}{G_{\mathrm{LPA}}}\right)$ |
|---|---|---|---|---|
| 4.161 | 7.745 | 8.150 | 1.873 | 1.971 |
| 6.19 dBi | 8.89 dBi | 9.11 dBi | 2.72 dB | 2.95 dB |

At the lower and upper frequency ends, the gain deteriorates slightly due to geometrical discontinuities. From simulations, the pyramidal reflector shows better performance. It is also easier to fabricate. The value of $p$ in Eq. (4.21) can be estimated from the reciprocal of the linear values listed in the third and fourth columns of Table 4.1.

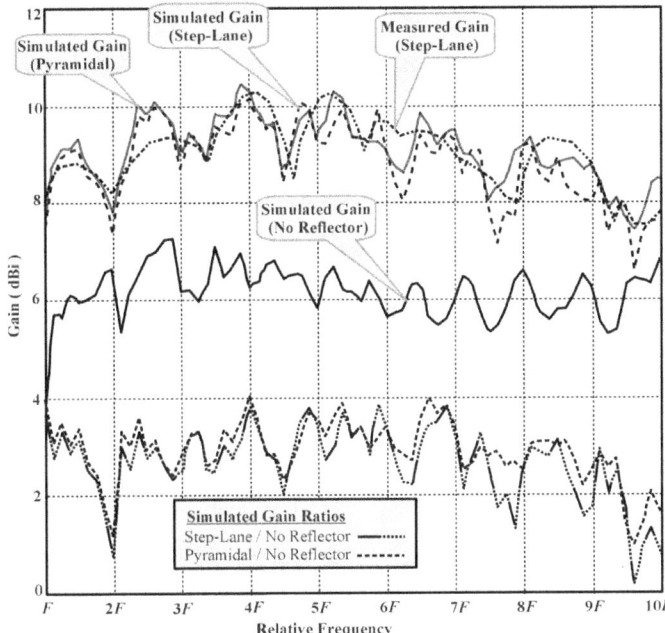

**FIGURE 4.12:** Simulated and measured antenna gains and gain ratios as functions of frequency. Contributions from the secondary reflector are included.

### 4.4.2.3 Log-Periodic Dipole Arrays (LPDA)

LPDA are used with paraboloid dishes as feeds or in antenna arrays. They consist of multiple half-wave dipoles (with different frequencies) arranged as shown in Fig. 4.13. The thickness $\Delta_n$, length $L_n$, and interspacing $d_n$, of the dipoles increases with distance from the apex. Similar to the other log-periodic structures, the scale factor $\tau$ is given by Eq. (4.40), where $N$ is the total number of dipoles, and $n$ stands for the $n^{th}$ dipole. The electrical connections between adjacent dipoles are reversed. As a special case, when $N = 1$, one obtains a half-wave dipole antenna. The spacing factor $\sigma$ is given by Eq. (4.41). From the antenna geometry, the apex angle $\alpha$ can be written as in Eq. (4.42). This is related with $\tau$ and $\sigma$ as shown in Eq. (4.43).

$$\tau = \frac{R_n}{R_{n+1}} = \frac{L_n}{L_{n+1}} = \frac{\Delta_n}{\Delta_{n+1}} < 1, \quad \text{where, } 1 \le n \le N \tag{4.40}$$

$$\sigma = \frac{d_n}{2L_n}, \quad \text{where, } \quad d_n = R_{n+1} - R_n \tag{4.41}$$

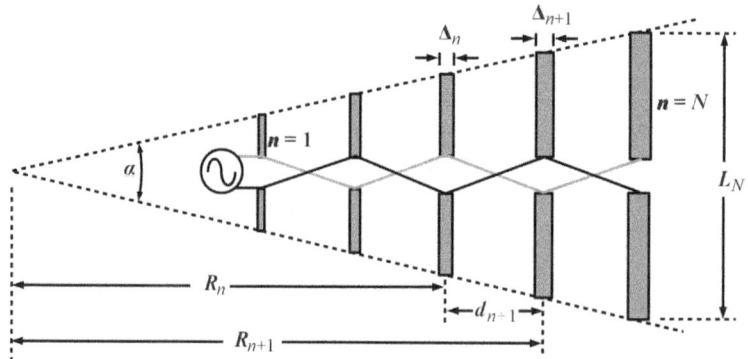

**FIGURE 4.13:** Geometrical structure and design parameters of a LPDA.

$$\tan\left(\frac{\alpha}{2}\right) = \frac{0.5L_n}{R_n} = \frac{0.5L_{n+1}}{R_{n+1}} \tag{4.42}$$

$$\alpha = \tan^{-1}\left(\frac{1-\tau}{4\sigma}\right) \tag{4.43}$$

To design a LPDA, use the gain-curve graph shown in Fig. 4.14. The curve selected must satisfy the gain requirements of the antenna. The characteristic impedance $Z_{of}$ of the feeder line is 100 Ohm. The ratio of a dipole length $L$ to its diameter $\Delta$ is 125. The optimized values of $\sigma$ lie on the dotted line bisecting the gain curves. The smallest dipole length is obtained from the highest frequency, as given in Eq. (4.44), where $v_{min}$ is the minimum frequency and $c$ is the speed of light. Using $\sigma$ as the length of this first dipole in Eq. (4.40), the lengths of the remaining dipoles (lower frequencies) are calculated. The value of $\alpha$ is obtained by substituting the values of $\sigma$ and $\tau$ in Eq. (4.43). The impedance $Z_{LPDA}$ of the LPDA is given by Eq. (4.45), where $\sigma$ is the mean spacing factor, $Z_a$ is the average characteristic impedance of a short dipole, and $Z_{of}$ is the feeder impedance.

$$L_1 = 0.5\frac{c}{v_{min}} \tag{4.44}$$

$$\left.\begin{array}{l} Z_{LPDA} = Z_{of}\left(1 + \dfrac{Z_{of}}{4\sigma' Z_a}\right)^{-1/2} \\[4mm] \text{where,} \, Z_a = 120\left[\ln\left(\dfrac{2L_n}{\Delta_n}\right) - 2.25\right], \text{and} \, \sigma' = \dfrac{\sigma}{\sqrt{\tau}} \end{array}\right\} \tag{4.45}$$

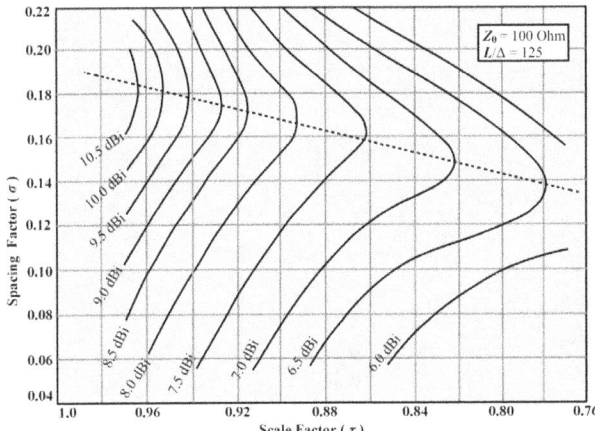

**FIGURE 4.14:** Gain variations of a LPDA in relation to number of dipoles and different spacings.

### 4.4.3 Horn Antenna

Horn antennas are operated at microwave frequencies. They can form narrow beams and can operate as single-aperture antennas. They are also used as antenna feeds for dish antenna systems. The horn antenna is a flared wave guide. Basically, there are two types of horns: (*i*) rectangular-aperture horn, and (*ii*) conical-aperture horn. The former has a rectangular aperture, while the later has a circular aperture. The flaring of a rectangular horn can be in the E-plane, the H-plane, or both. These are named, respectively, the *E-plane sectoral*, the *H-plane sectoral*, and *pyramidal*. These are shown in Fig. 4.15. Excitations are in the $TE_{10}$ mode.

Pyramidal horns are popular, because beam narrowing is possible in both directions. Thus pencil beams can be formed. Fig. 4.16 shows the flaring of a pyramidal horn in both planes. It is a combination of E-plane and H-plane sectoral horns. Note the top and side views in Figs. 4.16b and 4.16c, which resemble top and side views of an H-plane sectoral horn and an E-plane sectoral horn, respectively. The field equations can be obtained by combining the equations of the two horns. The top view of a E-plane sectoral horn is shown in Fig. 4.16c. Geometrical relationships are established using Eqs. (4.46) through (4.48). The electric field is given in Eq. (4.49), where $E_0$ is its peak amplitude. The side view shown in Fig. 4.16b resembles an H-plane sectoral horn. In this case, the relationships obtained are given in Eqs. (4.50) through (4.52). Eq. (4.53) gives the electric field, where $E_0$ is its peak amplitude.

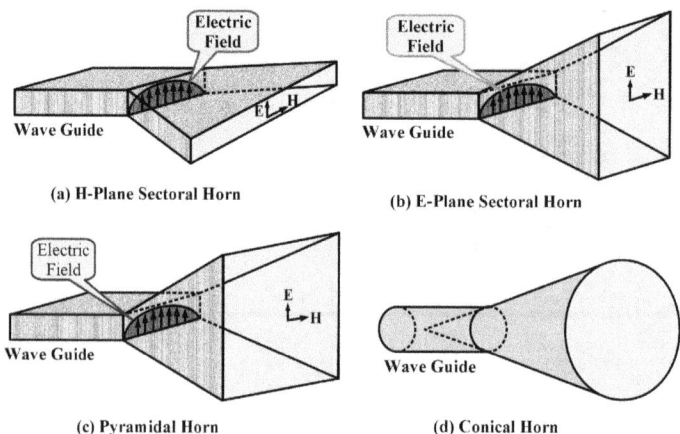

**FIGURE 4.15:** Common horn antennas. Aperture areas are shown in light gray. Some E and H fields across the aperture are shown corresponding to a $TE_{10}$-mode excitation. (a) H-plane sectoral horn. (b) E-plane sectoral horn. (c) Pyramidal horn. (b) Conical horn.

$$l_E^2 = R_2^2 + \left(\frac{B}{2}\right)^2 \tag{4.46}$$

$$\alpha_E = \tan^{-1}\left(\frac{B}{2R_2}\right) \tag{4.47}$$

$$R_E = (B-b)\sqrt{\left(\frac{l_E}{B}\right)^2 - \frac{1}{4}} \tag{4.48}$$

$$E_{ay} = E_0 \cos\left(\frac{\pi x}{a}\right) \exp\left[-j\left(\frac{\beta}{2R_2}\right)y^2\right] \tag{4.49}$$

$$l_H^2 = R_1^2 + \left(\frac{A}{2}\right)^2 \tag{4.50}$$

$$\alpha_H = \tan^{-1}\left(\frac{A}{2R_1}\right) \tag{4.51}$$

$$R_H = (A-a)\sqrt{\left(\frac{l_H}{A}\right)^2 - \frac{1}{4}} \tag{4.52}$$

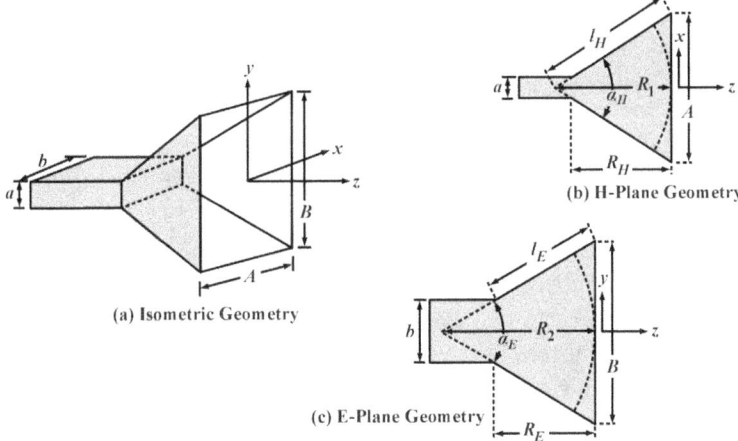

(a) Isometric Geometry

(b) H-Plane Geometry

(c) E-Plane Geometry

**FIGURE 4.16:** (a) Isometric geometry of a pyramidal horn showing flaring in both the E and H planes. (b) The H-plane geometry (side view) resembles the side view of an H-plane sectoral horn antenna. (c) The E-plane geometry (top view) resembles the top view of an E-plane sectoral horn antenna.

$$E_{ay} = E_0 \cos\left(\frac{\pi x}{A}\right) \exp\left[-j\left(\frac{\beta}{2R_1}\right)x^2\right] \qquad (4.53)$$

Combining Eqs. (4.49) and (4.53), respectively, for the E-plane and H-plane sectoral horns, the field expression for a pyramidal horn antenna is shown in Eq. (4.54). The directivity $G_D$ of the pyramidal horn is expressed in Eq. (4.55), where $A_e$ is its effective aperture area and $\eta_A$ is the aperture efficiency. For good designs, $\eta_A$ lies between 0.5 and 0.7.

$$E_{ay} = E_0 \cos\left(\frac{\pi x}{A}\right) \exp\left[-j\left(\frac{\beta}{2}\right)\left(\frac{x^2}{R_1} + \frac{y^2}{R_2}\right)\right] \qquad (4.54)$$

$$G_D = \left(\frac{4\pi}{\lambda^2}\right) A_e = \left(\frac{4\pi}{\lambda^2}\right)\left(A B \eta_A\right) \qquad (4.55)$$

## 4.5 DESIGNING SINGLE-DISH RADIO TELESCOPES

Before designing a single-dish radio telescope, its scope of usage must be defined. Form the type of objects which the telescope would be used to see, the following parameters should be known: (*i*) angular separation between two adjacent sources, (*ii*) source brightness, (*iii*) background sky brightness, (*iv*) frequency or wavelength, and (*v*) bandwidth. We can calculate the minimum diameter $D'$ (m) required to resolve the source using Eq. (4.56), where, $\lambda$ is the wavelength (m) and $\alpha_s$ is the angular separation between adjacent sources (rad). The aperture illumination efficiency $\eta_A$ of a dish antenna ranges between 0.4 and 0.5. Using $\eta_A \approx 0.45$, we calculate the actual dish diameter $D$ using Eq. (4.57). The effective aperture area $A_e$ (m$^2$) is calculated using Eq. (4.58). The receiver temperature $T_R$ of a fairly good system ranges from 35 to 50 K. Eq. (4.59) can be used to caluclate the system temperature $T_{\text{Sys}}$ for the two cases: (*i*) the antenna beam focussed on the source, and (*ii*) the antenna beam focussed on a nearby cold (empty) sky. Using these values, we can calculate the incremental temperature $\Delta T$ by using Eq. (4.60), where the system temperatures $(T_{\text{Sys}})_{\text{src}}$ and $(T_{\text{Sys}})_{\text{sky}}$, respectively, represent the cases (*i*) and (*ii*). If $\Delta T$ emerges as a positive quantity, the design will certainly succeed without any integration. Otherwise, we can calculate the sensitivities using Chapter 2 Eqs. (4.59) through (4.61). If $\Delta S_{\text{min}}$ is less than the source flux density, the design will succeed. For this, one has to carefully chose the integration time $\tau$, bandwidth $\Delta v$, and number of records $n$ for to be averaged.

$$D' = 1.22 \frac{\lambda}{\alpha_s} \tag{4.56}$$

$$D = D'\sqrt{\eta_A} \tag{4.57}$$

$$A_e = \frac{\pi D^2}{4}\eta_A \tag{4.58}$$

$$T_{Sys} = T_A + T_R \tag{4.59}$$

$$\Delta T = \left(T_{\text{Sys}}\right)_{\text{src}} - \left(T_{\text{Sys}}\right)_{\text{sky}} \tag{4.60}$$

## 4.6   APERTURE AND FAR FIELD

For any antenna, the spatial behavior of its far field is directly proportional to the Fourier transform of the spatial-frequency field at its aperture. Consider the geometry shown in Fig. 4.17. An aperture sheet holding the field distributions is located at the origin of a three-dimensional coordinate system. For simplicity, let the aperture plane behave like a PEM (perfect magnetic conductor)[1]. Hence the magnetic

---

[1] In general, the aperture surface may be neither a PEM nor a PEC (perfect electric conductor). Additional components are expected, though these, too, maintain a Fourier-transform relationship with the far field.

field is zero on its surface. Only the electric field exists. Let the electric field be $E_{ay}$. It is aligned with the y-axis as shown. The far electric field components $E_\theta$ and $E_\phi$ are related to $E_{ay}$ through Fourier transformations. These are given in Eqs. (4.61) and (4.62), where $\lambda$ represents the wavelength, $\beta = 2\pi/\lambda$, and $dx$ and $dy$ are infinitesimal lengths along the x- and y-axes, respectively.

$$E_\theta = \left( j\beta \frac{e^{-j\beta r}}{2\pi r} \sin\phi \right) E_{ay} \iint_{S_a} e^{j\beta(\sin\theta\cos\phi)x} e^{j\beta(\sin\theta\sin\phi)y} dx\, dy \qquad (4.61)$$

$$E_\phi = \left( j\beta \frac{e^{-j\beta r}}{2\pi r} \cos\theta\cos\phi \right) E_{ay} \iint_{S_a} e^{j\beta(\sin\theta\cos\phi)x} e^{j\beta(\sin\theta\sin\phi)y} dx\, dy \quad (4.62)$$

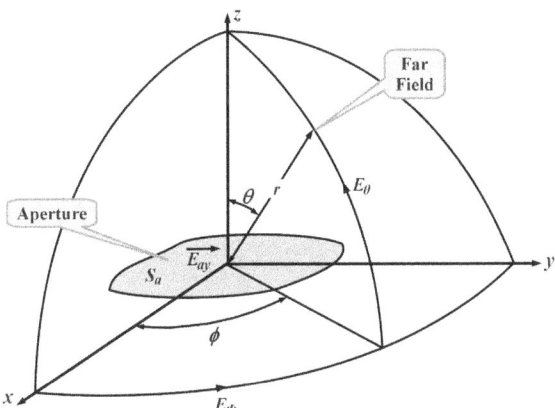

**FIGURE 4.17:** An antenna aperture positioned in the x-y plane of a three-dimensional coordinate. The E and H fields lie in the aperture plane. The far-field pattern is measured at a large distance r from the origin.

## 4.6.1 Uniform Rectangular Aperture and the Far Field

Consider the rectangular aperture shown in Fig. 4.18. As in the previous case, let the aperture plane behave like a PMC, and let the electric field $E_{ay}$ lie along the y-axis. Setting the limits of integration to the aperture size in Eqs. (4.61) and (4.62), we obtain $E_\theta$ and $E_\phi$, respectively, as shown in Eqs. (4.63) and (4.64).

$$E_\theta = E_{ay} \left[ \frac{j\beta \sin\phi e^{-j\beta r}}{2\pi r} \right] \int_{-\frac{L_x}{2}}^{\frac{L_x}{2}} e^{j\beta(\sin\theta\cos\phi)x} dx \int_{-\frac{L_y}{2}}^{\frac{L_y}{2}} e^{j\beta(\sin\theta\sin\phi)y} dy \qquad (4.63)$$

$$= E_{ay} L_x L_y \left[ j\beta \left( \frac{e^{-j\beta r}}{2\pi r} \right) \sin\phi \right] \left( \frac{\sin\left[ (\beta L_x/2)u \right]}{(\beta L_x/2)u} \right) \left( \frac{\sin\left[ (\beta L_y/2)v \right]}{(\beta L_y/2)v} \right)$$

where, $u = \sin\theta\cos\phi$ and $v = \sin\theta\sin\phi$

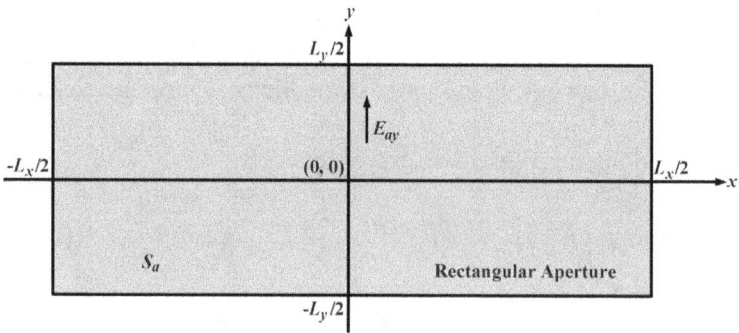

**FIGURE 4.18:** A rectangular aperture having a uniform electric field along the $y$-axis.

$$E_\phi = E_{ay}\left[\frac{j\beta e^{-j\beta r}\cos\theta\cos\phi}{2\pi r}\right]\int_{-\frac{L_x}{2}}^{\frac{L_x}{2}}e^{j\beta(\sin\theta\cos\phi)x}\,dx\int_{-\frac{L_y}{2}}^{\frac{L_y}{2}}e^{j\beta(\sin\theta\sin\phi)y}\,dy$$

$$= E_{ay}L_xL_y\left[j\beta\left(\frac{e^{-j\beta r}}{2\pi r}\right)\cos\theta\cos\phi\right]\left(\frac{\sin\left[(\beta L_x/2)u\right]}{(\beta L_x/2)u}\right)\left(\frac{\sin\left[(\beta L_y/2)v\right]}{(\beta L_y/2)v}\right)$$

(4.64)

where, $u = \sin\theta\cos\phi$ and $v = \sin\theta\sin$

There are two sine functions along the $x$- and $y$-axes that are responsible for the side lobes. $E_\theta$ and $E_\phi$ are directed along the $z$-axis ($\theta = 0$). Their peak values depend on $L_xL_y$. The field components $E_\theta$ and $E_\phi$ vanish at $\phi = 0$, $\pi$ and $\phi = \pi/2$, $3\pi/2$, respectively, because they contain sin$\phi$ and cos $\phi$. As $L_x \gg \lambda$, the main beam narrows in the $x$-$z$ plane (because of the sine functions). Similarly, as $L_y \gg \lambda$, the main beam narrows in the $y$-$z$ plane. In summary, ($i$) the beam peaks perpendicular to the aperture surface, and ($ii$) the directivity increases with an increase in the aperture size (with respect to $\lambda$). Hence, for a fixed aperture size, the directivity increases with the operating frequency.

We saw that sin$\phi$ and cos$\phi$ appeared in the expressions for $E_\theta$ and $E_\phi$ because the electric field was polarized. For a randomly polarized wave, these are replaced by unity. In this case, the normalized beam pattern $P_n(\theta,\phi)$ is given as Eq. (4.65). The beam solid angle $\Omega_A$ is obtained by integrating $P_n(\theta,\phi)$ over the entire hemisphere, as expressed in Eq. (4.66). Using this the directivity $G_D$ is expressed by Eq. (4.67). These results show that the effective aperture area of a uniformly illuminated rectangular aperture is identical to its physical aperture area ($L_xL_y$).

$$P_n(\theta,\phi) = \left(\frac{\sin(k_x u)}{k_x u}\right)^2 \left(\frac{\sin(k_y v)}{k_y v}\right)^2 \tag{4.65}$$

$$\text{where, } k_x = \frac{\beta L_x}{2} = \pi\left(\frac{L_x}{\lambda}\right),\ k_y = \frac{\beta L_y}{2} = \pi\left(\frac{L_y}{\lambda}\right)$$

$$\Omega_A = \int_0^{2\pi}\int_0^{\pi} P_n(\theta,\phi)\,d\Omega = \frac{\lambda^2}{L_x L_y} \tag{4.66}$$

$$G_D = \frac{4\pi}{\Omega_A} = \left(\frac{4\pi}{\lambda^2}\right)\left(L_x L_y\right) \tag{4.67}$$

### 4.6.2 Tapered Rectangular Aperture and the Far Field

Consider a tapered electric field $E_{ay}$ directed along the $y$-axis, as shown in Fig. 4.19. The variation of $E_{ay}$ along the $x$-axis is modeled by Eq. (4.68), where $E_0$ is the peak intensity. This type of electric field variation is seen as a waveguide operating in the $TE_{10}$ mode. The intensity of $E_{ay}$ is maximum at $x = 0$ and minimum at $x = \pm L_x/2$. The far electric fields $E_\theta$ and $E_\phi$ are given, respectively, by Eqs. (4.69) and (4.70).

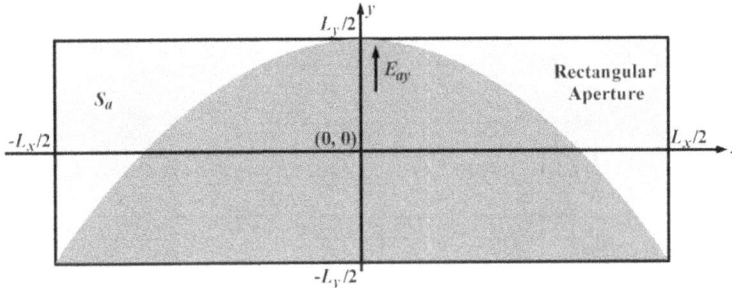

**FIGURE 4.19:** A rectangular aperture having a tapered electric field along the $y$-axis.

$$E_{ay} = E_0 \cos\left(\frac{\beta\lambda}{4L_x}x\right) = \frac{1}{2}\left[e^{j\beta x(\lambda/2L_x)} + e^{-j\beta x(\lambda/2L_x)}\right] \tag{4.68}$$

$$E_\theta = \frac{E_0}{2} L_x L_y \left(j\beta \frac{e^{-j\beta r}}{2\pi r}\sin\phi\right) \tag{4.69}$$

$$\times \left(\frac{\sin\left[(\beta L_x/2)u_1\right]}{(\beta L_x/2)u_1} + \frac{\sin\left[(\beta L_x/2)u_2\right]}{(\beta L_x/2)u_2}\right)\left(\frac{\sin\left[(\beta L_y/2)v\right]}{(\beta L_y/2)v}\right)$$

where, $u_1 = \sin\theta\cos\phi + \lambda/(4L_x)$,

$u_2 = \sin\theta\cos\phi - \lambda/(4L_x)$ and $v = \sin\theta\sin\phi$

$$E_\phi = \frac{E_0}{2} L_x L_y \left( j\beta\frac{e^{-j\beta r}}{2\pi r} \cos\theta\cos\phi \right)$$

$$\times \left( \frac{\sin\left[(\beta L_x/2)u_1\right]}{(\beta L_x/2)u_1} + \frac{\sin\left[(\beta L_x/2)u_2\right]}{(\beta L_x/2)u_2} \right)\left( \frac{\sin\left[(\beta L_y/2)v\right]}{(\beta L_y/2)v} \right) \right\}$$

(4.70)

where, $u_1 = \sin\theta\cos\phi + \lambda/(4L_x)$,

$u_2 = \sin\theta\cos\phi - \lambda/(4L_x)$ and $v = \sin\theta\sin\phi$

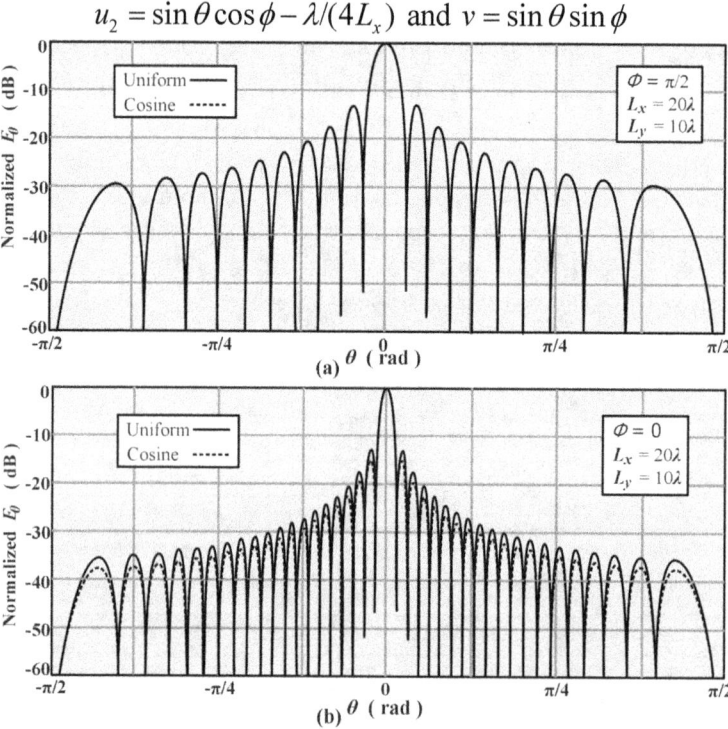

**FIGURE 4.20:** Normalized $E_\theta$ from the uniform and taper aperture illuminations plotted as functions of $\theta$ The side lobes in the x-z plane are reduced by taper illumination. (a) Plot of $E_\theta$ in the y-z plane. (b) Plot of $E_\theta$ in the x–z plane.

Fig. 4.20 shows various distributions of $E_\theta$ resulting from (*i*) a uniformly il-luminated aperture, (*ii*) a tapered cosine illumination in the y-z plane ($\theta = \pi/2$), and (*iii*) a tapered cosine illumination in the x-z plane ($\theta\theta = 0$). The normalized values of $E_\theta$ are converted to dB and plotted as functions of $\theta$ in two orthogonal planes. The effect of $\cos\phi$ has been eliminated by dividing the expressions by $\cos\phi$. The dimensions of the aperture chosen are $L_x = 20\lambda$ and $L_y = 10\lambda$. Note that for a tapered illumination, the side-lobe amplitudes in the x-z plane are reduced.

### 4.6.3 Uniform Circular Aperture and the Far Field

Parabolic dishes are good examples of circular aperture areas. Fig. 4.21 shows a circular aperture of radius $a$ placed in the x-y plane at the center of a spherical coordinate system. We will analyze the field components $E_\theta$ and $E_\phi$ assuming the aperture illumination to be uniform.

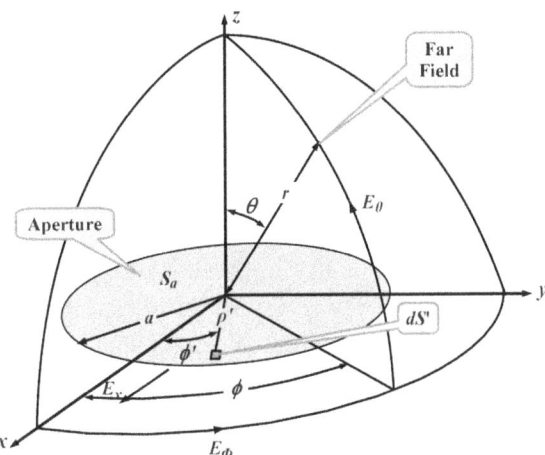

**FIGURE 4.21:** An uniformly illuminated circular aperture having an electric field along the x-axis is positioned in the x-y plane.

Let the electric field $E_x$ be directed along the x-axis. Let $dS'$ be an infinitesimal area on the aperture at a distance $\rho'$ from the origin subtending an angle $\phi'$ with the x-axis, as shown. Thus, $dS'$ can be expressed as a function of $d\rho'$, $d\phi'$, and $\rho'$ as in Eq. (4.71). The projection of $\rho'$ on $r$ is represented by $r'$ and is expressed in Eq. (4.72). Using the Fourier-transform relationship between the near and far fields, $E_\theta$ and $E_\phi$ are stated in Eqs. (4.73) and (4.74), respectively, where $J_1$ is a Bessel function of the first kind of order one.

$$dS' = \rho d\phi' \, d\rho' \qquad (4.71)$$

$$r' = \rho' \cos(\phi - \phi') \sin\theta \qquad (4.72)$$

$$E_\theta = \left( j\beta \frac{e^{-j\beta r}}{2\pi r} \cos\phi \right) \iint_{S_a} E_x e^{j\beta r'} dS'$$

$$= \left( j\beta \frac{e^{-j\beta r}}{2\pi r} \cos\phi \right) E_x \int_0^a \int_0^{2\pi} e^{j\beta \sin\theta \cos(\phi-\phi')\rho'} d\phi \rho' d\rho'$$

$$= E_x \left( j\beta \frac{e^{-j\beta r}}{2\pi r} \cos\phi \right) \left( \frac{2J_1\left(\beta a\sin\theta\right)}{\beta a \sin\theta} \right)$$

(4.73)

$$E_\phi = \left( j\beta \frac{e^{-j\beta r}}{2\pi r} \cos\theta \sin\phi \right) \iint_{S_a} E_x e^{j\beta r'} dS'$$

$$= \left( j\beta \frac{e^{-j\beta r}}{2\pi r} \cos\theta \sin\phi \right) E_x \int_0^a \int_0^{2\pi} e^{j\beta \sin\theta \cos(\phi-\phi')\rho'} d\phi \rho' d\rho'$$

$$= E_x \left( j\beta \frac{e^{-j\beta r}}{2\pi r} \cos\theta \sin\phi \right) \left( \frac{2J_1\left(\beta a\sin\theta\right)}{\beta a \sin\theta} \right)$$

(4.74)

### 4.6.4 Tapered Circular Aperture and the Far Field

Let the electric field intensity peak at the aperture origin and fall as one moves away in the x-y plane. Let the electric field variations of the aperture as a function of $\rho'$ be parabolic, as shown in Fig. 4.22. Thus the electric field intensity $E_a(\rho')$ is expressed in Eq. (4.75), where $E_0$ represents the peak intensity. Using the Fourier transform relationship between the near and far fields, $E_\theta$ and $E_\phi$ are given by Eqs. (4.76) and (4.77), respectively, where $J_{n+1}$ is a Bessel function of the first kind of order $n + 1$.

$$E_a\left(\rho'\right) = E_0 \left[ 1 - \left( \frac{\rho'}{a} \right)^2 \right]^n \quad \text{where, } n = 0, 1, 2, 3.$$

(4.75)

$$E_\theta = \left( j\beta \frac{e^{-j\beta r}}{2\pi r} \cos\phi \right) \iint_{S_a} E_0 \left[ 1 - \left( \frac{\rho'}{a} \right)^2 \right]^n e^{j\beta r'} dS'$$

$$= E_0 \left( j\beta \frac{e^{-j\beta r}}{2\pi r} \cos\phi \right) \left( \frac{\pi a^2}{n+1} \right) \left( \frac{2^{n+1}(n+1)! J_{n+1}\left(\beta a\sin\theta\right)}{\left(\beta a\sin\theta\right)^{n+1}} \right)$$

(4.76)

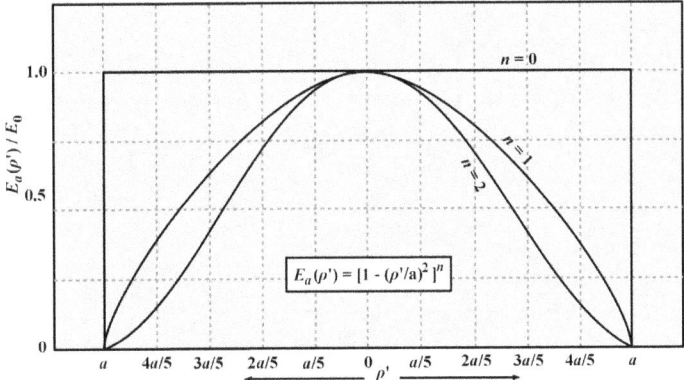

**FIGURE 4.22:** A general parabolic taper function.

$$\left. \begin{aligned} E_\phi &= \left( j\beta \frac{e^{-j\beta r}}{2\pi r} \cos\theta \sin\phi \right) \iint_{S_a} E_0 \left[ 1 - \left( \frac{\rho'}{a} \right)^2 \right]^n e^{j\beta r'} dS' \\ &= \left( j\beta \frac{e^{-j\beta r}}{2\pi r} \cos\theta \sin\phi \right) \left( \frac{\pi a^2}{n+1} \right) \left( \frac{2^{n+1}(n+1)! J_{n+1}(\beta a \sin\theta)}{(\beta a \sin\theta)^{n+1}} \right) \end{aligned} \right\} \quad (4.77)$$

### 4.6.5 General Far-Field Response of Circular Apertures

Let the aperture illumination be unpolarized. In general, the near electric field $E_{\text{Near}}$ is related to the far electric field $E_{\text{Far}}$ as stated in Eq. (4.78), where $K_A$ is some function described by the aperture geometry and $\mathcal{F}$ represents a Fourier transform (see discussion of the angular spectrum in Appendix B.17). Note that $K_A$ may be complex. Recall the $l,m$ (direction cosines) and $u,v$ coordinates described in Appendix A.10. Consider the antenna aperture at the origin of $u$, $v$ such that the near electric field is now $E_{\text{Near}}(u, v)$. Using $l,m$ coordinates, the far electric field at a fixed distance $r$ is represented by $E_{\text{Far}}(l, m)$. The two fields are related by Eq. (4.79). Some examples of aperture illumination and resulting far fields are shown in Fig. 4.23. The corresponding near-fields are listed in Table 4.2.

$$E_{\text{Far}}\left( \text{direction cosines} \right) = K_A \, \mathcal{F}\left[ E_{\text{Near}}(\text{rect.coords.}) \right] \quad (4.78)$$

$$E_{\text{Far}}(l,m) = \int_{-\infty}^{\infty} \int_{-\infty}^{\infty} E_{\text{Near}}(u,v) e^{-j2\pi l u} e^{-j2\pi m v} \, du \, dv \quad (4.79)$$

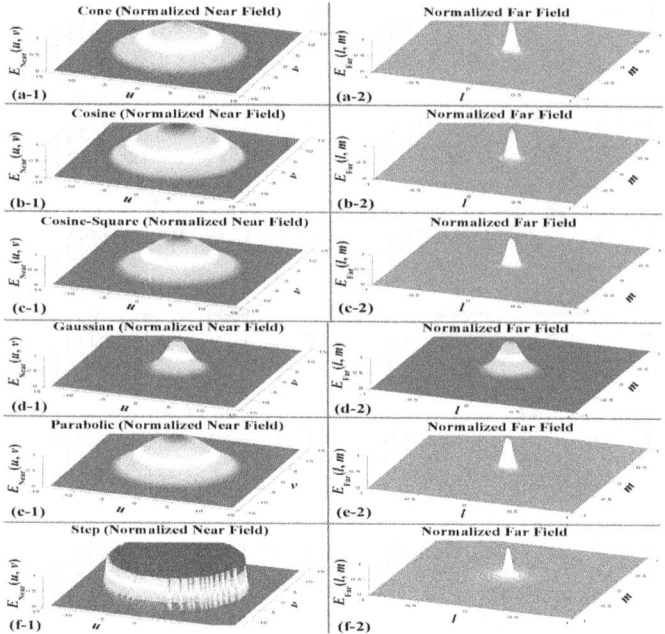

**FIGURE 4.23:** Near and far E-fields produced from a circular aperture with excitation (a) conical, (b) cosine, and (c) cosine squared. (Simulated by author.)

**Table 4.2:** Near-field excitations in Fig. 4.23.

| No. | Excitation | Electric Field across Aperture $E(u,v) = 0, \rho > a, \rho = \sqrt{u^2 + v^2}$ |
|---|---|---|
| a-1 | Cone | $E(u,v) = E_0 \left( \dfrac{a - \rho}{a} \right), \rho \le a; E_0 = 1, a = 10\lambda$ |
| b-1 | Cosine | $E(u,v) = E_0 \cos\left( \dfrac{\pi}{2a} \rho \right), \rho \le a; E_0 = 1, a = 10\lambda$ |
| c-1 | Cosine$^2$ | $E(u,v) = E_0 \left[ \cos\left( \dfrac{\pi}{2a} \rho \right) \right]^2, \rho \le a; E_0 = 1, a = 10\lambda$ |
| d-1 | Gaussian | $E(u,v) = E_0 \exp\left( \dfrac{-\rho^2}{k^2} \right), \rho \le a; E_0 = 1, k = 2.25, a = 10\lambda$ |
| e-1 | Parabolic | $E(u,v) = E_0 \left( 1 - \dfrac{\rho^2}{a^2} \right)^n, \rho \le a; E_0 = 1, n = 2, a = 10\lambda$ |
| f-1 | Step | $E(u,v) = E_0, \rho \le a; E_0 = 1, a = 10\lambda$ |

## 4.7  PHASED ANTENNA ARRAYS

Large antennas are difficult to construct. However, it may be possible to form a single large-aperture antenna using a number of small-aperture antennas. Such an arrangement is known as an *antenna array*. The small antennas are called *array elements*. They are scattered over a plane and are electrically connected together with phase adjustments. To reduce the design complexity, array elements having identical gain, beam shape, radiation, resistance, etc., are used. These are organized in a geometrical shape. One array element near the center is generally used as reference. Each element behaves as a point. The directivity of the effective array beam pattern can be changed by modifying the phases and amplitudes of the elements relative to the reference element. It may be possible to track a radio source in the sky without physically moving the elements by changing the relative phases. Because phases are frequency dependent, scanning-beam-type arrays possess narrow bandwidths.

### 4.7.1  Linear Antenna Array

Fig. 4.24a shows a linear antenna array where all the elements are equally spaced on a straight line. The array elements are identical. Signals received by each element are processed using a combination of a phase shifter and an attenuator. The processed signals are added and sent to a receiver. The beam pattern of an array formed by isotropic elements is known as an *array factor* $K_{AF}$. The beam pattern of an array of nonisotropic elements is a product of an individual element pattern with $K_{AF}$. Hence, analysis of an array is done in two steps: (*i*) find the array factor $K_{AF}$ by replacing the actual array elements with isotropic radiators (Fig. 4.24b); (*ii*) multiply this $K_{AF}$ with the pattern of an individual element.

As shown in Fig. 4.24b, $N$ isotropic elements are located in a straight line. The gap between the elements is $d$. The wave-fronts reaching the array from a distant radio source at an angle $\theta$ with respect to the zenith, as shown. Let the element $N$ be the reference antenna. The wave front touching it (dotted lines) is taken as the reference. The signal reaching the adder from the antenna $N$ will be $A_N e^{j0}$, provided the attenuators and phase shifters are set to zero. Similarly, signals contributed by elements $(N-1)$, $(N-2)$, ..., 2, 1 will be $A_{N-1}e^{j\beta d \sin\theta}$, $A_{N-2}e^{j2\beta d \sin\theta}$, ..., $A_2 e^{j(N-2)\beta d \sin\theta}$, $A_1 e^{j(N-1)\beta d \sin\theta}$, respectively. Because the elements are assumed identical, the magnitudes are also the same ($|A_n| = A_1$). In general, the signal from the $n^{th}$ antenna is given by Eq. (4.80), where $\lambda$ is the wavelength, $\beta = 2\pi/\lambda$, and $\psi = \beta d\sin\theta$. Thus the adder output is $K_{AF}$, *which is* a function of $\theta$, as shown in Eq. (4.81). The normalized variation of $K_{AF}$ as a function of $\theta$ for three differ-

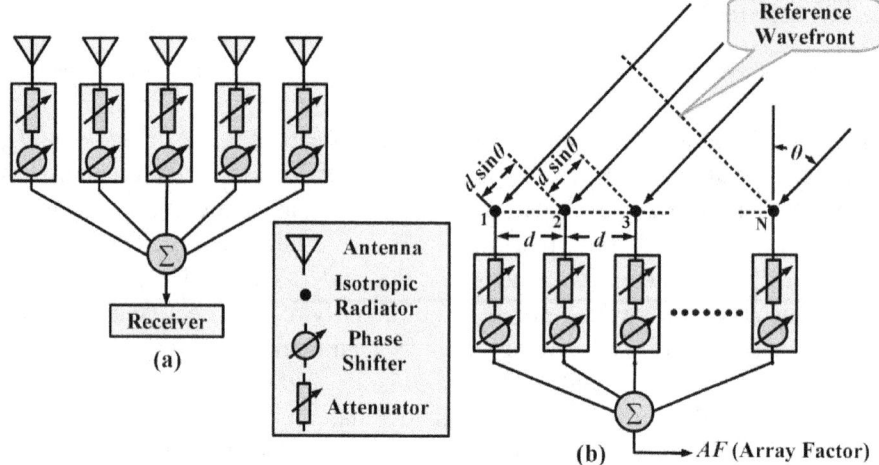

**FIGURE 4.24:** (a) Array formed by antennas having identical patterns and characteristics. (b) Equally spaced array of isotropic antennas receiving signals from a distant source inclined at an angle $\theta$ with respect to the zenith.

ent values of $d$ and $N = 10$ are shown in Fig. 4.25. From the antenna reciprocity theorem (Appendix B.14), we conclude that this receiving pattern also holds for a radiating array. When these isotropic elements are replaced by identical antennas with their major lobes towards the zenith, we get maximum directivity towards the zenith. The minor lobes are also reduced.

$$A_n = A_1 e^{j(N-n)\beta d \cos\theta} = A_1 e^{j(N-n)\psi}, \text{ where } \psi = \beta d \sin\theta \tag{4.80}$$

$$K_{AF}(\theta) = A_1 \sum_{n=1}^{N} e^{j(N-n)\psi} = A_1 \frac{1 - e^{jN\psi}}{1 - e^{j\psi}} = A_1 e^{j(N-1)\psi/2} \left[ \frac{\sin(N\psi/2)}{\sin(\psi/2)} \right] \tag{4.81}$$

We now consider an array of isotropic elements with fixed spacing, but with different attenuation and phase shifts applied to the individuals. Let the attenuation applied to the $n^{\text{th}}$ antenna be $\alpha_n$, and let the phase shift be $\kappa_n$. The array factor $K_{AF}$ is given by Eq. (4.82), where $0 \leq \alpha_n \leq 1$, $0 \leq \kappa_n \leq 2\pi$, and $A_0$ is the maximum signal amplitude (for all elements). An analytical solution to this equation is not possible, but we can find a numberical solution by using a computer. For a nonuniformly spaced array, $K_{AF}$ is given by Eq. (4.83), where $d_n$ is the spacing between the $(n + 1)^{\text{th}}$ and $n^{\text{th}}$ elements.

$$\left. \begin{array}{c} K_{AF}(\theta) = A_0 \sum_{n=1}^{N} \alpha_n e^{j[\beta(N-n)\psi + \kappa_n]} \\[2mm] \text{where, } 0 \leq \alpha_n \leq 1, \ 0 \leq \kappa_n \leq 2\pi \end{array} \right\} \tag{4.82}$$

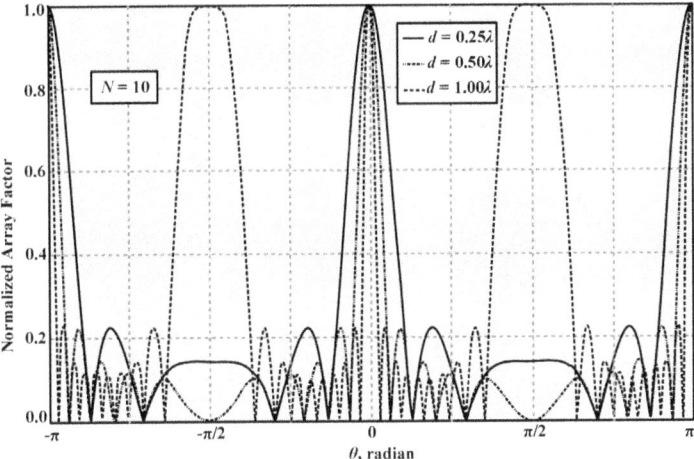

**FIGURE 4.25:** A grating response: Normalized variation of the array factor with respect to $\theta$ for different spacings between the isotropic radiators ($N = 10$).

$$K_{AF}(\theta) = A_0 \sum_{n=1}^{N} \alpha_n e^{j\left[\beta(N-n)\psi_n + \kappa_n\right]} \left.\rule{0pt}{24pt}\right\}$$

$$\text{where, } 0 \leq \alpha_n \leq 1, \, 0 \leq \kappa_n \leq 2\pi, \, \psi_n = d_n \sin\theta$$

(4.83)

### 4.7.2 Two-Dimensional Planar Antenna Array

A linear array can control the beam pattern only in the plane containing the elements and the direction at which the antenna is pointed (the zenith). Perpendicular to this plane, the beam pattern resembles that of an antenna element. In radio astronomy, a pencil beam is preferred for scanning the source. For narrowing the beam on the other plane, a two-dimensional array can be used.

Consider an array consisting of $M \times N$ elements, as shown in Fig. 4.26. The array factor $K_{AF}$ now depends on both $\theta$ and $\phi$. Let $d_m$ and $d_n$ be the spacings along the $x$- and $y$-axes, where $m$ and $n$ represent the elements. For simplicity we begin with a uniform array with no attenuation or phase shifts applied. Here, $d_m$ and $d_n$ are constants. Thus $K_{AF}$ is given by Eq. (4.84), where $\psi_m = d_m \sin\theta\cos\phi$ and $\psi_n = d_m \sin\theta\cos\phi$. Applying the attenuation and phase shifts to the elements, we get the generalized case, where $K_{AF}$ is given by Eq. (4.85). The factors $\alpha_{mn}$ and $\kappa_{mn}$ are, respectively, the attenuation and phase shifts applied at the element outputs. Fig. 4.27 shows the normalized $K_{AF}$ of a two-dimensional array with $M = 5$ and $N = 10$ for the planes $\phi = 0°$ and $\phi = 90°$.

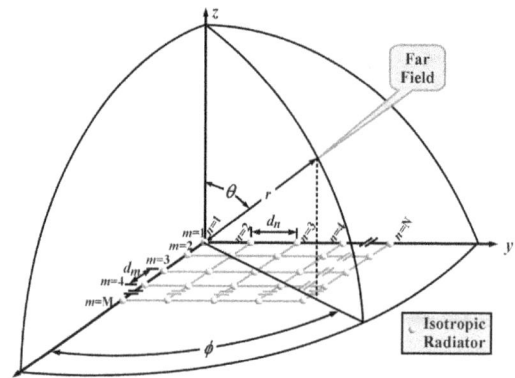

**FIGURE 4.26:** A two-dimensional array of isotropic antennas.

$$K_{AF}(\theta,\phi) = A_0 \sum_{m=1}^{M} e^{j\beta(M-m)\psi_m} \sum_{n=1}^{N} e^{j\beta(N-n)\psi_n} \Bigg\}$$

$$\text{where, } \psi_m = d_m \sin\theta \cos\phi, \ \psi_n = d_n \sin\theta \sin\phi \Bigg]$$

(4.84)

$$K_{AF}(\theta,\phi) = A_0 \sum_{n=1}^{N} \sum_{m=1}^{M} \alpha_{mn} e^{\kappa_{mn}} e^{j\beta[(M-m)\psi_m+(N-n)\psi_n]} \Bigg\}$$

$$\text{where, } 0 \le \alpha_{mn} \le 1, \ 0 \le \kappa_{mn} \le 2\pi \Bigg]$$

(4.85)

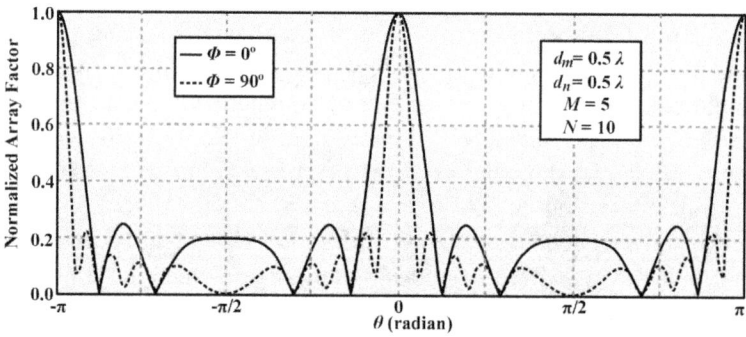

**FIGURE 4.27:** A grating response: Normalized variation of the array factor with respect to $\theta(\phi = 90°)$ for different spacings between the isotropic radiators for $M = 5$ and $N = 10$.

### 4.7.3 Practical Phased Arrays

If the array elements are closely packed, mutual couplings take place between the elements. They affect the beam shape or the array factor. The mutual couplings can be measured and included in the analysis. When dual polarized antenna elements are used, the polarization coupling must also be considered. Fixed-beam arrays with wide bandwidths are used in solar radio spectrography, where the transient signals are important. The elements of such arrays are usually log-periodic antennas. The beam pattern, however, changes with frequency, but a reasonable gain is achieved at all frequencies. Fig. 4.28 shows one such radio spectrography array constructed by the Indian Institute of Astrophysics. It is located about 100 km North of Bangalore at Gauribidanur. It operates between 30 to 150 MHz. The East-West beam-width is 60°, with the peak directed towards the zenith. Solar observations are possible at about ±2 hours around the local noon.

**FIGURE 4.28:** The Gauribidanur Solar Array (GRASS). Eight LPDA are stacked along a north-south axis with their E-plane resting along an east-west axis. The interelement gap is 7 m along the north-south axis.

## 4.8 ARRAYS AS A SINGLE ANTENNA

In an analogy to optical telescopes using concave mirrors, a dish antenna focusses the incoming parallel rays on the feed. If the antenna beam is narrower than the extent of the source, the later can be scanned. The corresponding signal power or intensity may be plotted as a function of the angular scans to form an image. The basic idea behind arrays is to synthesize a large antenna using a number of small antennas. This is due to the practical limitations of constructing a single large antenna that can satisfy the aperture size required for obtaining a good resolution and signal-to-noise ratio. The basic principles are discussed here.

### 4.8.1 Filled Array

Fig. 4.29a shows an aperture strip having an area $A$. An approximate aperture strip is formed from four parabolic dish antennas placed side by side. Hence $A$ may be approximated as the sum of the individual apertures, as in Eq. (4.86).

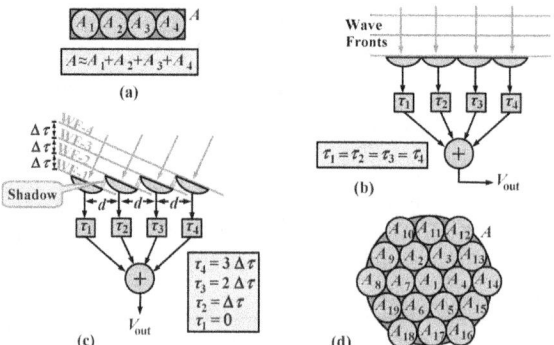

**FIGURE 4.29:** (a) Four circular apertures synthesize a rectangular aperture. (b) Four dish antennas are pointing to a source at the zenith. They are hit by same wave fronts, hence, the compensating delays must be the same. (c) Four antennas track an inclined source. They are hit by different wave fronts. Hence, the compensating delays are different. (d) Synthesizing a large circular aperture using small circular apertures.

$$A \approx A_1 + A_2 + A_3 + A_4 = \sum_{i=1}^{4} A_i \qquad (4.86)$$

The disadvantages are visible from Fig. 4.29b. Time delays have been introduced to phase the resulting signals from the same wave fronts (WF) produced from a distant radio source. This arrangement is efficient as long as the source is at the zenith. In this case, the introduced time delays must be identical (preferably zero). This is shown in Eq. (4.87).

$$\tau_1 = \tau_2 = \tau_3 = \tau_4 \qquad (4.87)$$

When the source is inclined from the zenith (Fig. 4.29c), the wave fronts are different at different antennas. Because the antennas are equally spaced ($d$), the wave front WF-1 first reaches the fourth antenna (right). After a time delay $\Delta\tau$, it reaches the third antenna, and so on. Hence, the instrumental delays ($\tau_1$, $\tau_2$, $\tau_3$, $\tau_4$) are adjusted suitably before the individual antenna signals are summed, as in Eq. (4.88).

$$\left.\begin{array}{r} \tau_4 = 3\Delta\tau \\ \tau_3 = 2\Delta\tau \\ \tau_2 = \Delta\tau \\ \tau_1 = 0 \end{array}\right\} \qquad (4.88)$$

When the dishes are inclined, shadows are cast on the neighboring antennas because the antennas are closely packed. This in turn reduces the effective aperture area. Using the same technique, a large circular aperture may be synthesized, as shown in Fig. 4.29d. Filled arrays are not common.

### 4.8.2 Grating Array

The array factor $K_{AF}$ of a phased array with interspacing $d$ has several crests and troughs at various angles. Fig. 4.30a shows a system for tracking a distant radio source on the sky using a one-dimensional phased array. The source is inclined by an angle $\theta$ from the zenith. The rays are parallel near the antennas. No shadow is cast from adjacent antennas because $d$ is much larger than the individual antenna diameters. Any wave front first approaches the right antenna, then after a delay of $\Delta\tau = d\sin\theta/c$, it approaches the next antenna, and so on. Here, $c$ is the propagation speed of the waves. Fig. 4.30b compares the normalized power pattern of a single dish with the same of an array of isotropic elements. The array response resembles a grating. The angular separation between two neighboring crests is known as the *grating interval*. Fig. 4.30c shows the pattern of the phased array of dish elements. The grating interval depends upon both the interspacing $d$ and the wavelength. Under the major lobe of an elemental antenna beam, more than one crest of the grating may appear. This may upset the performance because unwanted signals may enter through these lobes.

The power output $P_{out}$ of the phased array is given by Eq. (4.89), where $K_{scale}$ is a proportionality constant, $V_1$, $V_2$,... are the voltage outputs from individual elements, and $n_a$ is the number of antennas. Hence, the number of cross multiples $n_c$ using the outputs from different antenna pairs is given by the combination factor $^{n_a}C_2$, as shown in Eq. (4.90). The number of self-multiples is $n_a$.

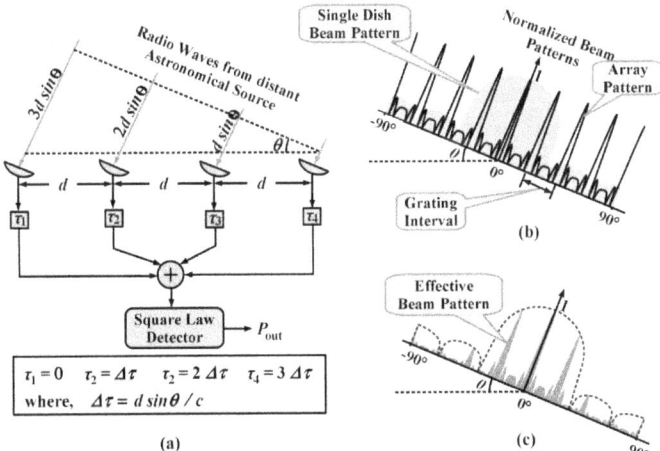

**FIGURE 4.30:** A grating array tracking a radio source and its response. (a) Signals are added after correcting the phases by appropriate delays. (b) Normalized power patterns of a dish and the array of isotropic point sources. (c) Effective response.

$$P_{\text{out}} = K_{\text{scale}} \left( V_1 + V_2 + V_3 + \ldots \right)^2 = K_{\text{scale}} \left( \sum_{i=1}^{n_a} V_i \right)^2 \tag{4.89}$$

$$n_c = {}^{n_a}C_2 = \frac{n_a \left( n_a - 1 \right)}{2} \tag{4.90}$$

| BOX 4.1 | Low-Frequency Array (LOFAR) |
|---|---|

**FIGURE 4.31:** Group of prototype LOFAR Low-Band Antennas (LBA) from the ITS (Initial Test Station) radio telescope in Exloo, Netherlands. Image credit: Wikimedia Commons

**FIGURE 4.32:** Single LOFAR High-Band Antenna (HBA) Image credit: Wikimedia Commons

The LOFAR (LOw Frequency ARray) telescope consists of a phased array of ~$10^4$ fixed dipole antennas designed by the ASTRON Netherlands Institute for Radio Astronomy. Observing frequencies range between 10-240 MHz. Dipole antennas are distributed over the northern part of the Netherlands and in four other European countries.

The LOFAR array comprises two antenna types: (*i*) high-band antennas (HBA) that operate between 110-250 MHz, and (*ii*) low-band antennas (LBA) that operate between 10 – 90 MHz. Fig. 4.31 shows a group of prototype LBAs, each 1.0 m in height. A single HBA is shown in Fig. 4.32. The antennas are arranged in aperture array stations, with thrity-six stations in the Netherlands, five stations in Germany, and one station each in Sweden, the UK, and France. The eighteen core array stations in the Netherlands are arranged in a square array. Ninety-six LBAs and forth-eight HBA tiles (with 4 × 4 elements spaced at 1.25 m) comprise each core and remote station field. Part of the LOFAR core near Exloo, the Netherlands, is shown in Figure 4.33. Array stations outside the Netherlands have ninety-six LBAs and ninety-six HBA tiles each. The maximum European LOFAR baseline is roughly 1,500 km. Electronic beam-steering is achieved by introducing phase shifts between the antennas. Digitized telescope data are processed on an IBM Blue Gene/Q supercomputer at the University of Groningen in the Netherlands.

## Key science projects of LOFAR include

(1) Epoch of Reionization (EoR)

Reionization occured with the birth of the first stars in the early universe, following a period of expansion and cooling after recombination, known as the Dark Ages. Earlier polarization measurements by the Wilkinson Microwave Anisotropy Probe (WMAP) suggest that the onset of reionization may have occured in multiple phases, ending at a redshift of z~6. LOFAR will probe this epoch by mapping the 21 cm line at a range of redshifts between z = 11.4 (115 MHz) and z = 6 (180 MHz).

(2) Cosmic Rays

The source of high-energy cosmic rays (HECRs) with energies between $10^{15}$ - $10^{20.5}$ eV is currently unknown. LOFAR will be used to study HECRs from radiation produced by secondary particle showers that result when HECRs collide with the Earth's atmosphere.

(3) Cosmic Magnetism

Local magnetic fields in the Milky Way are studied with observations of polarized synchrotron radiation. LOFAR enables the study of very small Faraday rotations of distant polarized sources due to weak magnetic fields.

(4) Solar Physics and Space Weather

Solar imaging may be conducted at twenty simultaneous frequencies to study the evolution of active solar regions. Individual stations may serve as stand-alone solar spectrometers to measure the intensity of solar bursts as a function of time over the available frequencies. LOFAR will also enable continuous mapping of the solar wind.

(5) Large-Sky Surveys

Low-frequency, large-sky surveys of z ≥ 6 radio galaxies may be used to investigate the formation of massive galaxies and galaxy clusters. Surveys of starburst galaxies may serve as probes of star formation in the early universe.

(6) Transient Sources

Transient emissions can be monitored by LOFAR, including flares, stars, astrophysical jets in active galactic nuclei (AGN), gamma ray bursters (GRBs), pulsars, and new transient events.

FIGURE 4.33 Part of the LOFAR core "superterp" near Exloo, Netherlands. Image credit: LOFAR / ASTRON

## REVIEW QUESTIONS

1. A telescope has an objective diameter of 2 m. If the wavelength is 500 nm, find its angular resolution. [***Hint***: use Eq. (4.1)]

2. The dish diameter of an antenna is 45 meters. Find its angular resolution at the frequencies (*i*) 150 MHz, (*ii*) 233 MHz, (*iii*) 327 MHz, (iv) 610 MHz, and (*v*) 1420 MHz.

   [***Hint***: first calculate the wavelengths using $\lambda = c/v$ and then use Eq. (4.1)]

3. Using a diagram, describe the common antenna-feed mountings and focusing arrangements used in single-dish radio telescopes. [***Hint***: see Fig. 4.3]

4. Explain the terms: (*i*) spillover, (*ii*) illumination leakage, (*iii*) side-lobe pickup, and (*iv*) back-lobe pickup. How do these parameters change with changes in dish inclination from the zenith? Explain using a diagram.

   [***Hint***: see Fig. 4.4]

5. Explain the various efficiencies applicable to a dish antenna system: (*i*) efficiency of the antenna feed, (*ii*) spillover efficiency, (*iii*) dish leakage efficiency, (*iv*) surface smoothness efficiency, (*v*) illumination efficiency, and (*vi*) polarization efficiency.

6. What is meant by aperture efficiency of a dish antenna system? Show its dependencies on the other efficiencies mentioned in problem no. 5 using an equation. [***Hint***: see Eq. (4.6)]

7. How does the effective gain depend on aperture efficiency? Using an equation, calculate the effective gain of a dish having a diameter of 45 meters and an aperture efficiency of 0.4 at a frequency of 1420 MHz.

   [***Hint***: see Eq. (4.7)]

8. Under what conditions is reduced aperture illumination preferable? Tabulate its advantages and disadvantages.

9. Explain the differences between a nonplanar and a planar log-periodic antenna.

10. Explain the differences between a log-periodic dipole array and a log-periodic antenna.

11. Design a log-periodic dipole array having a gain of 7.5 dBi and a 1:4 bandwidth. The minimum frequency is 200 MHz.

    [***Hint***: Use the graph in Fig. 4.14, and calculate $\tau$ and $\alpha$. Calculate the longest half-wave dipole using the given frequency and then proceed.]

12. Describe the differences between the E-plane and the H-plane sectoral horn antennas. [***Hint***: See Fig. 4.15]

13. Describe the differences between pyramidal and conical horns.

    [***Hint***: See Fig. 4.15]

14. What is the effective aperture area of a pyramidal horn antenna, given the lengths of the two sides are A and B, and the aperture efficiency is $\eta_A$?

    [***Hint***: See Eq. (4.55)]

15. Calculate the diameter of a dish antenna for resolving 1.5° at 150 MHz. Assume the aperture efficiency as 0.4. [***Hint***: See Eqs. (4.56) and (4.57).]

16. In a single-dish telescope design, the difference between on-source and off-source (cold-sky) temperatures (incremental temperature) is found to be zero. What are the possible ways to make it positive?

    [***Hint***: Reduce $T_{sys}$ or increase $D$.]

17. The incremental temperature is found to be negative in a single-dish radio telescope. It is not possible to reduce $T_{sys}$ or to increase $D$. What can you do to make the telescope work? [***Hint***: Use time integration.]

18. In general terms, how is the far electric field related to the aperture electric field? Give an equation. [***Hint***: See Eqs. (4.78) and (4.79)]

19. A one-dimensional antenna array uses ten isotropic radiators as its elements. If the separation between the elements is $d$, what is the array factor?

    [***Hint***: See Eq. (4.81)]

20. Describe the merits and demerits of a filled array.

21. Explain the grating array with a diagram.

22. How many cross and self multiplications can be obtained from a grating array consisting of thirty elements? [***Hint***: use Eq. (4.90)]

# *INTERFEROMETRY AND RADIO ARRAYS*

## 5.1 INTRODUCTION

Radio wavelengths are much larger than optical wavelengths. Thus single-dish radio telescopes have poor angular resolution as compared to optical telescopes. The resolution can be increased using antenna arrays with large spacings. An optical technique called *interferometry,* where two or more telescopes are combined, was extended to radio astronomy to improve the resolution. Interferometry uses antenna arrays, and we will begin with the basic principles of optics.

## 5.2 PRINCIPLES OF OPTICAL INTERFEROMETERS

Consider the geometry shown in Fig. 5.1a, where a distant monochromatic point source produces parallel rays near the vicinity of a convex lens of diameter $D$. The lens is shadowed by a screen that has two small holes of diameter $\Delta$. The holes act as apertures. Depending on the distance $d$ *between the holes*, an interference occurs on the screen. Fig. 5.1b shows the interference fringe pattern. If one of the holes is closed, the fringe vanishes, and the image pattern now takes the shape of the fringe envelope. If the apertures are made infinitely small ($\Delta \rightarrow 0$), the fringe becomes uniform (Fig. 5.1c). Note that there is a large difference between the maxima and minima, which results in a good contrast between the two.

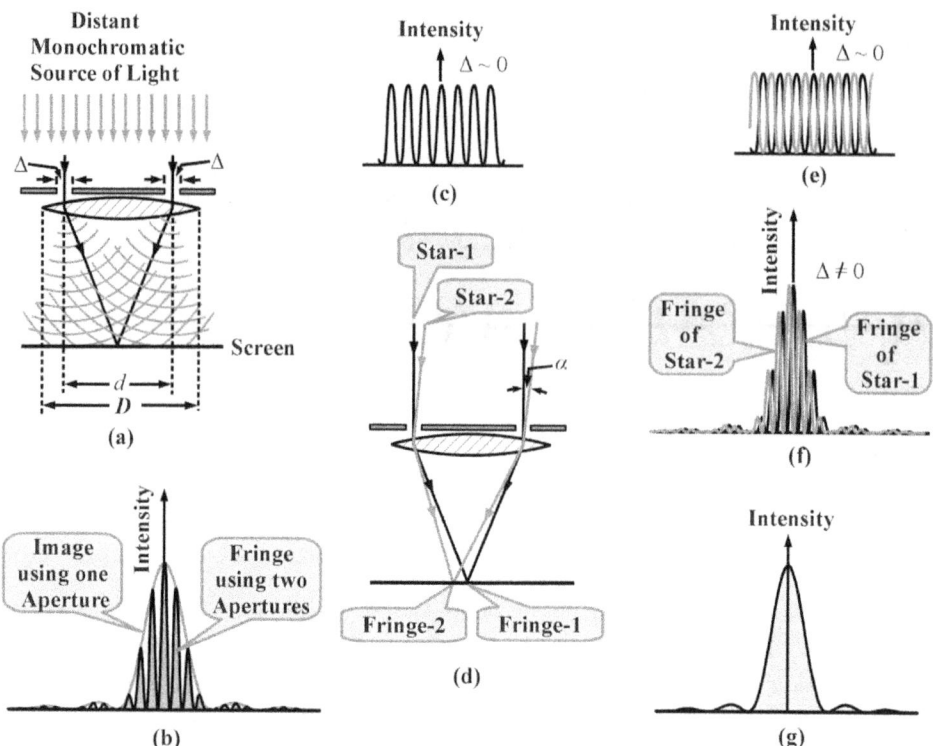

**FIGURE 5.1** Formation of fringes in an interferometer.

Now consider two stars (point sources) of equal intensity and separated by an angle $\alpha$, as shown in Fig. 5.1d. If the stars are monochromatic and have the same wavelength $\lambda$, two fringes, one corresponding to each, are formed. If both the stars are aligned ($\alpha = 0$), the two fringe patterns overlap each other, and the resulting intensity is doubled. The fringes cancel out if $\alpha = \alpha''$, where $\alpha''$ is given by Eq. (5.1). This is because the maxima of one fringe pattern coincides with the minima of the other, as shown in Figs. 5.1e and 5.1f. As $\Delta \to 0$, the light intensity tends to be more uniform on the screen. For finite aperture diameters ($\Delta \neq 0$), slightly blurred envelopes of the images are seen (Fig. 5.1f). If $\alpha$ is increased further, the fringes reappear. In general, the fringes are clearly visible when $\alpha = 2n\alpha''$, where $n$ is an integer. The fringes disappear when $\alpha = (2n + 1)\alpha''$.

$$\alpha'' = \frac{\lambda}{2d} \tag{5.1}$$

### 5.2.1 Interferometer Resolution

Applying the Rayleigh criterion, we find the resolution of a dual-aperture interferometer is $\alpha''$ (rad). In other words, it is the angular separation of two monochromatic point sources at which the corresponding fringe patterns are mutually displaced by half a fringe width (Figs. 5.1e and 5.1f). This is shown in Eq. (5.1). If the apertures are on the edges of the lenses ($d = D$), then $\alpha'' = \alpha'/2$, where $\alpha'$ is the resolution of the lens aperture (Chapter 4, Eq. (4.1)). Hence the resolution is doubled, but the aperture size is reduced. However, this is not a problem when two separate lenses are used with an arrangement of mirrors to combine their outputs. Increasing the distance $d$ of separation between the two apertures increase the angular resolution. This concept can be easily applied to radio arrays because they use separate antennas as elements. The two lenses can be replaced by two array elements.

### 5.2.2 Visibility (Fringe Contrast)

The fringe pattern can be seen as long as the contrast between the bright and dark areas of the fringe are good. It depends on the intensity differences between the fringe maxima and fringe minima. A dimensionless quantity known as *visibility* $\mathcal{V}$ is defined in Eq. (5.2), where $I_{max}$ and $I_{min}$, respectively, are two neighboring maxima and minima within the fringe. Visibility quantifies the fringe contrast. The image possesses maximum contrast if $I_{min} = 0$ . Under these conditions, $\mathcal{V} = 1$. No fringe is visible when $I_{min} = I_{max}$, which is indicated by $\mathcal{V} = 0$.

$$\mathcal{V} = \frac{I_{max} - I_{min}}{I_{max} + I_{min}} \tag{5.2}$$

### 5.2.3 Coherence Length

Reconsider looking at Fig. 5.1a. Irrespective of the source distance or wavelength $\lambda$, the brightest spot of the fringe always appears at the center of the image. However, the distance between a maxima and its neighboring minima depends on $\lambda$. If the source is polychromatic, a number of frequency components reach the screen, and their phases differ depending on their wavelengths. Each of these interferes with the others, which effectively reduces the fringe contrast $\mathcal{V}$. The path difference between two waves that results in $\mathcal{V} = 0$ is known as the *coherence length* $l_c$. It is given by Eq. (5.3), where $c$ is the speed of light, $n_r$ is the refractive index of the medium, $\Delta v$ is the bandwidth of light, $\lambda$ is the wavelength, and $\Delta\lambda$ is the waveband of the source (also known as the *spectral width*).

$$l_c = \frac{c}{n_r \Delta v} \approx \left( \frac{1.4}{\pi n_r} \right) \frac{\lambda^2}{\Delta \lambda} \tag{5.3}$$

The refractive index is given by $n_r = \sqrt{\varepsilon_r \mu_r}$, where $\varepsilon_r$ and $\mu_r$, respectively, represent the relative permittivity and relative permeability of the medium.

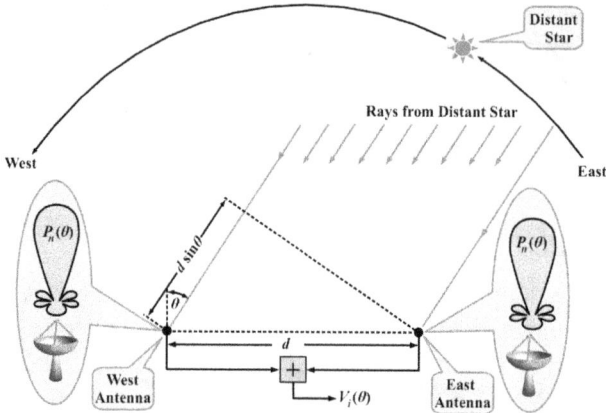

**FIGURE 5.2:** An additive radio interferometer. Two antennas with identical characteristics are aligned along the east-west line, separated by a distance $d$. The main beams are towards the zenith. The outputs of the antennas are added to form the interferometer output.

## 5.3   ADDITIVE RADIO INTERFEROMETER

Fig. 5.2 shows two identical antennas positioned along the east-west line under an open sky. The distance $d$ between the two is known as the *baseline*. In terms of wavelength $\lambda$, the baseline is given as $d_\lambda = d/\lambda$. The antenna beams are directed towards the zenith. Because the antenna output voltages are added, this arrangement is sometimes known as an *additive interferometer*. Note that the arrangement resembles a two-element grating array. Assume the instrumental phase delay is zero when the instantaneous antenna outputs are added. Consider a single radio star in the sky. Let the experiment be conducted on the Earth's equator. Hence the star rises in the east, passes through the zenith, and sets in the west. Let $\theta$ be the angle between the star and the zenith. We will consider for analysis the normalized beam patterns in a plane containing the east-west line and the zenith.

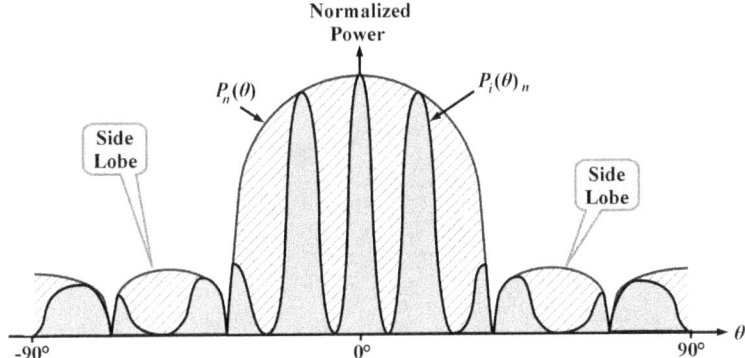

**FIGURE 5.3:** Fringe pattern of the additive radio interferometer shown in Fig. 5.2.

$$\tau = \frac{d \sin \theta}{c} \qquad (5.4)$$

We now attempt to establish a relationship between the interferometer output power and the source location ($\theta$). If the star is on the eastern sky, the wave fronts first reach the east antenna, and after a time delay $\tau$ they reach the west antenna. Eq. (5.4) expresses $\tau$ as a function of $d$ and $\theta$, where $c$ is the speed at which the wave propagates. When the source rises, $\tau$ reaches a maximum given by $\tau = d/c$. Similarly, when the source sets, $\tau = -d/c$. At the zenith, both antennas simultaneously receive the same wave fronts. In the adder, the two signals add up constructively or destructively depending on their relative phases. Hence, the adder output voltage $V_i$ is a function of $\theta$. The antenna beam patterns also scale the incoming signals. When the power output from the adder is plotted against $\theta$, a modulation of the antenna beam pattern is seen over the fringe. This is shown in Fig. 5.3, where $P_n(\theta)$ is the normalized power pattern of the antennas and $P_i(\theta)_n$ is the normalized power output from the adder. Note that the fringe pattern $P_i(\theta)_n$ is not equally spaced along $\theta$. Because the relative delay increases with the magnitude of $\theta$, the spacings also increase and get maximized at $\theta = \pm 90°$. When the star is at the zenith ($\theta = 0$), $\tau = 0$ and the adder output power peaks.

The additive interferometer can be improved. The antennas can track the source on the sky, and thus the modulation of the fringe by the beam pattern is avoided. Before adding the signals, the phases can be corrected by introducing artificial time delays so as to maximize the adder output. The artificial delay is also known as *instrumental time delay*. LNA (low-noise amplifiers) can be used after the antennas for maintaining a good SNR (signal-to-noise ratio).

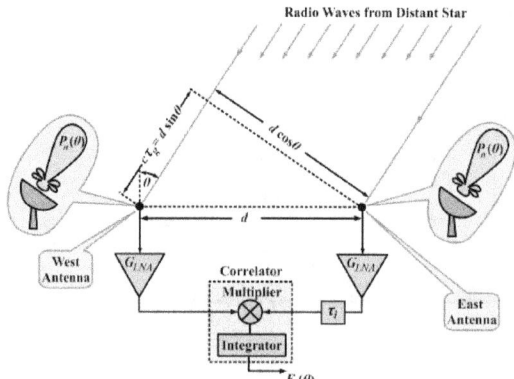

**FIGURE 5.4:** A multiplicative radio interferometer consisting of a pair of tracking antennas, correlator, and an adjustable time delay. Here, $c$ is the speed of electromagnetic waves.

## 5.4 MULTIPLICATIVE RADIO INTERFEROMETER

Fig. 5.4 shows a popular interferometer. The antennas keep tracking the source with time, and a time-varying instrumental delay $\tau_i$ is introduced at the antenna path that first receives the wave fronts. Unlike an additive interferometer, here, the signals are multiplied. Two LNAs with similar gains $G_{LNA}$ are connected to the antennas for signal amplification. Assuming no instrumental delay ($\tau_i = 0$), the multiplier output $F$ (before the integrator) is given by Eq. (5.5), where n is the frequency, $\tau_g = d \sin \theta / c$ is the geometrical delay between the antennas, and $t$ is the time. A tentative plot of the multiplier output is shown in Fig. 5.5.

$$F = 2\sin\left(2\pi v t\right)\sin\left\{2\pi v\left(t - \tau_g\right)\right\} \tag{5.5}$$

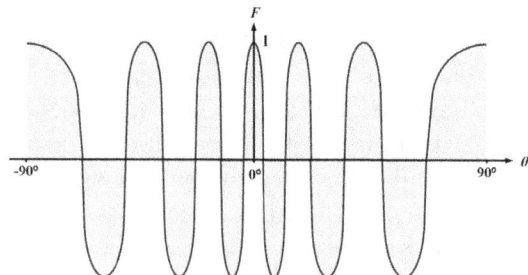

**FIGURE 5.5** Normalized squared voltage output $F$ from the multiplier for the interferometer shown in Fig. 5.4 ($\tau_i = 0$).

## 5.4.1 A Simplified Analysis

With reference to Fig. 5.4, for a point source, let the antennas produce voltages of equal magnitude. These are fed to the correlator consisting of a multiplier followed by an integrator. Consider one of the voltages $V(t)$ as reference. Let the other voltage lag behind by a time delay $\tau = \tau_g - \tau_i$, and represent the voltage as $V(1 - \tau)$. This delay is determined from the geometry of the baseline $d$ and the source angle $\theta$. Let integration be performed over $2T$ seconds, which happens to be the time constant. Due to the multiplication of voltages, the correlator output $r(\tau)$ is proportional to the power (square of the voltage). It is given in Eq. (5.6). The amplifier noise has been ignored here.

$$r(\tau) = \frac{1}{2T} \int_{-T}^{T} V(t) V(t - \tau)\, dt \qquad (5.6)$$

Generally, the time constant $2T$ varies from several milliseconds to few seconds. It should be chosen much higher than the reciprocal of the bandwidth $\Delta v$, where $v$ is the frequency. The upper limit of the integration time is the duration of the observation, which can be hours. Thus, for a very large integration, we may express $r(\tau)$ as shown in Eq. (5.7).

$$r(\tau) = \lim_{T \to \infty} \frac{1}{2T} \int_{-T}^{T} V(t) V(t - \tau)\, dt \qquad (5.7)$$

According to the Wiener-Khinchin relationship, the power spectrum (squared amplitude of the voltage spectrum) is the Fourier transform of the autocorrelation function of the signal. It applies to signals that are either deterministic or statistical by nature. Let $H(v)$ be the voltage spectrum. Then $|H(v)|^2$ represents the power spectrum. The correlator output $r(\tau)$ is related to $|H(v)|^2$ as shown in Eqs. (5.8) and (5.9), where we integrate over frequency and time, respectively.

$$|H(v)|^2 = \int_{-\infty}^{\infty} r(\tau) \exp(-j2\pi v\tau)\, d\tau \qquad (5.8)$$

$$r(\tau) = \int_{-\infty}^{\infty} |H(v)|^2 \exp(j2\pi v\tau)\, dv \qquad (5.9)$$

Note that for a wide-band noise source, $H(v)$ resembles the passband shape of the receiving system. Thus we may say that the interferometer output as a function of time delay $\tau$ is the Fourier transform of the cosmic signal's power spectrum, whose bandwidth is identical to the passband of the amplifiers.

### 5.4.2 Fringe-Stopping

Let the antennas be tracking a source on the sky. The propagation delay $d \sin\theta$ keeps changing with the Earth's rotation. If we compensate the geometrical delay $\tau_g = d \sin\theta/c$ by providing an equal instrumental delay $\tau_i$, such that $\tau = \tau_g - \tau_i = 0$, then the signals entering the correlator are in phase. The interferometer will thus give a peak output. This point of observation on the sky is called the *fringe-stopping center*. Because modern interferometers compensate the geometrical delay at the IF (intermediate frequency) instead of the RF (radio frequency), the visibility at any other direction changes slowly due to the rotation of the baseline with respect to the source. The time constant of the integrator should be short enough to observe these variations. If the delay is compensated in the RF stages, fringe-stopping is possible in any direction.

### 5.4.3 Simplified Analysis for a Gaussian Passband Response

Assume the amplifier passbands are Gaussian-shaped, centered at a frequency $\nu_0$, and have a bandwidth factor $\sigma$. Hence their bandwidth at half maxima is $\sqrt{8 \ln 2}\,\sigma$. The power spectrum obtained using these can be expressed as shown in Eq. (5.10). Substituting in Eq. (5.9) and integrating, we get $r(\tau)$ as given by Eq. (5.11). This is shown in Fig. 5.6a. Note that $r(\tau)$ is a cosine wave modulated by a Gaussian envelope. The correlator response $r(\tau)$ for a rectangular passband is shown in Fig. 5.6b.

$$|H(\nu)| = \frac{1}{2\sigma\sqrt{2\pi}}\left[\exp\left\{-\frac{(\nu - \nu_0)^2}{2\sigma^2}\right\} + \exp\left\{-\frac{(\nu + \nu_0)^2}{2\sigma^2}\right\}\right] \quad (5.10)$$

$$r(\tau) = \exp\left(-2\pi^2\tau^2\sigma^2\right)\cos\left(2\pi\nu_0\tau\right) \quad (5.11)$$

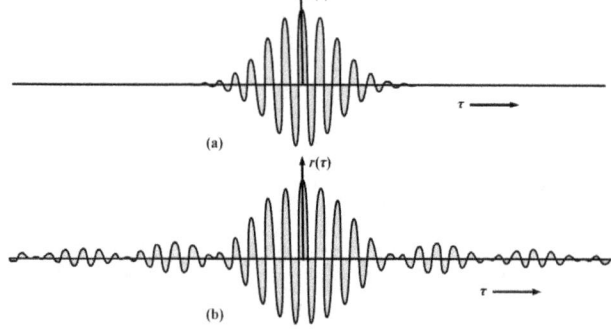

**FIGURE 5.6:** Response of the interferometer shown in Fig. 5.4 for a point radio source using (a) Gaussian and (b) rectangular passbands.

Let us set the instrumental delay $\tau_i = 0$ by using some electronic circuits, such that the delay is based on the angle $\theta$, wavelength $\lambda$, and baseline distance $d$. The correlator response may be expressed as Eq. (5.12). Note that wide bandwidths (large $\sigma$) and large baselines (large $d$) result in narrow fringe envelopes.

$$r(\tau_g) = \exp\left\{-2\left(\frac{\pi d\sigma}{c}\sin\theta\right)^2\right\}\cos\left(\frac{2\pi v_0 d}{c}\sin\theta\right) \qquad (5.12)$$

The ultimate aim of using a correlator is to image the astronomical sources in two dimensions (right ascension and declination), which will be explained later. For the present, remember that the fringe envelope is considered a nuisance for most mapping applications, except in VLBI (very long baseline interferometry). Generally, one uses the fringe where the amplitude is optimum, which is done by making $\tau_i$ equal to $\tau_g$ periodically with the antenna movement such that $\tau \to 0$. For practical purposes, $\tau_i$ is adjusted in time steps on the order of $v_0^{-1}$. This maintains a cosine correlator response.

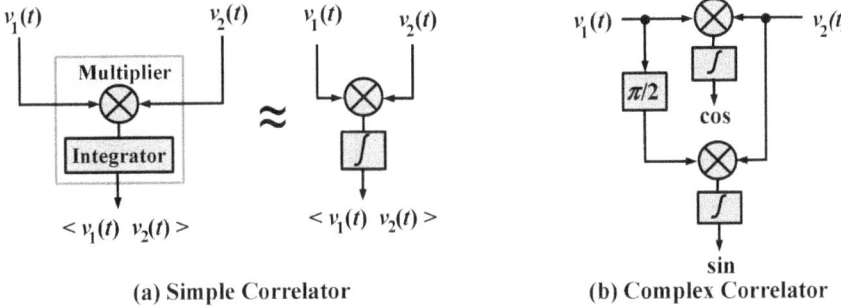

(a) Simple Correlator          (b) Complex Correlator

**FIGURE 5.7:** Block representations of correlator. (a) A simple correlator. (b) A complex correlator.

## 5.5   THE COMPLEX CORRELATOR

Fig. 5.7a symbolizes the basic correlator. The two signal voltages $V_1(t)$ and $V_2(t)$ are its inputs. The output is represented as $r(t) = < V_1(t)\ V_2(t) >$. Consider the two inputs as cosine waves of the same frequency $v$ but with different phase shifts $a_1$ and $a_2$. Let their amplitudes be $A_1$ and $A_2$. These inputs are expressed in Eq. (5.13). The product $P_{12}(t)$ of the two signals (before integration) is given by Eq. (5.14). The integrated output is expressed in Eq. (5.15).

$$\left.\begin{array}{l} V_1(t) = A_1\cos\left(2\pi vt + \alpha_1\right) \\ V_2(t) = A_2\cos\left(2\pi vt + \alpha_2\right) \end{array}\right\} \qquad (5.13)$$

$$P_{12}(t) = \frac{1}{2}A_1 A_2 \left\{ \cos\left(4\pi vt + \alpha_1 + \alpha_2\right) + \cos\left(\alpha_1 - \alpha_2\right) \right\} \qquad (5.14)$$

$$r(\tau) = <V_1(t)V_2(t)> = \frac{1}{2}A_1 A_2 \cos\left(\alpha_1 - \alpha_2\right) \qquad (5.15)$$

Thus we see that correlator output $r(\tau)$ depends on the phase difference between the two signals. The output is a maximum when the phase difference becomes zero.

Fig. 5.7b shows a symbolized complex correlator. Let the input signals be complex, as shown in Eq. (5.16), where $A_1$ and $A_2$ represent the amplitudes, and $\alpha_1$ and $\alpha_2$ represent the phases of the inputs $V_1(t)$ and $V_2(t)$, respectively. The complex correlator gives the product of one signal with the complex conjugate of the other, as in Eq. (5.17). Complex correlations thus produce both real and imaginary outputs. The real output (shown as the cosine) depends only on the phase difference. The imaginary output (shown as the sine) has an added phase difference of $\pi/2$ in one of the two signals.

$$\left.\begin{array}{l} V_1(t) = A_1\, e^{j(2\pi vt + \alpha_1)} \\[4pt] V_2(t) = A_2\, e^{j(2\pi vt + \alpha_2)} \end{array}\right\} \qquad (5.16)$$

$$\left.\begin{array}{l} V_1(t)V_2^*(t) = A_1 A_2\, e^{j(2\pi vt + \alpha_1)}\, e^{-j(2\pi vt + \alpha_2)} \\[4pt] \qquad = A_1 A_2 \left[\cos\left(\alpha_1 - \alpha_2\right) + j\sin\left(\alpha_1 - \alpha_2\right)\right] \end{array}\right\} \qquad (5.17)$$

The complex correlator gives information about both amplitude and phase. Because the two channels are independent, the noise immunity of the complex correlator is $\sqrt{2}$ times better.

## 5.6  CORRELATOR ARRAYS

Like phased arrays, correlator arrays also possess certain advantages over single-dish radio telescopes. They, too, comprise a number of antenna elements. Fig. 5.8 shows a four-element correlator array. Here, simple correlators (multiplier-integrator) are used. Four self-correlations ($<V_1^2>$, $<V_2^2>$, $<V_3^2>$, and $<V_2^4>$) and six cross-correlations ($<V_1V_2>$, $<V_1V_3>$, $<V_1V_4>$, $<V_2V_3>$, $<V_2V_4>$, and $<V_3V_4>$) are produced. In general, for $n_a$ antennas, the number of cross-correlations $n_c$ is given by Eq. (5.18). The number of self-correlations is $n_a$.

$$n_c = n_a \frac{(n_a - 1)}{2} \qquad (5.18)$$

If the self-correlations are ignored, the instantaneous sensitivity of the array is reduced by a factor $\kappa_c$, as given by Eq. (5.19). This loss becomes negligible if $n_a \gg 1$.

$$\kappa_c = 1 - \frac{1}{n_a} \qquad (5.19)$$

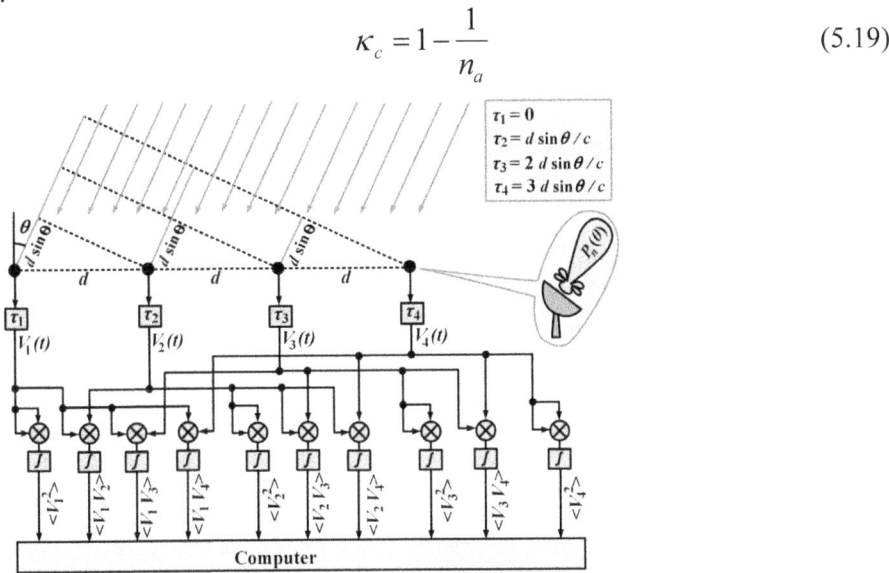

**FIGURE 5.8:** A four-antenna correlator array.

In principle, the correlated outputs can be combined to generate crests similar to phased arrays. If the output of one of the antennas is delayed by a phase shift, corresponding phase changes are seen in the fringes obtained using that antenna as one element. A beam-scanning effect may be achieved by combining the cross-correlations with appropriate phase variations using a computer. Unlike the phased array where the response is determined from the narrow beam it forms, the correlator array responds to the whole field of view of the individual antennas. Hence, correlator arrays receive data more efficiently than phased arrays.

## 5.6.1 Cross-Correlator Arrays versus Phased Arrays

A cross-correlator array generates only the cross terms from two different antennas. Eq. (5.19) shows that the sensitivity of the reduction is negligible for an array composed of many antennas. Thus from a large array, radio maps can be constructed using the cross terms alone, without affecting their quality. A comparison between the cross-correlator array with the phased array is listed below:

- The response of a point source using a cross-correlator array is the same as that of a phased array, except for the absence of the self-correlator terms in the former.

- If the source is extended, the cross-correlator gives complete information, whereas the phased array catches signals at equal phase intervals within the main beam and side lobes.

- Because the self-correlator terms are absent, the cross-correlator output can be positive or negative, whereas the phased-array output is always positive because it uses a square-law detector.

- The signal-to-noise ratio obtained from cross-correlation is better than that of a phased array. The reason is that while the correlator gathers coherent signals, incoherent noise from different antennas is diminished. Hence the cross terms are more important than the self terms. A phased array adds both noise and signals, so the noise cancellation is less.

## 5.7 CROSS-CORRELATORS USING DIFFERENT RECEIVERS

We have discussed simple correlator systems employing delay lines in the RF (radio frequency). We now discuss correlators using complex heterodyne receivers with delay applied in the IF (intermediate frequency). Phase changes taking place within the receivers significantly affect the visibilities.

Fig. 5.9a shows a model of a cross-correlator using a heterodyne receiver. There are two heterodyne receivers fed from a common LO (local oscillator) having a frequency $V_0$. Each receiver consists of a mixer followed by an IF bandpass filter and an amplifier. The changing geometrical time delay $\tau_g$ is continuously compensated after IF conversion by providing a suitable instrumental time delay $\tau_i$ based on the antenna movements. Let the selected RF bandwidth be the same as $\Delta v_{IF}$ but centered at a frequency $V_0$ (the RF and IF bandwidths are same). This is shown in Fig. 5.9b. The input signal spectra is $v_0 \pm 0.5\Delta v_{IF}$. Each sideband from the mixer output has a bandwidth equal to $\Delta V_{IF}$. The lower and upper sidebands, respectively, are centered at $v_{LO} - v_0$ and $v_{LO} + v_0$. When the incoming RF at the antennas is $v_{LO} + v_0$ or $v_{LO} - v_0$, the IF at the correlator input is $v$.

### 5.7.1 Single-Sideband Response

Let $\phi_m$ be the phase change in the signal path from antenna $m$ to the correlator input resulting from $\tau_g$ and the local oscillator. Let $\phi_n$ be the corresponding phase change in the signal path from antenna $n$ to the correlator (including the instrumental delay $\tau_i$). With reference to the $u,v,w$ coordinate system shown in Appendix A.9, we express the visibility $\mathcal{V}(u, \mathrm{v})$ by its magnitude and phase form in the $u$–$v$ plane as

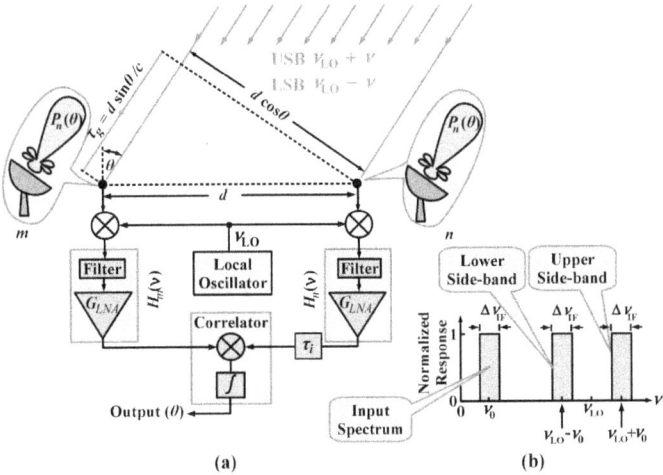

**FIGURE 5.9:** (a) A heterodyne correlator receiving system using two antennas of an array. The geometric delay $\tau_g$ is nullified continuously by the instrumental delay $\tau_i$ while tracking the source. $H_m(v)$ and $H_n(v)$ are the overall bandpass characteristics of the amplifiers and filters in the signal path. (b) Bandpass characteristics of the input signal and mixer outputs with reference to LO frequency $V_{LO}$ and input center frequency $V_0$.

in Eq. (5.20), where $\phi_v$ is the phase of the visibility. Note that the $u$–$v$ plane is perpendicular to the direction of the source.

$$\mathcal{V}(u,v) = |\mathcal{V}|e^{j\phi_v} \tag{5.20}$$

Within a passband $dv$, the infinitesimal correlator response $dr$ can be expressed as Eq. (5.21), where $A_e$ is the effective aperture area of any one antenna (same for both antennas) towards the source, and $H_m(v)$ and $H_n(v)$ are the response of the filter amplifier combinations. Note that $H_n^*(v)$ is the complex conjugate of $H_n(v)$. The correlator response over the entire system passband is the integration of $dr$ over the entire bandwidth and is given by Eq. (5.22).

$$dr = \text{Re}\left[ A_e |\mathcal{V}| H_m(v) H_n^*(v) e^{j(\phi_n - \phi_m - \phi_v)} \right] \tag{5.21}$$

$$r = \text{Re}\left[ A_e |\mathcal{V}| \int_{-\infty}^{\infty} H_m(v) H_n^*(v) e^{j(\phi_n - \phi_m - \phi_v)} \, dv \right] \tag{5.22}$$

### 5.7.1.1 Upper-Sideband Reception with Single IF Conversion

Let the filters in Fig. 5.9a allow only the upper-sideband frequencies $v_{LO} + v_0$ to pass. Due to the geometric delay $\tau_g$, signals from antenna $m$ undergo a phase shift of $2\pi(v_{LO} + v)\tau_g$ when they reach the mixer. Let $\phi_m$ be the negative phase contribution from the LO to the signal from antenna $m$ in the mixer. Similarly, let $\phi_n$ be the negative phase contribution from the LO to the signal from antenna $n$ in the mixer. The overall phase changes $\phi_m$ and $\phi_n$, respectively, for the paths from $m$ and $n$ antennas to the mixer, are expressed in Eqs. (5.23) and (5.24).

$$\phi_m = -2\pi\left(v_{\text{LO}} - v\right)\tau_g - \theta_m \tag{5.23}$$

$$\phi_n = -2\pi v\tau_i - \theta_n \tag{5.24}$$

The upper-sideband correlator output $r_{\text{usb}}$ is given in Eq. (5.25), where $\Delta\tau$ is given as $\Delta\tau = \tau_g - \tau_i$. It is the difference between geometric and instrumental time delays.

$$r_{\text{usb}} = \text{Re}\left[ A_e | \mathcal{V} | e^{j\left[2\pi v_{\text{LO}}\tau_g + (\phi_n - \phi_m - \phi_v)\right]} \right.$$
$$\left. \times \int_{-\infty}^{\infty} H_m(v) H_n^*(v) e^{j2\pi v\Delta\tau} dv \right] \tag{5.25}$$

Let $\Delta\tau$ be small but nonzero and continuously maintained. If the source is sufficiently close to the center of the observing field, then the real part of the integral in Eq. (5.25) becomes half of the Fourier transform of the cross power spectrum $H_m(v)H_n(v)$. Because $\Delta\tau = \tau_g - \tau_i$ is nonzero, it introduces a linear phase change across the band. Let the magnitudes of $H_m(v)$ and $H_n(v)$ be identical and flat across the bandwidth as shown in Eq. (5.26), where $\Delta v_{\text{IF}}$ represents the IF filter bandwidth (see Fig. 5.9b).

$$|H_m(\mathcal{V})| = |H_m(\mathcal{V})| = \begin{cases} H_0, & |v - v_0| < \dfrac{\Delta v_{\text{IF}}}{2} \\ 0, & |v - v_0| > \dfrac{\Delta v_{\text{IF}}}{2} \end{cases} \tag{5.26}$$

We define $G_{mn} = |G_{mn}| e^{j\phi_G}$ as the instrumental gain factor, where $\phi_G$ is the phase difference between the two signal paths produced by the filter-amplifier combinations. Note that the LO phase contributions ($\phi_m$ and $\phi_n$) are not included in $\phi_G$ because they appear in the upper and lower sidebands with opposite signs. For a nonzero $\Delta\tau$ we obtain Eq. (5.27).

$$|G_{mn}(\Delta\tau)| e^{j(2\pi v_0\Delta\tau + \phi_G)} = A_e \int_{-\infty}^{\infty} H_m(v) H_n^*(v) e^{j2\pi v\Delta\tau} dv \tag{5.27}$$

Substituting Eq. (5.27) in Eq. (5.25) we obtain the correlator output $r_{sub}$ for the upper sideband as given in Eq. (5.28).

$$r_{usb} = |\mathcal{V}| |G_{mn}(\Delta\tau)| \cos\left[2\pi\left(\nu_{LO}\,\tau_g + \nu_0\Delta\tau\right) + \left(\theta_m - \theta_n\right) - \phi_v + \phi_G\right] \quad (5.28)$$

The $2\pi\nu_0\tau_g$ term inside the cosine function produces a quasisinusoidal variation of $r_{sub}$ with the motion of the source. The overall phase depends on (*i*) the phase responses of the signal channels, (*ii*) the delay error $\Delta\tau$, (*iii*) the phase of the visibility function, and (*iv*) the relative phases of the LO signals. Because the geometrical delay $\tau_g$ appears at the RF, whereas the compensating instrumental delay $\tau_i$ is applied in IF, the output oscillates. The oscillation frequency $\nu_{fringe}$ at the output is the *natural fringe frequency*. It is expressed in Eq. (5.29). Because the frequencies at RF and IF differ by $\nu_{LO}$, the contribution to $\nu_{fringe}$ comes from $\nu_{LO}$ as the source changes its position in the sky.

$$\nu_{fringe} = \nu_{LO}\frac{d\tau_g}{dt} \quad (5.29)$$

*5.7.1.2 Lower-Sideband Reception with Single IF Conversion*
In this case, the filters are designed to allow the lower sideband to pass alone. Hence, increasing the phase of the signal at the RF decreases the phase at the IF. The signs of the phases $\phi_m$, $\phi_n$, and the visibility phase $\phi_v$, will be the opposite from the upper-sideband case. We accordingly modify the phase Eqs. (5.23) and (5.24) and rewrite them as Eqs. (5.30) and (5.31), respectively.

$$\phi_m = 2\pi\left(\nu_{LO} - \nu\right)\tau_g + \theta_m \quad (5.30)$$

$$\phi_n = -2\pi\nu\tau_i + \theta_n \quad (5.31)$$

The correlator output $r_{lsb}$ for the lower sideband can be expressed as Eq. (5.32). After introducing the instrumental gain $G_{mn}$, we rewrite Eq. (5.32) as expressed in Eq. (5.33).

$$r_{lsb} = \text{Re}\left[A_e|\mathcal{V}|e^{-j\left\{2\pi\nu_{LO}\tau_g + \left(\phi_n - \phi_m - \phi_v\right)\right\}}\right. \quad (5.32)$$

$$\left. \times \int_{-\infty}^{\infty} H_m(\nu)H_n^*(\nu)e^{j2\pi\nu\Delta\tau}d\nu\right]$$

$$r_{lsb} = |\mathcal{V}||G_{mn}(\Delta\tau)|\cos\left[2\pi\left(\nu_{LO}\,\tau_g - \nu_0\Delta\tau\right) + \left(\theta_m - \theta_n\right) - \phi_v - \phi_G\right] \quad (5.33)$$

*5.7.1.3 Lower-Sideband Reception with Multiple IF Conversions*
For technical reasons, it may sometimes be advantageous to upconvert or down-convert the signals more than once. These conversions are done at several stages of the receiving system using different mixers and LO combinations. For a signal that is downconverted once, the lower-sideband spectrum gets flipped. The frequencies at the lower end of the RF appear at the higher end of the IF. If the down conversion is done once more, the spectrum flips back. Hence, if the number of down conversions are odd, the final IF has a reversed spectrum. The lower-sideband correlator output $r_{even(lsb)}$ for an even number of IF conversions is shown in Eq. (5.34), where $\Sigma \nu_{LO}$ represents the sum of the LO frequencies. The lower-sideband correlator output $r_{odd(lsb)}$ for an odd number of IF conversions is given in Eq. (5.35).

$$r_{even(lsb)} = |\mathcal{V}| |G_{mn}(\Delta \tau)| \times \left. \begin{array}{c} \\ \cos\left\{2\pi\left(\sum \nu_{LO} \tau_g + \nu_0 \Delta \tau\right) + \left(\theta_m - \theta_n\right) - \phi_v + \phi_G\right\} \end{array} \right\} \tag{5.34}$$

$$r_{odd(lsb)} = |\mathcal{V}| |G_{mn}(\Delta \tau)| \times \left. \begin{array}{c} \\ \cos\left\{2\pi\left(\sum \nu_{LO} \tau_g - \nu_0 \Delta \tau\right) + \left(\theta_m - \theta_n\right) - \phi_v - \phi_G\right\} \end{array} \right\} \tag{5.35}$$

*5.7.1.4 Fringe-Stopping in Single-Sideband Correlators*
The basic concept of fringe-stopping has been explained in Subsection 5.5.4.2. From Eq. (5.29) we know that the correlator also produces a natural fringe frequency component $\nu_{fringe}$ if heterodyne receivers are used. To stop these fringe oscillations, a continuous phase change is applied to one of the LO. As understood from Eqs. (5.28) and (5.33), these fringe oscillations can be eliminated if we are able to vary $\theta_m - \theta_n$ at a rate that maintains a constant modulo $2\pi$ to the term $[2\pi\nu_{LO}\tau_g + (\theta_m - \theta_n)]$. This is achieved either by adding $2\pi\nu_{fringe}$ to $\theta_m$ or subtracting it from $\theta_n$.

To understand the effect of fringe frequency in more detail, consider the geometry shown in Fig. 5.10. Let two antennas A and B be located at different longitudes and/or latitudes. Relative to a radio source, the correlator baseline AB is shown for three different time instants $t_1, t_2, t_3$, where $t_1 < t_2 < t_3$. As shown, on a $u,v,w$ coordinate system (see Appendix A.9) the baseline AB and the source distance (along $w$) changes with the Earth's rotation. The path length of the waves reaching the antennas keeps changing. The angular frequency $\omega_e$ of the Earth's rotation equals $7.29115 \times 10^{-5}$ rad sec. This is obtained as $\omega_e = dH/dt$, where $H$ is the hour angle (see the equatorial coordinate systems in Appendix A.3). The fringe frequency at the correlator output is given by Eq. (5.36), where $\Delta$ is the source declination.

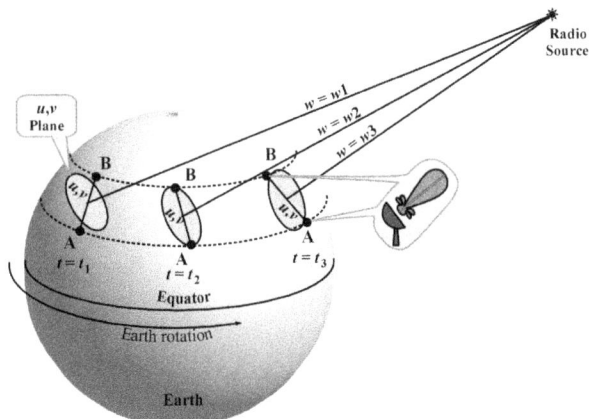

**FIGURE 5.10:** With respect to a radio source, the positions of two antennas A and B (at different latitudes) forming an interferometer baseline are shown for three different instants of time. The position of the antennas with respect to the *u-v* plane and the length of the *w* coordinate changes with the Earth's rotation.

$$\frac{dw}{dt} = \frac{dw}{dH}\frac{dH}{dt} = \frac{dw}{dH}\omega_e = -\omega_e u \cos\delta \qquad (5.36)$$

For fringe-stopping, we evaluate $d\tau_g/dt$. The $w$ coordinate is given by Eq. (5.37), where $c$ is the speed of light. Using Eqs. (5.36) and (5.37), the value of $d\tau_g/dt$ can be expressed as Eq. (5.38). Fringe-stopping is usually done by introducing a continuous phase change to one of the local oscillators.

$$w = c\tau_g \qquad (5.37)$$

$$\frac{d\tau_g}{dt} = -\frac{\omega_e u \cos\delta}{c} \qquad (5.38)$$

### 5.7.1.5 Complex Correlators in Single-Sideband Systems

Complex correlators are preferred in fringe-stopping because they produce real and imaginary parts (Fig. 5.7b). One of the signal inputs to the mixer is given a phase shift of $\pi/2$ using an electronic quadrature phase shifter. The cosine output $r_{real}$ and sine output $r_{imag}$ are considered real and imaginary, respectively. The phase information is gathered from these two. Under fringe-stopping conditions, $r_{real}$ and $r_{imag}$ are, respectively, shown in Eqs. (5.39) and (5.40).

$$r_{real} = \mathrm{Re}[\mathcal{V}]\int_{-\infty}^{\infty} H_m(\nu)H_n^*(\nu)d\nu \qquad (5.39)$$

$$r_{imag} = \mathrm{Im}[\mathcal{V}]\int_{-\infty}^{\infty} H_m(\nu)H_n^*(\nu)d\nu \qquad (5.40)$$

Mathematically, two conditions are maintained: (*i*) $\Delta\tau = 0$ and (*ii*) $2\pi\nu_{LO}\tau_i + (\theta_m - \theta_n) = 0$. Hence, correlator outputs $r_{real}$ and $r_{imag}$ represent the real and imaginary parts of $G_{mn}\mathcal{V}(u, v)$, respectively.

### 5.7.2 Double-Sideband Response

Fig. 5.11a shows the spectra of the input and sidebands for $\nu_0 = 1.5\nu_{IF}$. The bandpass filters block everything except the two sideband signals. Hence the double-sideband correlator response $r_{dsb(cos)}$ is the sum of Eqs. (5.28) and (5.33). This is given in Eq. (5.41).

$$\begin{aligned} r_{dsb(cos)} = r_{usb} + r_{lsb} = 2\,|\,\mathcal{V}\,||\,G_{mn}\left(\Delta\tau\right)|\cos\left\{2\pi\nu_0\Delta\tau + \phi_G\right\} \\ \times\cos\left\{2\pi\nu_{LO}\,\tau_g + \left(\theta_m - \theta_n\right) - \phi_\nu\right\} \end{aligned} \qquad (5.41)$$

Note that a similar result is obtained from the cosine (real) output of a complex correlator. In that case, the sine (imaginary) output is given by Eq. (5.42).

$$\begin{aligned} r_{dsb(sin)} = 2\,|\,\mathcal{V}\,||\,G_{mn}\left(\Delta\tau\right)|\sin\left\{2\pi\nu_0\Delta\tau + \phi_G\right\} \\ \times\cos\left\{2\pi\nu_{LO}\,\tau_g + \left(\theta_m - \theta_n\right) - \phi_\nu\right\} \end{aligned} \qquad (5.42)$$

In Eq. (5.41), the cosine function operating on $2\pi\nu_{LO}\tau_g + (\theta_m - \theta_n) - \phi_\nu$ is independent of $\Delta\tau$ or $\phi_G$. It rapidly modulates the fringe amplitude. This can be seen if $\tau_i$ is held constant but $\Delta\tau$ keeps changing.

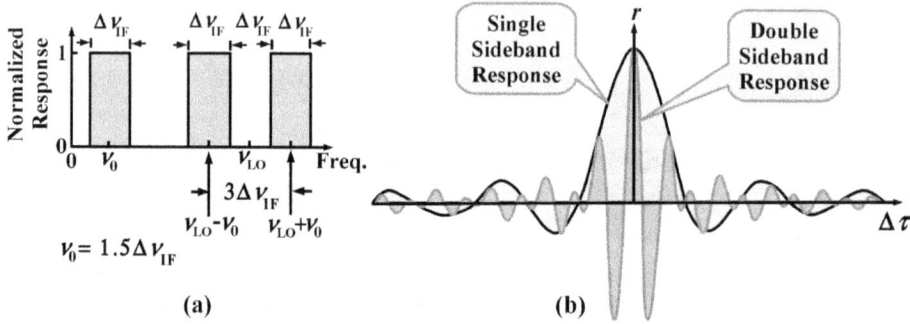

**FIGURE 5.11:** An example of fringe amplitude variation as a function of $\Delta\tau$ for double- and single-sideband correlator systems. (a) Spectra of the input and sidebands for $\nu_0 = 1.5\nu_{IF}$. (b) Amplitude variation of the fringes as a function of $\Delta\tau$.

Fig. 5.11b shows an example of fringe amplitude variation as a function of $\Delta\tau$, where the center frequency $\nu_0$ of the received signal is taken as $1.5\Delta\nu_{IF}$, and the instrumental delay $\tau_i$ is kept constant. If the term $2\pi v_0 \Delta \tau + \phi_G$ is adjusted to maximize the real output, the imaginary output vanishes, and vice versa. Hence, in continuum observations where both the sidebands are of equal strength, the complex correlator does not give any extra sensitivity.

### 5.7.2.1 Double-Sideband Reception with Multiple IF Conversions

Fig. 5.12a shows a simple system consisting of two frequency conversions. The interferometer has two IF stages. Next to the antennas, double-sideband receivers are used. The second conversion near the correlator has only an upper sideband. Two compensating delays ($\tau_{i1}$ and $\tau_{i2}$) are used. The filter-amplifier responses are shown in Figs. 5.12b and 5.12c. The signal phases $\theta_m$ and $\theta_n$ at the correlator input are given by Eqs. (5.43) and (5.44), respectively.

$$\phi_m = \mp 2\pi \left( v_1 \pm v_2 \pm v \right) \tau_g \mp \theta_{m1} - \theta_{m2} \tag{5.43}$$

$$\phi_n = -2\pi \left\{ \left( v_2 + v \right) \tau_{i1} + v \tau_{i2} \right\} \mp \theta_{m1} - \theta_{m2} \tag{5.44}$$

Note that upper signs in the symbols $\pm$ represent the upper-sideband conversions at both first and second IF for each antenna. The lower signs represent the same for the lower sideband. The correlator responses of the upper and lower sidebands are given in Eqs. (5.45) and (5.46), respectively. The complete double-sideband response is given in Eq. (5.47), where $\Delta \tau = \tau_g - \tau_{i1} - \tau_{i2}$.

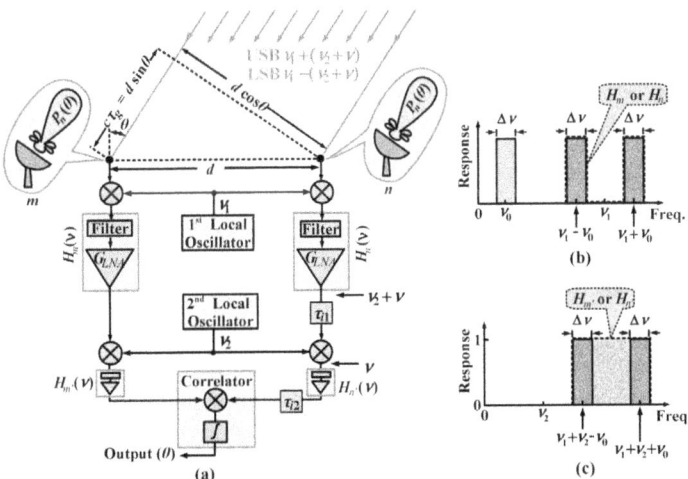

**FIGURE 5.12:** Interferometer with two-stage frequency conversions. (a) The frequency conversion next to the antennas uses a double-sideband receiver, whereas the second conversion (near the correlator) uses the upper sideband alone. Two compensating delays ($\tau_{i1}$ and $\tau_{i2}$) are included to study their effect in mixing. (a) Response of $H_m$ or $H_n$. (c) Response of $H_m$ or $H_n$.

$$r_{usb} = |\mathcal{V}|\,|G_{mn}(\Delta\tau)|\cos\left[2\pi\nu_1\tau_g + 2\pi\nu_2\left(\tau_g - \tau_{i1}\right)\right.$$
$$\left. + 2\pi\nu_0\Delta\tau + \left(\theta_{m1} - \theta_{n1}\right) + \left(\theta_{m2} - \theta_{n2}\right) - \phi_\nu + \phi_G\right] \tag{5.45}$$

$$r_{lsb} = |\mathcal{V}|\,|G_{mn}(\Delta\tau)|\cos\left[2\pi\nu_1\tau_g - 2\pi\nu_2\left(\tau_g - \tau_{i1}\right)\right.$$
$$\left. - 2\pi\nu_0\Delta\tau + \left(\theta_{m1} - \theta_{n1}\right) - \left(\theta_{m2} - \theta_{n2}\right) - \phi_\nu - \phi_G\right] \tag{5.46}$$

$$r_{dsb} = r_{usb} + r_{lsb}$$
$$= 2|\mathcal{V}|\,|G_{mn}(\Delta\tau)|\cos\left[2\pi\left\{\nu_2\left(\tau_{i1} - \tau_g\right) - \nu_0\Delta\tau\right\}\right. \tag{5.47}$$
$$\left. - \left(\theta_{m2} - \theta_{n2}\right) - \phi_G\right]\cos\left\{\nu_1\tau_g + \left(\theta_{m1} - \theta_{n1}\right) - \phi_\nu\right\}$$

The first cosine term modulates the fringe amplitude. The fringe phase is determined by the second cosine term, which depends only on the phase of the first LO. Hence, the phase shift should be applied to the first LO. The effects of $\tau_{i1}$ and $\tau_{i2}$ are seen when one is held constant and the other is varied as explained below:

(i) Let $\tau_{i1}$ be a variable that compensates for the delay, and let $\tau_{i2} = 0$. Hence in Eq. (5.47), $\tau_{i1} - \tau_g$ must be ideally zero, and $\phi_G$ should be minimized. It follows that $\theta_{m2}$ and $\theta_{n2}$ must be equalized to maximize the fringe amplitude. This is similar to the single-sideband conversion expressed in Eqs. (5.41) and (5.42).

(ii) Let $\tau_{i2}$ be a variable that compensates for the delay, and let $\tau_{i1} = 0$. This is preferred in large arrays with digital delay compensation. With changing $\tau_g$, a continuously varying phase shift is applied to $\theta_{m2}$ and $\theta_{n2}$. This is to maintain the first cosine term in Eq. (5.47) near unity. The fringe phase is not affected, but its amplitude varies.

### 5.7.2.2 Fringe-Stopping in Double-Sideband Correlators

For the correlator system shown in Fig. 5.12, let $\tau_{i1} = 0$, and let $\tau_{i2}$ be changed continuously to maintain $\Delta\tau_g = 0$. When the source moves over the sky, the path difference between the signals received at antennas $m$ and $n$ changes, resulting in fringe frequency. Let the signal frequency received by antenna $n$ be $\nu_{nRF}$. With respect to the wave front at antenna $n$, the antenna $m$ moves away as a function of time. Thus, there is a relative Doppler shift occurring in antenna $m$ with respect to antenna $n$. This results from the changing geometrical delay $\tau_g$ with the source position as a function of time. Mathematically, the RF frequency $\nu_{mRF}$ at antenna $m$ is related to the RF frequency $\nu_{nRF}$ of antenna $n$, as expressed in Eq. (5.48). For the upper-sideband signals received using antenna $m$, the correlator input frequency $v_{mCRR(in)}$ is given in Eq. (5.49), where $v_1$ and $v_2$ are the LO frequencies.

$$V_{mRF} = V_{nRF}\left(1 - \frac{d\tau_g}{dt}\right) \tag{5.48}$$

$$V_{mCRR(in)} = V_{nRF}\left(1 - \frac{d\tau_g}{dt}\right) - v_1 - v_2 \tag{5.49}$$

For fringe-stopping, the frequency obtained at antenna $n$ must be suitably decreased so that input signals to the correlator have identical frequencies. For this, the two LO frequencies, which are mixed with the signals from antenna $n$, are multiplied by $(1 + d\tau_g/dt)$. Mathematically, we add $2\pi(d\tau_g/dt)v_1$ to $\theta_{n1}$ and $2\pi(d\tau_g/dt)v_2$ to $\theta_{n2}$. These are the required LO phase-changing rates for maintaining the two cosine terms in Eq. (5.47) at fixed values. Simultaneously, the antenna $n$ signal is delayed by $\tau_{i2}$ at a frequency $v_{nRF}$ Because the delay $\tau_{i2}$ is adjusted continuously, the signal frequency gets reduced by a factor $(1 - d\tau_g/dt)$. Hence the correlator input frequency $v_{nCRR(in)}$ from the path of antenna $n$ can be expressed as in Eq. (5.50).

$$V_{nCRR(in)} = \left\{ V_{nRF} - \left(v_1 + v_2\right)\left(1 + \frac{d\tau_g}{dt}\right)\right\}\left(1 - \frac{d\tau_g}{dt}\right) \tag{5.50}$$

Note that if we neglect the second-order terms of $d\tau_g/dt$, Eq. (5.50) reduces to Eq. (5.48). If the signs of both $v_{nRF}$ and $v_1$ are reversed, then Eqs. (5.50) and (5.48) can also be applied to the lower sideband, thus making both the correlator input frequencies identical. The net result is that the fringes are stopped in both sidebands.

### 5.7.3 Signal-to-Noise Ratio

The SNR (signal-to-noise ratio) is considered at the final output. The power contribution from the source alone is considered to be the signal. The power contribution from the rest of the system is considered to be noise. These contributions depend on: (*i*) antenna temperature, (*ii*) system temperature, (*iii*) receiver system bandwidth, and (*iv*) integration bandwidth after multiplication.

In the following, we assume the source is at the center of the fringe pattern. We may thus eliminate the effect of the delay in our derivations. We also assume identical phase responses of the signal channels. The chosen correlator response is therefore the peak fringe amplitude, which is the modulus of the visibility ($|\mathcal{V}|$).

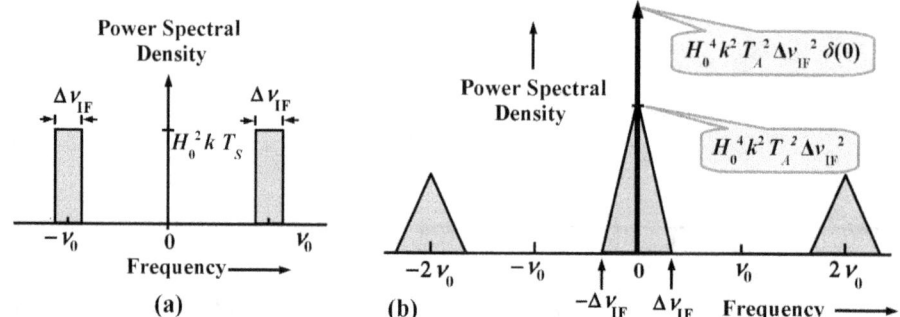

**FIGURE 5.13:** (a) Input spectrum of a correlator with a rectangular passband $\Delta v_{\mathrm{IF}}$. (b) The complete spectrum generated from the time-lag product (before integration) of two signals, including noise bands. Only frequencies close to zero are able to pass out from the integrator. The required signal spectrum appears as a delta function at zero frequency (shown by a thick arrow). It is assumed $T_A \ll T_S$, and  is the Boltzmann constant.

### 5.7.3.1 Signal Spectra near the Correlator

Assume that the antenna temperatures are less than the respective system temperatures. Fig. 5.13a shows the input spectra of a correlator with a rectangular passband $\Delta v_{\mathrm{IF}}$. Fig. 5.13b shows the complete spectra of the time-lag product of two signals (before integration). Only those components close to zero frequency remain after integration. The required signal spectrum is a delta function located at zero frequency. It is shown as a thick arrow. Because the input spectra $|H_m(v)|^2$ and $|H_n(v)|^2$ contain both negative and positive frequencies and are symmetric about the origin, the output noise spectrum can be considered proportional to either (i) the convolution between $|H_m(v)|^2$ and $|H_n(v)|^2$, or (ii) the cross-correlation between $|H_m(v)|^2$ and $|H_n(v)|^2$.

### 5.7.3.2 SNR in Analog Systems

Refer back to the system shown in Fig. 5.12. Let $T_{Am}$ and $T_{An}$ be the antenna temperatures of antennas $m$ and $n$, respectively. If the system temperatures are $T_{Sm}$ and $T_{Sn}$, respectively, for the paths associated with the $m$ and $n$ antennas, then the ratio of the rms signal-to-noise voltages $\mathcal{R}_{\mathrm{sn}}$ is given by Eq. (5.51), where $H_m(v)$ and $H_n(v)$ are, respectively, the responses of the systems associated with the $m$ and $n$ antennas, and $2\Delta v_{\mathrm{LF}}$ is the equivalent integration bandwidth of the correlator. Because the correlator output voltage is a result of the product between two antenna voltages, $\mathcal{R}_{\mathrm{sn}}$ actually represents the signal-to-noise power ratio (SNR) of the signal and system noise.

$$\mathcal{R}_{sn} = \frac{\sqrt{T_{Am}T_{An}}}{\sqrt{(T_{Am}+T_{Sm})(T_{An}+T_{Sn})+T_{Am}T_{An}}}$$
$$\times \frac{\int_{-\infty}^{\infty} H_m(v)H_n(v)dv}{\sqrt{2\Delta v_{LF}\int_{-\infty}^{\infty}|H_m(v)|^2|H_m(v)|^2\,dv}} \Bigg\} \qquad (5.51)$$

In general, $H_m(v)$ and $H_n(v)$ have identical rectangular bandwidths $\Delta v_{IF}$. Unless the antenna temperatures are greater than their respective system temperatures, detection is not possible. With these considerations we evaluate Eq. (5.51) and express it as Eq. (5.52).

$$\mathcal{R}_{sn} = \sqrt{\frac{T_{Am}T_{An}}{T_{Sm}T_{Sn}}}\sqrt{\frac{\Delta v_{IF}}{\Delta v_{LF}}} \qquad (5.52)$$

Let $\tau_a$ be the data-averaging time of the correlator. The averaging in the time domain may be described as the convolution of the time signal with a rectangular function of the unit area having a width $\tau_a$. Hence the equivalent bandwidth $2\Delta v_{LF}$ (including both positive and negative frequencies) is related to $\tau_a$ as given in Eq. (5.53).

$$2\Delta v_{LF} = \int_{-\infty}^{\infty}\frac{\sin^2(\pi\tau_a v)}{(\pi\tau_a v)^2}dv = \frac{1}{\tau_a} \qquad (5.53)$$

From Eqs. (5.52) and (5.53), the SNR ($\mathcal{R}_{sn}$) is given as shown in Eq. (5.54). If the antennas are single polarized, then each antenna receives only half the total flux density $S$. Then the received power $p_r$ for each antenna is given by Eq. (5.55), where $A_e$ is the effective aperture area of the antenna.

$$\mathcal{R}_{sn} = \sqrt{2\Delta v_{IF}\tau_a\left(\frac{T_{Am}T_{An}}{T_{Sm}T_{Sn}}\right)} \qquad (5.54)$$

$$p_r = k_B T_A = \frac{1}{2}A_e S \qquad (5.55)$$

Let the antennas be identical in nature. Using Eqs. (5.54) and Eq. (5.55), $\mathcal{R}_{sn}$ is expressed in Eq. (5.56). Note that the SNR can be improved by increasing the factor $\sqrt{\Delta v_{IF} \tau_a}$. Thus larger bandwidths ($\Delta v_{IF}$) and larger integration times ($\tau_a$) improve the sensitivity by improving the SNR.

$$\mathcal{R}_{sn} = \frac{A_e S}{k_B T_S} \sqrt{\frac{\Delta v_{IF} \tau_a}{2}} \qquad (5.56)$$

### 5.7.3.3 SNR in Digital Systems

The quantization process in digitizing the signals adds a little more noise to the system and is known as *quantization noise*. To include the quantization loss, an efficiency factor $\eta_Q$ has been introduced. The final expression for SNR ($\mathcal{R}_{sn}$) is given in Eq. (5.57). Here, $0 \le \eta_Q \le 1$. In terms of the antenna temperature, the SNR is given by Eq. (5.58).

$$\mathcal{R}_{sn} = \frac{A_e S \eta_Q}{k_B T_S} \sqrt{\frac{\Delta v_{IF} \tau_a}{2}} \qquad (5.57)$$

$$\mathcal{R}_{sn} = \frac{T_A \eta_Q}{T_S} \sqrt{\frac{\Delta v_{IF} \tau_a}{2}} \qquad (5.58)$$

Note that in deriving the above expressions, no delay was introduced between the two signals reaching the correlator. Also the phase responses of the signal channels are assumed identical. Hence the source must be in the center of the fringe pattern, and the response is for the peak fringe amplitude.

### 5.7.4 RMS Noise at Complex Correlator Output

Consider a simple correlator. We now substitute $\mathcal{R}_{sn} = 1$ and replace $S$ by $\sigma$ in Eq. (5.58). After simplification, we express $\sigma$ as shown in Eq. (5.59). We may visualize $\sigma$ as the flux density (in Watt m$^{-2}$ Hz$^{-1}$) of an unresolved source located at the phase reference point, and which gives a peak fringe amplitude equal to the noise from the system. Here, $T_S$ represents the system temperature in K, $k_B$ is the Boltzmann constant, $\Delta v_{IF}$ represents the effective rectangular IF bandwidth, $\tau_a$ is the data-averaging time, $A_e$ is the effective aperture area of each antenna, and $\eta_Q$ is the efficiency factor for the quantization loss.

$$\sigma = \frac{\sqrt{2} k_B T_S}{A_e \eta_Q} \frac{1}{\sqrt{\Delta v_{IF} \tau_a}} \qquad (5.59)$$

In a complex correlator under the fringe-stopping condition, the output cosine (real) and sine (imaginary) signal components form Hilbert transform pairs. However, this is not true for the noise components. Hence, the real and imaginary noise outputs are uncorrelated.

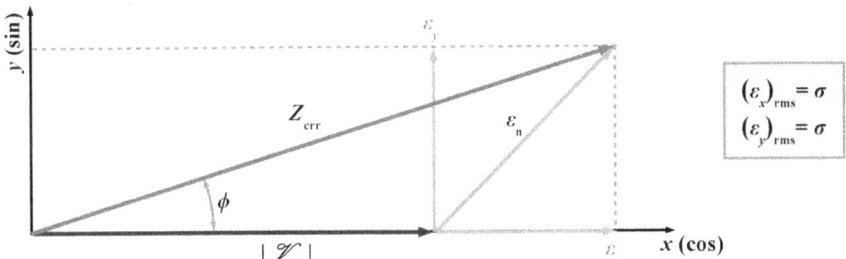

**FIGURE 5.14** The modulus of the noise-free visibility $|\mathcal{V}|$ is real. The noise $\varepsilon_n$ contains both real and imaginary components of rms amplitude $\sigma$. It causes a phase deviation $\phi$. The sum of the noise-free visibility and the noise is represented by $Z_{\text{crr}}$ which is the correlator output.

Let $|\mathcal{V}|$ be the modulus of the noise-free visibility. It is the square-root of thre sum of squares of the real (cosine) and imaginary (sine) outputs of a complex correlator. Let $\varepsilon_n$ be the noise. The correlator output $Z_{\text{crr}}$ contains both the signal $V$ and the noise $\varepsilon_n$. Using a vector diagram, with $|\mathcal{V}|$ as a reference, we visualize the effect of the noise on the correlator output. This is shown in Fig. 5.14. The phase deviation caused in the correlator output $Z_{\text{crr}}$ by the noise is $\phi$. The real and imaginary components of $\varepsilon_n$ are represented by $\varepsilon_x$ and $\varepsilon_y$, respectively. The corresponding rms values are represented by the real $(\varepsilon_x)_{\text{rms}}$ and the imaginary $(\varepsilon_y)_{\text{rms}}$, respectively. Each of these is equal to $\sigma$, which is the uncertainty in the measurement. The noise $\varepsilon_n$ has an rms amplitude $\varepsilon_{\text{rms}} = \sqrt{2}\sigma$. The derivation is given in Eq. (5.60).

$$\varepsilon_{\text{rms}} = \sqrt{\left(Z_{\text{crr}} \cdot Z_{\text{crr}}\right) - \left(Z_{\text{crr}}\right)^2} = \sqrt{\left(\varepsilon_n \cdot \varepsilon_n\right)} = \sqrt{2}\,\sigma \qquad (5.60)$$

When the antennas are pointed to a blank sky, the correlator output is due to noise alone, which can be measured. Pointing the antennas towards an unresolved source, the correlator output becomes $Z_{\text{crr}}$, which is the sum of the visibility and the noise. The measurement of $Z_{\text{crr}}$ must be under the fringe-stopping condition so that the correlator output is maximum. This can be done if the source position is known accurately. Note that the source must be unresolved, and both the real and the imaginary outputs of the complex correlator are required for computing $Z_{\text{crr}}$.

## 5.8 DIGITAL CORRELATOR SYSTEMS

Though digital systems have technological limitations in the context of high-speed sampling, they are advantageous to analog systems in many respects. They offer better delay control and give highly accurate timing pulses to the system. For these reasons, most of the digital systems of today stand on cutting-edge technology.

Analog signals are sampled at the Nyquist rate (or higher) and converted to discrete form without any information loss. These discrete values are then approximated to the nearest available digital value through a process called *quantization*. The number of quantized values $N_{qnt}$ comes from the number of bits $N_{bits}$ used for representing a discrete value. These are related as $N_{qnt} = 2^{N_{bits}}$. Because $N_{qnt}$ is a finite quantity, a distortion occurs in the digitized signal known as *quantization noise*. Hence, the digital correlator output is an approximation of the linear correlation functions $r(\tau)$ and $R(\tau)$.

### 5.8.1 Baseband Sampling

If the power spectrum of a signal is restricted within a band of frequencies, the signal is said to be *band limited*. If this band starts at zero frequency and ends at some fixed upper frequency (low-pass), it is known as a *baseband*.

According to the sampling theorem, a continuous signal can be discritized without loss of information if the sampling frequency is twice or more than the maximum frequency content of the signal. The smallest sampling rate at which the information content of the signal remains intact (after sampling) is known as the *Nyquist rate*. It is given in Eq. (5.61), where $\Delta v$ is the bandwidth of the baseband signal.

$$\text{Nyquist rate} \quad v_s = 2\Delta v \qquad (5.61)$$

Fig. 5.15 illustrates the sampling theorem, the Nyquist rate, and the reconstruction of signals. A baseband signal $x(t)$ having a maximum frequency component of $\Delta v$ is sampled by multiplying it with an equally spaced train of pulses (delta functions). The separation between the pulses is $\tau_s$, which is known as the *sampling rate*. The resulting sampled signal $x_s(t)$ is given in Eq. (5.62), where $n$ is an integer.

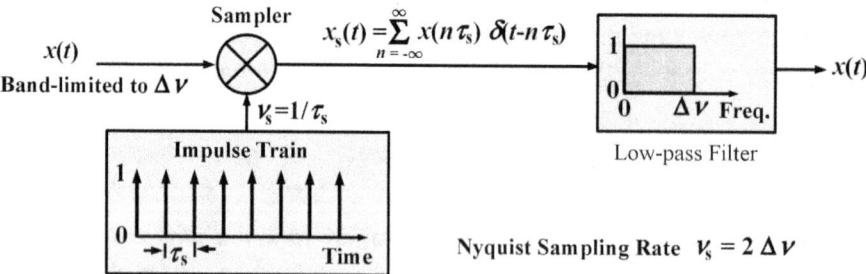

**FIGURE 5.15:** Discrete sampling and reconstruction of signals based on the sampling theorem.

$$x_s(t) = \sum_{n=-\infty}^{\infty} x(n\tau_s)\delta(t - n\tau_s) \tag{5.62}$$

The analog signal can be reproduced from $x_s(t)$ by passing it through a low-pass filter having a cut-off frequency $\Delta v$. Sampling at rates higher or lower than the Nyquist rate are known as *over-* and *under sampling*, respectively. A majority of radio telescope receivers convert the final IF into a baseband, which is then sampled at a rate slightly higher than the Nyquist rate and converted to digital. For a rectangular passband as shown in Fig. 5.15, the auto correlation function $R_\infty(\tau)$ is given by Eq. (5.63). The subscript symbol $\infty$ indicates infinite quantization levels (states before quantization).

$$R_\infty(\tau) = \frac{\sin(2\pi\Delta v\tau)}{2\pi\Delta v\tau} \tag{5.63}$$

### 5.8.2 Bandpass Sampling

Bandpass spectra can also be sampled at the Nyquist rate. Here, the analog signals lie within a frequency range from $n\Delta v$ to $(n+1)\Delta v$, where $n$ is an integer. In other words, the signal spectrum starts at some integral multiple of $\Delta v$ instead of zero frequency. Because the signal bandwidth is $\Delta v$, the Nyquist rate remains the same as for $2\Delta v$. The necessary condition for bandpass sampling at the Nyquist rate is that the lower and upper limits of the spectral band must be integral multiples of the bandwidth. Based on the Wiener-Khinchin relation, the autocorrelation $R_\infty(\tau)$ is given in Eq. (5.64).

$$R_\infty(\tau) = \frac{\sin(2\pi\Delta v\tau)}{2\pi\Delta v\tau}\cos\left\{2\pi\left(n + \frac{1}{2}\right)\Delta v\tau\right\} \tag{5.64}$$

Note that $R_\infty(\tau)$ is zero when the time interval $\tau$ becomes an integral multiple of $1/(2\Delta v)$. For a rectangular-shaped passband, the successive samples obtained at the Nyquist rate are uncorrelated. The central frequency of the signal spectrum may be adjusted suitably for conforming it to bandpass sampling. This will also minimize the sampling rate without any information loss. If the spectrum fails to conform, a slightly greater hypothetical bandwidth may be selected. The center frequency of this hypothetical band should be adjusted for proper bandpass sampling. The sampling will now be at slightly higher Nyquist rate (defined by the new hypothetical bandwidth).

### 5.8.3 Correlation of Sampled Signals without Quantization

Assume that quantization has not been done after sampling. In the following, we investigate what happens when sampling is performed at Nyquist and non-Nyquist rates.

### 5.8.3.1 Sampled at Nyquist Rates

Let $x_s(t)$ and $y_s(t)$ be the sampled versions of two band-limited signals $x(t)$ and $y(t)$, respectively. The samplings are done at Nyquist rates. These are fed to a correlator consisting of a multiplier and an integrator. Let the signals be in phase (time delay between them is zero). The normalized cross-correlation coefficient $r_{nrm}$ is shown below in Eq. (5.65).

$$r_{nrm} = \frac{< x_s(t)\, y_s(t) >}{\sqrt{<\left[x_s(t)\right]^2 ><\left[y_s(t)\right]^2 >}} \qquad (5.65)$$

The symbol $<>$ stands for the expected value of the contents enclosed. It is the mean value of a large number of samples. This is like integrating $x(t)y(t)$ over a large amount of time. Assuming $x(t)$ and $y(t)$ possess equal variance ($\sigma^2$), we express the numerator of Eq. (5.65) as shown in Eq. (5.66).

$$<x(t)y(t)> = r_{nrm}\, \sigma^2 \qquad (5.66)$$

Note that $<x(t)y(t)>$ is the analog correlation between $x(t)$ and $y(t)$. The correlator output $r_\infty$ (of unquantized signals $x_s(t)$ and $y_s(t)$) is effectively $<x(t)y(t)>$. Let $x_i$ and $y_i$, respectively, represent the $i^{th}$ samples in $x_s$ and $y_s$. Let $N_{Nq}$ be the number of samples averaged. The subscript Nq indicates that samplings are done at the Nyquist rate. The digital correlator output $r_\infty$ is mathematically shown in Eq. (5.67).

$$r_\infty = \frac{1}{N_{Nq}} \sum_{i=1}^{N_N} x_i\, y_i \qquad (5.67)$$

The value of $r_\infty$ shown here is only for a fixed number of samples $N_{Nq}$. To represent the analog function $<x(t)y(t)>$ more accurately, one may increase the blocks of $r_\infty$. Because the numbers of $x_i$ and $y_i$ are large, we assume they have the same Gaussian statistical nature as $x(t)$ and $y(t)$. We may thus reduce Eq. (5.66) to Eq. (5.68).

$$<r_\infty> = \rho\sigma^2, \quad \text{since} <r_\infty> = <x(t)\,y(t)> \qquad (5.68)$$

It is clear that the a digital correlator output having infinite quantization levels is equivalent to a linear measure of the normalized cross-correlation $r_{nrm}$. The variance $\sigma_\infty^2$ of the correlator output is expressed in Eq. (5.69). The signal-to-noise ratio $\mathcal{R}_{sn\infty}$ is shown in Eq. (5.70), where the approximation holds for $r_{nrm} \ll 1$.

$$\left.\begin{aligned} \sigma_\infty^2 &= \langle r_\infty^2\rangle - \langle r_\infty\rangle^2 \\ &= \frac{\sigma^4}{N_N}\left(1+r_{nrm}^2\right) \end{aligned}\right\} \tag{5.69}$$

$$\left.\begin{aligned} \mathcal{R}_{sn\infty} &= \frac{\langle r_\infty\rangle}{\sigma_\infty} = \frac{r_{nrm}\sqrt{N_{Nq}}}{\left(1+r_{nrm}^2\right)} \\ &= r_{nrm}\sqrt{N_{Nq}} \end{aligned}\right\} \tag{5.70}$$

Practically, $N_{Nq}$ ranges within $10^6$ to $10^{12}$, which makes the approximation good. The threshold for detection of the signal is found from evaluating Eq. (5.70) with $\mathcal{R}_{sn\infty} = 1$. This shows that $r_{nrm}$ lies between $10^{-12}$ and $10^{-6}$. The SNR of Eq. (5.70) can be expressed in terms of bandwidth, as in Eq. (5.71).

$$\mathcal{R}_{sn\infty} = r_{nrm}\sqrt{2\Delta\nu\tau} \tag{5.71}$$

### 5.8.3.2 Sampled at Non-Nyquist Rates

Let the sampling frequency be scaled as $\beta$ times the Nyquist rate, where $\beta$ is dimensionless. Then the number of samples is $N = \beta N_{Nq}$, and the sampling interval is $\tau_s = 1/(2\beta\Delta\nu)$. The samples are spaced in time by $q\tau_s$, where $q$ is an integer. Substituting these into Eq. (5.63), the autocorrelation function $R_\infty$ is obtained in Eq. (5.72). The variance $\sigma_\infty^2$ of the correlator output signal and the SNR are given in Eqs. (5.73) and (5.74), respectively.

$$R_\infty\left(q\tau_s\right) = \frac{\sin\left(\pi q/\beta\right)}{\pi q} \tag{5.72}$$

$$\sigma_\infty^2 = \frac{\sigma^4}{N}\left\{1+2\sum_{q=1}^\infty R_\infty^2\left(q\tau_s\right)\right\} \tag{5.73}$$

$$\mathcal{R}_{sn\infty} = \frac{r_{nrm}\sqrt{\beta N_{Nq}}}{\sqrt{1+2\sum_{q=1}^\infty R_\infty^2\left(q\tau_s\right)}} \tag{5.74}$$

Let us first consider under sampling, where $0<\beta<1$. This makes $R_\infty = 0$. Hence the denominator in Eq. (5.74) becomes unity. This causes the sensitivity to drop.

Now consider over sampling, where $\beta > 1$. This results in $\sum_{q=1}^{\infty} R_{\infty}^2(q\tau_s) = (\beta - 1)/2$.
Here, the sensitivity is the same as is achieved at the Nyquist rate.

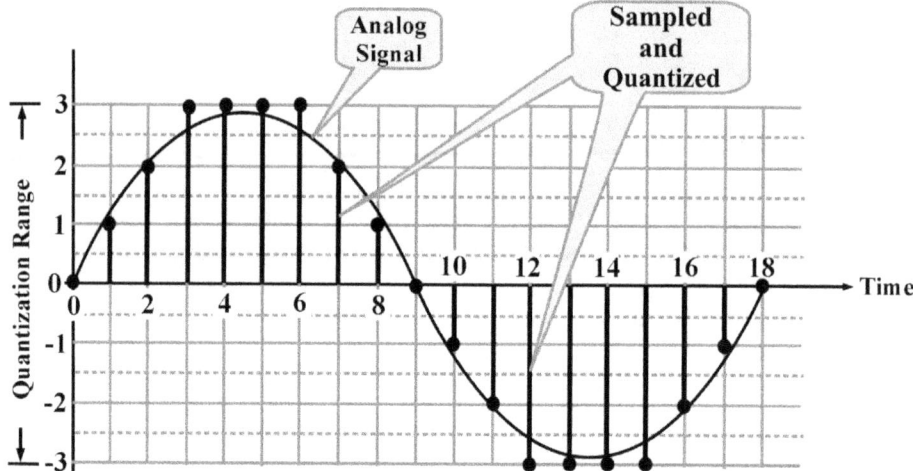

**FIGURE 5.16** A sinusoidal signal is sampled and quantized (same as quantization after sampling). The number of quantization levels are seven. The sampled signal values are fixed at the nearest available quantization levels.

### 5.8.4 Correlation of Sampled Signals with Quantization

Let a sinusoidal signal be sampled and then quantized. The results are the same as would be produced by quantizing first and then sampling. Fig. 5.16 shows a case of quantization with seven levels. The sampled values are fixed at the nearest available quantization level. The quantized values may or may not be close to the signal value. However, as the number of quantization levels are increased, the quantized values fall closer to the signal. When the number of quantization levels becomes infinity, the quantized signal values exactly match the signal values at those time instants. We now discuss the correlation results of sampled but finitely quantized signals. Our focus will be on three important points: (*i*) the relationship between the measured correlation and the normalized cross-correlation coefficient $r_{nrm}$, (*ii*) sensitivity losses, and (*iii*) the extent to which the lost sensitivity may be restored by oversampling. Numerical subscripts denoting the number of quantized levels have been extensively used. For example, $r_{n2}$ represents the normalized cross-correlation coefficient obtained using two-level quantized signals at the correlator input.

### 5.8.4.1 Two-Level Quantizations

Consider a system that has only two quantization levels. Let them be represented by +1 and −1. Only one bit is sufficient to digitally represent +1 and −1. The signals reaching the correlator will be either +1 or −1. In this case, the two-level quantized normalized correlation coefficient $r_{n2}$ is related to $r_{nrm}$ (unquantized) as in Eq. (5.75). This is also known as the *Van Vleck relationship*.

$$r_{n2} = \frac{2}{\pi} \sin^{-1}(r_{nrm}) \tag{5.75}$$

Let $\hat{x}_i$ and $\hat{y}_i$ represent the quantized states of $x_i$ and $y_i$. The subscript $i$ is an integer that represents the sample point. The correlator output $r_2$ produced by two-level quantized inputs is given in Eq. (5.76), where $N$ is the total number of samples sent to the correlator. The mean square value of $r_2$ is $<r_2^2>$. It is shown in Eq. (5.77).

$$r_2 = \frac{1}{N} \sum_{i=1}^{N} \hat{x}_i \hat{y}_i \tag{5.76}$$

$$<r_2^2> = \frac{1}{N^2} \sum_{i=1}^{N} <[\hat{x}_i]^2 [\hat{y}_i]^2> + \frac{1}{N^2} \sum_{i=1}^{N} \sum_{i \neq k} <\hat{x}_i \hat{y}_i \hat{x}_k \hat{y}_k> \tag{5.77}$$

Eq. (5.78) shows the variance $\sigma_2^2$ of the two-level quantized correlator output, where $q$ is an integer and $R_2$ is the autocorrelation. The points in time where the samples reside is given by $q\tau_s$. The approximation is good if $r_{n2} \ll 1$. In terms of $R_\infty$ (unquantized auto correlation), $R_2$ is given by Eq. (5.79). The SNR $\mathcal{R}_{sn2}$ of a two-level quantized correlator output is shown in Eq. (5.80).

$$\left. \begin{aligned} \sigma_2^2 &= <r_2^2> - <r_2>^2 \\ &= \frac{1}{N} \left\{ 1 + 2 \sum_{q=1}^{\infty} R_2^2(q\tau_s) \right\} \quad \text{if } r_{nrm} \ll 1 \end{aligned} \right\} \tag{5.78}$$

$$R_2(q\tau_s) = \frac{2}{\pi} \sin^{-1} \{ R_\infty(q\tau_s) \} \tag{5.79}$$

$$\mathcal{R}_{sn2} = \frac{<r_2>}{\sigma_2} = \frac{2 r_{nrm} \sqrt{N}}{\pi \sqrt{1 + 2 \sum_{q=1}^{\infty} R_2^2(q\tau_s)}} \tag{5.80}$$

Substituting $N = \beta N_{\mathrm{Nq}}$, we evaluate the ratio of SNR $\eta_2$ or relative sensitivity, which is the ratio of the SNR of a two-level quantized correlator to that of an unquantized correlator, as shown in Eq. (5.81). Recall that $N_{\mathrm{Nq}}$ is the number of samples at the Nyquist rate. The factor $\beta$ is for studying under- and over sampling cases. Similarly, the expression for $R_2$ from Eq. (5.79) gets modified as shown in Eq. (5.82). Note that $R_2$ is zero where $R_\infty$ is zero on the time axis $q\tau_s$.

$$\eta_2 = \frac{\mathcal{R}_{\mathrm{sn2}}}{\mathcal{R}_{\mathrm{sn\infty}}} = \frac{2\sqrt{\beta}}{\pi\sqrt{1 + 2\sum_{q=1}^{\infty} R_2^2 (q\tau_s)}} \tag{5.81}$$

$$R_2(q\tau_s) = \frac{2}{\pi}\sin^{-1}\left\{\frac{\beta\sin(\pi q / \beta)}{\pi q}\right\} \tag{5.82}$$

Sampling at Nyquist rates or less gives $\eta_2 = 0.64$. In other words, for $\beta = 1, \frac{1}{2}, \frac{1}{3}, \dots$, we obtain $\sum_{q=1}^{\infty} R_2^2(q\tau_s) = 0$, for which $\eta_2 = 0.64$. Sampling above the Nyquist rates gives higher values of $\eta_2$. For example, if $\beta = 2$ we obtain $\eta_2 = 0.74$. Dependency on the bandpass shape comes in when $\beta \geq 2$.

### 5.8.4.2 Four-Level Quantizations
We now consider a two-bit system that gives $2^2 = 4$ quantization levels for representing the data. Hence, the data accuracy is better than the two-level quantization (single-bit) process. Let $-n, -1, +1, +n$ represent the quantization states from lowest to highest. Inside the correlator, the product of two samples can take the values $+1, -1, +n, -n, +n^2$ and $-n^2$. In this case, the four-level quantized normalized correlation coefficient $r_{n4}$ as a function of $r_{\mathrm{nrm}}$ is given in Eq. (5.83), where $r_4$ is the four-level quantized correlator output. The probability $\Phi$ that the unquantized level is restricted within $\pm v_0$ is given in Eq. (5.84). The quantity $E$ is shown in Eq. (5.85). Recall that $\sigma^2$ is the variance of the correlator output with unquantized inputs.

$$\begin{aligned} r_{n4} &= \frac{\langle r_4 \rangle}{\Phi + n^2(1+\Phi)} \\ &= r_{\mathrm{nrm}} \frac{2\left[(n+1)E+1\right]^2}{\pi\left[\Phi + n^2(1-\Phi)\right]} \text{ if } r_{\mathrm{nrm}} \ll 1 \end{aligned} \tag{5.83}$$

Interferometry and Radio Arrays • **163**

$$\Phi = \frac{1}{\sigma\sqrt{2\pi}}\int_{-v_0}^{v_0}\exp\left(\frac{-x^2}{2\sigma^2}\right)dx = \mathrm{erf}\left(\frac{v_0}{\sigma\sqrt{2}}\right) \tag{5.84}$$

$$E = \exp\left(-v_0^2/2\sigma^2\right) \tag{5.85}$$

Let $\hat{x}_i$ and $\hat{y}_i$ be the quantized values of $x_i$ and $y_i$. The subscript $i$ is an integer that represents the sample point. The mean square value of $r_4$ is shown in Eq. (5.86), where $N$ represents the number of samples given as correlator input.

$$<r_4^2> = \frac{1}{N^2}\sum_{i=1}^{N}<\left[\hat{x}_i\right]^2\left[\hat{y}_i\right]^2> + \frac{1}{N^2}\sum_{i=1}^{N}\sum_{i\neq k}<\hat{x}_i\,\hat{y}_i\,\hat{x}_k\,\hat{y}_k> \tag{5.86}$$

Eq. (5.87) expresses the variance $\sigma_4^2$ of a four-level quantized correlator output. The variance as a function of autocorrelation function $R_4$ of the four-level quantized correlator is shown in Eq. (5.88).

$$\left.\begin{aligned}\sigma_4^2 &= <r_4^2>-<r_4>^2 \\ &= <r_4^2>-\rho_4^2\left\{\Phi+n^2\left(1+\Phi\right)\right\}^2\end{aligned}\right\} \tag{5.87}$$

$$\sigma_4^2 = \frac{1}{N}\left[\Phi+n^2\left(1-\Phi\right)\right]^2\left\{1+2\sum_{q=1}^{\infty}R_4^2\left(q\tau_s\right)-r_{n4}^2\right\} \tag{5.88}$$

Assuming $r_{\mathrm{nrm}} \ll 1$, the SNR $\mathcal{R}_{\mathrm{sn4}}$ of a four-level quantized correlator is given by Eq. (5.89). The ratio of SNR $\eta_4$ or relative sensitivity given by $\mathcal{R}_{\mathrm{sn4}}/\mathcal{R}_{\mathrm{sn}\infty}$ is shown in Eq. (5.90), where $\beta$ controls the Nyquist rate. Letting $\beta = 1$, the sampling rate becomes the Nyquist rate.

$$\mathcal{R}_{\mathrm{sn4}} = \frac{<r_4>}{\sigma_4} = \frac{2\rho\left\{(n-1)E+1\right\}^2\sqrt{N}}{\pi\left\{\Phi+n^2\left(1-\Phi\right)\right\}\sqrt{1+2\sum_{q=1}^{\infty}R_4^2\left(q\tau_s\right)}} \tag{5.89}$$

$$\eta_4 = \frac{\mathcal{R}_{\mathrm{sn4}}}{\mathcal{R}_{\mathrm{sn}\infty}} = \frac{2\left\{(n-1)E+1\right\}^2\sqrt{\beta}}{\pi\left\{\Phi+n^2\left(1-\Phi\right)\right\}\sqrt{1+2\sum_{q=1}^{\infty}R_4^2\left(q\tau_s\right)}} \tag{5.90}$$

Recall that the minimum and maximum values of the quantization states in our four-level quantized correlator are $-n$ and $+n$, respectively. Changing the level $n$ causes the optimized sensitivity to change. For example, when $n = 3$, the optimized sensitivity occurs with $v_0 = 0.996\sigma$, and when $n = 4$ it occurs with $v_0 = 0.942\sigma$. Fig. 5.17 shows the relative sensitivities ($\eta_4$) of a four-level quantized correlator for different values of $n$.

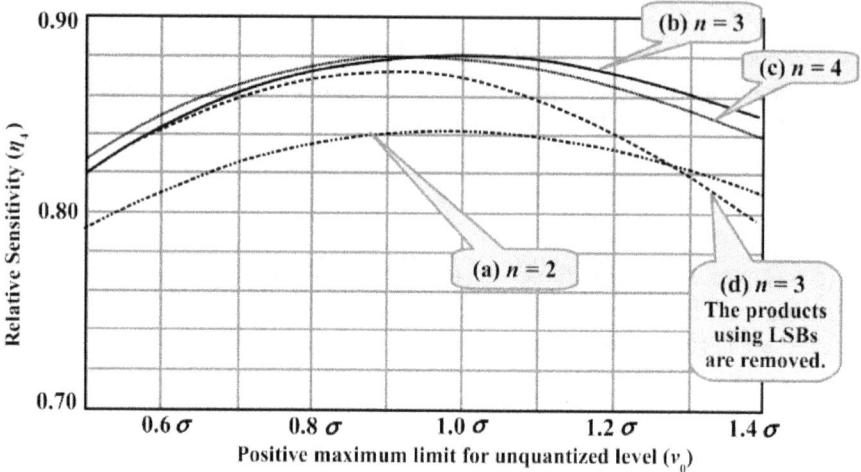

FIGURE 5.17: Relative sensitivities ($\eta_4$) of a four-level quantized correlator for different values of $n$. (a) Two-bit system with $n = 2$. (b) Two-bit system with $n = 3$. (c) Two-bit system with $n = 4$. (d) Two-bit system with $n = 3$, but with low-level products deleted.

### 5.8.4.3 Other-Level Quantizations

Until now, we have considered only even-numbered quantization levels. Sometimes an odd number of quantization levels is preferred for certain advantages. We now describe a three-level quantization where one of the quantization levels is zero. The variance $\sigma_3^2$ of the correlator output $r_3$ s shown in Eq. (5.91), where $R_3$ and $r_{n3}$ represent the auto correlation and the normalized quantized correlation coefficient, respectively.

$$\left. \begin{aligned} \sigma_3^2 &= <r_3^2> - <r_3>^2 \\ &= \frac{1}{N}(1+\Phi)^2 \left\{ 1 + 2\sum_{q=1}^{\infty} R_3\left(q\tau_s\right) - r_{n3}^2 \right\} \end{aligned} \right\} \tag{5.91}$$

Neglecting $r_{n3}^2$, the relative sensitivity $\eta_3$ in terms of the SNR of a three-bit quantized correlator $\mathcal{R}_{sn3}$ to the SNR of a nonquantized correlator $\mathcal{R}_{sn\infty}$ is expressed by Eq. (5.92), where $\Phi$ is given in Eq. (5.84). Sensitivity does not improve very much for quantization levels greater than four.

$$\eta_3 = \frac{\mathcal{R}_{\text{sn3}}}{\mathcal{R}_{\text{sn}\infty}} = \frac{<r_3>}{\sigma_3 \, \mathcal{R}_{\text{sn}\infty}} = \frac{2\sqrt{\beta}E^2}{\pi(1+\Phi)\sqrt{1+2\sum_{q=1}^{\infty}R_3^2(q\tau_s)}} \qquad (5.92)$$

### 5.8.5 Principle of Digital Time Delay

We know that the geometrical delay $\tau_g$ in a correlator system is compensated by generating an instrumental delay $\tau_i$. Digital delay systems generally multiply the basic sampling period $\tau_s$ with integers for controlling the delay period. Shift registers or parallel-to-serial conversion schemes may be used for this purpose. In a shift register, the first cell is assigned 1 and the remaining cells are assigned 0. Note that a time $\tau_s$ is required for entering the 1 in the first cell. The sampling pulses with period $\tau_s$ can be applied to time the movement of the digits. For example, in a four-cell register, after a period of $4\tau_s$ the 1 appears at the output. Currently, digital integrated circuits are available that are software controlled and generate customized delays.

### 5.8.6 Principle of Digital Quadrature Phase Shift

The sine component in a complex correlator is obtained by applying a 90° phase shift to one of the signals before multiplying them (see Fig. 5.7b). Because the Hilbert transform of cosine signals gives sine signals, it may be applied to one of the inputs of the analog correlator to generate the imaginary part of the visibility. Hilbert transforms using digital circuits give limited accuracy because of the limited number of samples. This is like multiplying a rectangular window with the input time-domain signal that introduces ripples in the spectrum due to the sine function resulting from the window. Hence the SNR is reduced by a few percent. The summation process in digital convolution increases the number of bits in the data. To reduce the complexity of the correlator system, the lower-order bits may be discarded. This results in further quantization loss. Effectively, the imaginary part of the correlator output suffers from spectral distortion and reduced SNR, as compared to the real part. When designing broad-band systems, these are more serious considerations when the data rate is so hight that there is scope only for simple processing.

### 5.8.7 XF Correlator

XF is a correlator symbol that indicates that the correlation (X) is done in the time domain, followed by a Fourier transformation (F) to the frequency domain. This is also known as a *lag correlator*. The input bit streams from each antenna pair are correlated first (preferably in real time), and then the complex output is converted to a cross-power spectrum by a FFT (fast Fourier transform) operation.

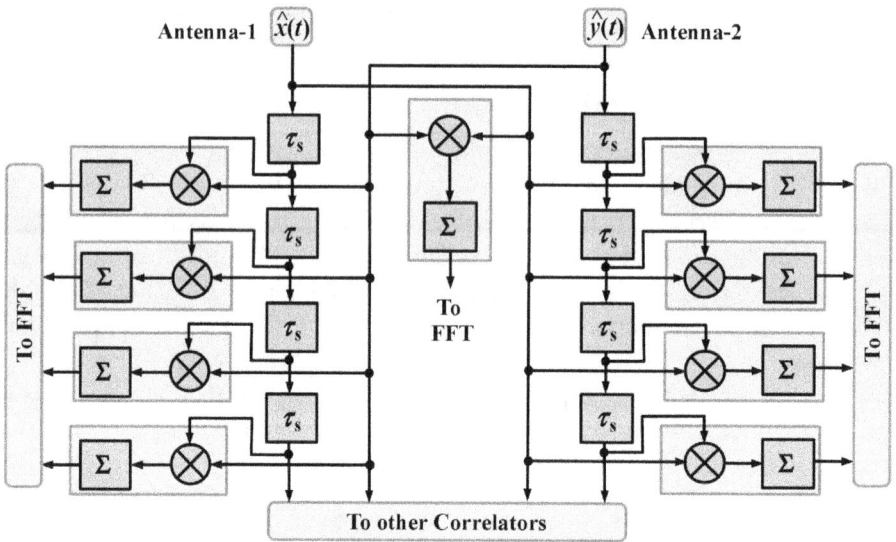

**FIGURE 5.18:** A XF (lag correlator) using two antennas. Delay is applied in integral multiples of the sampling period $\tau_s$ An FFT is applied to the correlator output. The left, middle, and right sets of outputs are, respectively, the correlation measures between (*i*) $\hat{y}$ and delayed $\hat{x}$, (*ii*) $\hat{y}$ and $\hat{x}$, and (*iii*) $\hat{x}$ and delayed $\hat{y}$.

Fig. 5.18 shows a simplified XF correlator. The instrumental delay is provided in integer multiples of the sampling time $\hat{o}_s$. The correlator output undergoes an FFT operation. The left set of outputs are the correlations between $\hat{y}$ and delayed $\hat{x}$. The middle output is the correlation between $\hat{y}$ and $\hat{x}$ without any delay. The right set of outputs are the correlations between $\hat{x}$ and delayed $\hat{y}$. A shift register type of memory structure that can store input samples in blocks and allow them to be read at the input rate of the correlator can be used. The memory units are used in pairs such that one is filled with data at the Nyquist rate, while the other is readable at the maximum data rate. These correlators are also known as *recirculating correlators*.

### 5.8.8 FX Correlator

FX is a correlator symbol that indicates that the Fourier transform (F) is performed first and then the correlation (X) is performed. The input bit streams from both antennas are converted to frequency spectra by a real time FFT. Each of the frequency channels are then multiplied in complex form with the complex conjugate of the corresponding frequency channels from another antenna to produce the cross-power spectrum. The number of FFT operations involved is proportional to the number of antennas and number of polarizations. The total number of computations is comparatively less than an equivalent XF correlator, especially if the number of antennas is large. Hence it is economical.

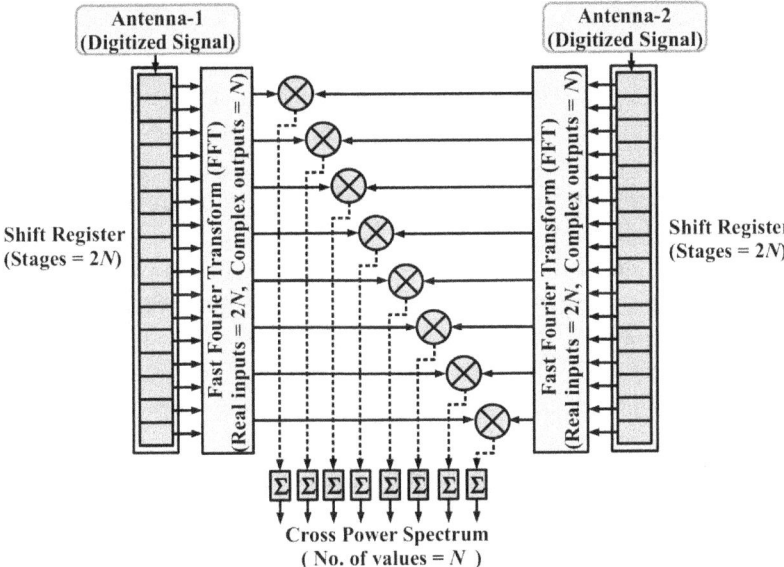

**FIGURE 5.19:** A FX correlator using two antennas. Digitized signals are fed into the shift registers, whose output undergoes FFT at intervals of 2*N* sample periods. Correlations are made between one signal with the complex conjugate of the other. For an array of $n_a$ antennas, each of the FFT outputs are split ($n_d$ – 1) ways for combining with the complex amplitudes resulting from the other antennas.

A simple FX correlator using two antennas is shown in Fig. 5.19. The digital signals are taken into shift registers, the output of which undergoes FFT at intervals of 2*N* sample periods, where *N* stands for the number of frequency channels. Correlations are performed between one of the signals with the complex conjugate of the other signal. For an array consisting of $n_a$ antennas, each FFT output is split in ($n_d$ – 1) ways and combined with the complex amplitudes from other antennas.

## 5.9 VERY LONG BASELINE INTERFEROMETRY (VLBI)

The VLBI or *very long baseline interferometry* developed from the requirements of higher resolution images of astronomical objects at radio frequencies. Recall that for a single wavelength $\lambda$, the angular resolution of the interferometer is $\lambda/(2d \cos\theta)$ (see Fig. 5.4), where *d* is the distance between the two antennas forming the interferometer. Because radio wavelengths are very large, the highest resolution one can achieve is by separating the two antennas by a distance equal to the diameter of Earth. In other words, the antennas would be placed on opposite sides of the Earth.

The VLBI is a technique by which data from radio arrays all over the world are calibrated in time and phase and then combined. Due to the unavailability of phase-stable fiber links between individual antennas and a correlator for distances larger than 200 km, it is not possible to operate all the telescope arrays together and obtain a single data set. It is, however, possible to avoid the transmission of phase-stable LO signals by using atomic clocks together with extremely phase-stable oscillators. Individual interferometer data marked with precise times are recorded independently. These are collectively matched and processed.

VLBI uses digital data recording. The antenna output data contains variations of the electric field strength marked with accurate timings. Mutual correlation functions can be generated from this digitized data. Phase jumps in LO affect the IF signal by destroying the coherence. Hence, the LO used must be extremely stable in phase. It also enables the extremely precise time marks required to align the signals from different telescopes.

Different local oscillator systems, including quartz oscillators, rubidium clocks, and hydrogen masers, have been tried. Hydrogen masers provide the best frequency and phase stability. Special digital correlators are operated on the data. The different data sets are aligned in time. The geometric delays and local oscillator offsets are compensated suitably. The differential Doppler shifts due to the rotation of the Earth are corrected. The clock-rate offsets are corrected, and finally a correlation function for each pair of observation sites is generated. From these operations, the data generated are like those from any conventional interferometer.

| BOX 5.1 | Space VLBI |
|---------|------------|

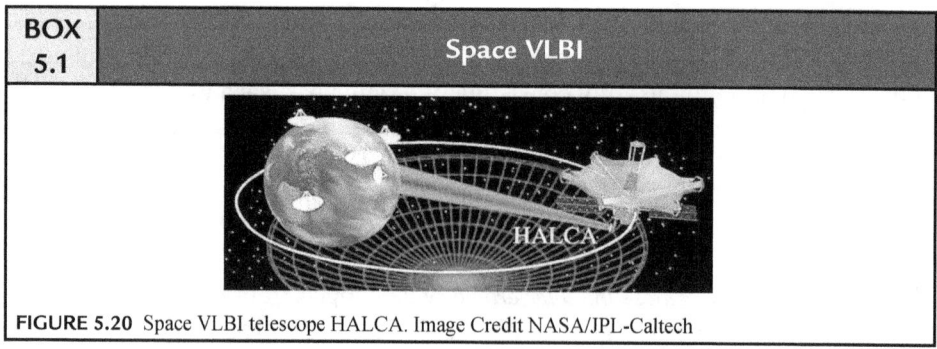

FIGURE 5.20 Space VLBI telescope HALCA. Image Credit NASA/JPL-Caltech

Space Very Long Baseline Interferometry (SVLBI) combines space- and ground-based radio telescopes to form an interferometer. The first SVLBI radio telescope HALCA (Highly Advanced Laboratory for Communication and Astronomy), depicted in Figure 5.20, was launched in 1997 and operated until 2003. The Japanese-built 8-m telescope was placed in a highly elliptical orbit with respective perigee-apogee altitudes ranging between 560–21400 km, enabling wide coverage of the (u,v) plane. HALCA studied hydroxyl masers, quasars, and pulsars, while operating at 1.6 and 5.0 GHz.

Spektr-R is currently the only operational SVBLI instrument and is part of the RadioAstron project initiated by the Astro Space Center (ASC) of the P. N. Lebedev Physical Institute of the Russian Academy of Science. The 3660 kg satellite was launched in 2011 from the Bikaner Cosmodrome. The Spektr-R deployable 10 m parabolic reflector is formed from twenty-seven carbon-fiber panels that open like flower petals. The science payload includes two separate hydrogen-maser frequency standards, feed arms, a focal container, an instrumentation module, a communication system, and a high-gain antenna.

The orbital semi-major axis of the telescope is on the order of the Earth-Moon separation with perigee-apogee altitudes ranging between 10,000–390,000 km and an orbital period between eight and nine days. Satellite observing frequencies are 0.327, 1.665, 4.830, and 18.392–25.112 GHz corresponding to the P, L, C, and K bands, respectively. Operational bandwidths are 4 MHz (P-band) and 16 MHz (L-, C-, and K-band). A prime-focus concentric feed arrangement enables simultaneous observation of either two frequencies or two circular polarizations. A single frequency band may be observed with two polarizations or two frequency bands may be observed simultaneously with opposite polarization.

The RadioAstron collaboration includes ground-based VLBI networks located in Australia, Chile, China, Europe, India, Japan, Korea, Mexico, Russia, South Africa, Ukraine, and the USA. Data recorders are located at the Pushchino tracking station and at the Russian radio telescopes. The RadioAstron science goals include the study of the M-87 galaxy and other extragalactic sources with an angular resolution up to $7 \times 10^{-6}$ arc seconds at the highest frequency and the longest baseline.

## REVIEW QUESTIONS

1. Two parabolic dish antennas of 2 m diameter are separated by 20 m. They form an interferometer. If the operating frequency is 150 MHz, calculate the angular resolution. [***Hint:*** use Eq. (5.1)]

2. An interferometer forms a fringe pattern on the screen due to light from a distant star. The maximum intensity measured is found to be twice the minimum intensity. Calculate the visibility. [***Hint:*** use Eq. (5.2)]

3. An optical interferometer uses a wave bandwidth of 1 nm. Find the coherence length at a wavelength of 500 nm. Compare this with a radio interferometer working at 610 MHz having a bandwidth of 32 MHz.
[*Hint:* use Eq. (5.3)]

4. Explain the additive radio interferometer and its fringe pattern using suitable diagrams. [*Hint:* see Figs. 5.2 and 5.3]

5. Using diagrams to explain a two-element multiplicative radio interferometer and its fringe pattern. [*Hint:* see Figs. 5.4 and 5.5]

6. What is meant by the fringe-stopping center of a radio interferometer?

7. What can be the maximum integration time of the correlator? Can it be increased infinitely for a moving source?

8. Give diagrammatic representations of (*i*) a simple correlator, and (*ii*) a complex correlator. Explain these using equations.
[*Hint:* use Eqs. (5.13) through (5.17) with Fig. 5.7]

9. Give the number of cross-correlator outputs obtained from an array of thirty antennas. [*Hint:* use Eq. (5.18)]

10. For a correlator array of thrity antennas, what would be the percentage loss in power if the self terms were eliminated? [*Hint:* use Eq. (5.19)]

11. Super-heterodyne receiving systems are used in a multiplicative interferometer, and the delay correction is done in IF. Explain the term *natural fringe frequency using* an equation. [*Hint:* see Eq. (5.29).

12. If the number of IF stages is *n,* and if only the LSB are chosen, for what values of *n* does the spectrum flip? [*Hint:* see Section 7.1.3]

13. If $r_{usb}$ and $r_{lsb}$ are, respectively, the upper- and lower-sideband correlator responses of a single IF conversion receiver, what is the correlator response if both sidebands are used? Is there any gain in sensitivity?
[*Hint:* See Section 7.2]

14. For problem 13, draw tentative correlator responses as functions of a nonzero $\Delta\tau$, where $\Delta\tau = \tau_g - \tau_i$ (the difference between geometrical and instrumental). [*Hint:* See Fig. 5.11]

15. Tentatively draw the complete noise spectra that will be available within a period equal to a single integration period of the correlator if integration is not performed. [*Hint:* See Fig. 5.13]

16. An analog interferometer consisting of two antennas A and B are used to observe a radio source. If the system temperatures using A and B are, respectively, 35 K and 40 K, and their respective antenna temperatures are 45 K and 43 K, find the SNR. Assume the integration bandwidth and IF system bandwidth are 32 MHz. [*Hint:* Use Eq. (5.52)]

17. An interferometer shows a system temperature of 40 K and an antenna temperature of 50 K. If the IF bandwidth is 32 MHz and the integration time is 16 seconds, find the SNR for (*i*) an analog system, and (*ii*) a digital system with quantization efficiency 70%. [*Hint:* Use Eq. (5.58)]

18. Calculate the rms noise at the correlator output of an interferometer, given the diameter of the dishes is 45 m, the aperture efficiency is 40%, the system temperature is 40 K, the integration time is 16 seconds, the IF bandwidth is 32 MHz, and the quantization efficiency is 70%.

    [*Hint:* Calculate the effective aperture area and use Eqs. (5.59) and (5.60)]

19. Explain the sampling theorem for discritizing analog signals without loss of information.

20. Explain the bandpass sampling technique. [*Hint:* See Section 8.2]

21. If $x_s$ and $y_s$ are the instantaneous signals from the two antennas of an interferometer, give an expression for the normalized cross-correlation coefficient. [*Hint:* See Eq. (5.65)]

22. Explain the basic principle behind generating a digital time delay. What is the minimum delay period that can be produced? [*Hint:* See Section 8.5]

23. Explain the basic technique used for generation of a digital quadrature phase-shifted output from a digital correlator. [*Hint:* See Section 8.6]

24. Use a diagram to explain the XF (lag) correlator. [*Hint:* Use Fig. 5.18]

25. Use a diagram to explain the FX correlator. [*Hint:* Use Fig. 5.19]

26. Explain the VLBI.

CHAPTER 6

# RECEIVING SYSTEMS

## 6.1 INTRODUCTION

In radio telescopes, antennas are followed by receiving systems that further process the signals. Except for transient emissions from the Sun and Jupiter, in general the signals are very weak (micro- and milli-Jansky). Hence the system requirements are (*i*) high sensitivity, (*ii*) high gain, and (*iii*) high stability. The antenna temperature plays a major role in determining the receiver requirements. As shown in Fig. 1.1 of Chapter 1, the dynamic range of radio emissions and spectra of the astronomical sources are very large. The corresponding antenna temperatures using a 3m² aperture antenna are also shown. Due to the different spectral ranges of different sources, it is difficult to design a single receiving unit for all sources. Similarly, a single-frequency-band antenna may not be sufficient to accommodate them all. Dish antenna systems can employ changeable feeds for different frequencies on the same reflector. These are brought to the focus of the dish one at a time based on the observation frequency. To support antenna feeds for various frequencies, different sets of receiving systems can be used.

### 6.1.1 Categorizing the Receiving Systems

In general, observing different radio sources may require different antenna structures and different types of receiving systems. Interferometers are used for spectral and continuum observations, whereas for receiving radio bursts, a spectrum-analyzer-based system is preferable. A good frequency resolution is required for spectral-line observations. For pulsar studies, phased arrays may be used with very high time resolution. Highly sensitive receivers with high system stability are required for observing distant radio galaxies. Receiver designs are categorized broadly based on (*i*) frequency bands, and (*ii*) sensitivity. They may further be subcategorized based on the observation types, namely (*i*) continuum, (*ii*) spectral, and (*iii*) radio burst.

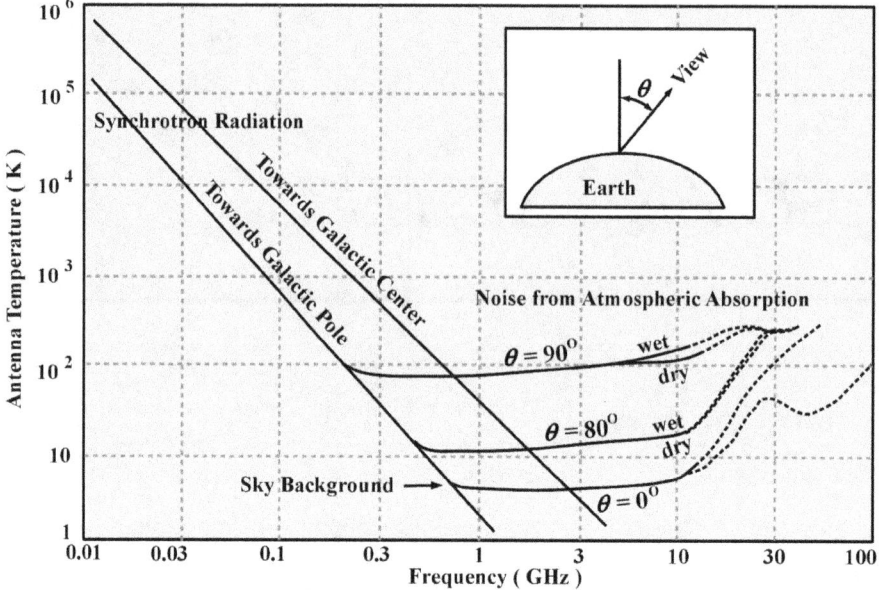

**FIGURE 6.1:** Antenna temperatures (for a beam less than a few degrees) obtained by pointing at various directions in the sky, plotted as a function of frequency.

### 6.1.2 Atmosphere on Receiving Sensitivity

We assume that the readers have gone through Sections 2.8, 2.9, and 4.5, and have a good knowledge of antenna temperature $T_A$, system temperature $T_{Sys}$, sensitivity, minimum detectable temperature $\Delta T_{min}$, etc. However sensitive a radio telescope might be, its capabilities are further limited by the Earth's atmosphere. Depending on the direction $\theta$ of the telescope with respect to the local zenith, atmospheric noise may limit the threshold of observation for frequencies above 0.2 GHz. This is shown in Fig. 6.1. Below this frequency, the noise from the galactic center and the galactic pole dominates. The atmospheric noise also depends on the weather conditions (wet or dry) above 3 GHz. The lowest noise levels are obtained by observing any source at the zenith.

## 6.2   BASIC RECEIVING SYSTEM (RADIOMETER)

A radio astronomy receiver is also known as a *radiometer*. It measures the RF (radio frequency) power delivered by the antenna system. The elements of the receiver are shown in Fig. 6.2. The RF amplifier block consists of an LNA (low-noise

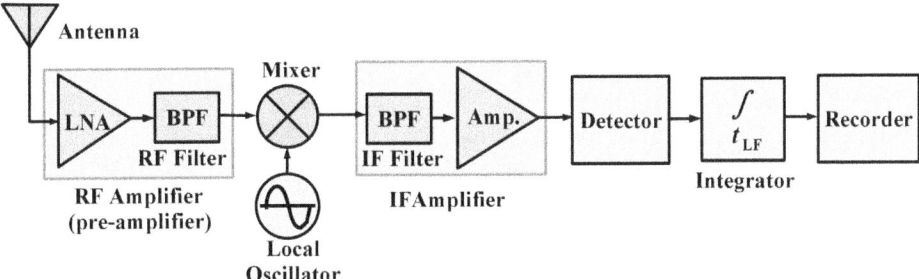

FIGURE 6.2: Functional blocks of a simple radiometer.

amplifier) and a BPF (bandpass filter, also known as an RF filter). It amplifies the signals from the antenna within the selected bandwidth and passes them to the mixer. The LO (local oscillator) frequency is mixed with it, and an IF (intermediate frequency) is generated. The image frequencies are rejected by the BPF (IF filter) and amplified. This is passed to the detector, followed by integration and recording. The important parameters of the receiver are defined below:

1. **RF Center Frequency ($\nu_0$):**   This is the center frequency of the RF band chosen for observation. It is the same as the center frequency of the RF filter. It is usually in the MHz range but can run up to the GHz range for modern telescopes.

2. **Gain ($G_{RCV}$):**   The objective of the receiver is to amplify the very minor power changes (on the order of $10^{-17}$ Watt) resulting from a radio astronomical source and make them capable of being recorded. The complete gain $G_{RCV}$ of a receiving system is defined by Eq. (6.1).

$$G_{Rcv} = \frac{\text{Power output to recorder}}{\text{Power input to receiver}} \tag{6.1}$$

$G_{RCV}$ is the product of gains of individual stages between the antenna and the recorder. It can be on the order of $10^{14}$ or 140 dB.

3. **Bandwidth ($\Delta\nu_{HF}$):**   This is the rectangular observation bandwidth of the receiver. Because the RF and IF bandwidths are identical, $\Delta\nu_{HF}$ represents both. It is usually in the range of MHz.

4. **Integration Time ($t_{LF}$):**   The detector output keeps fluctuating with time. To smooth out the large-scale fluctuations before recording, integration is applied over a fixed period of time. The effective time of integration represented by $t_{LF}$ is usually in the range of seconds.

5. **Receiver Temperature ($T_R$):** This is the effective noise temperature seen at the input terminals of a receiver. If a resistor equal to the input resistance of the receiver is heated, it will generate noise. Over a bandwidth $\Delta v_{HF}$, the temperature at which this resistor generates the same amount of noise power as that of the receiver is defined as the receiver temperature. It is measured in K. The noise power $W_R$ (Watt) of a receiver is given by Eq. (6.2), where $k_B$ is the Boltzmann constant, $T_R$ is in Kelvins, and $\Delta v_{HF}$ is in Hz.

$$W_R = k_B T_R \Delta v_{HF} \tag{6.2}$$

The receiver temperature is also related to the noise factor $F$ of the receiver as expressed in Eq. (6.3), where $T_{R(Phy)}$ is the physical temperature of the receiver in K (also see Eq. (2.58) of Chapter 2). The noise factor when converted to dB is known as the *noise figure*.

$$T_R = (F\text{-}1) \times T_{R(Phy)} \text{ K} \tag{6.3}$$

Because the receiver is composed of several cascaded stages, $T_R$ can be expressed as a function of the noise contribution from each stage. This is shown in Eq. (6.4), where $G_1, G_2,...$ are, respectively, the gains of the $1^{st}, 2^{nd},...$ stages; $T_1, T_2,...$ are, respectively, the individual temperatures generated at the inputs of the $1^{st}, 2^{nd},...$ stages; and $\varepsilon_T$ is the efficiency of the transmission line joining the receiver to the antenna.

$$T_R = \frac{1}{\varepsilon_T}\left( T_1 + \frac{T_2}{G_1} + \frac{T_3}{G_1 G_2} + .... \right) \tag{6.4}$$

Note that $\varepsilon_T$ plays a major role in increasing the temperature. The noise temperature contribution from the first amplifier stage is the largest, while those from the later stages are much smaller because they get divided by the gains of their previous stages.

## 6.3   SENSITIVITY AND STABILITY

The radiometer is a highly sensitive instrument that measures very weak signals. Its properties should remain unchanged over the period of a measurement that requires high stability. These factors also depend on the chosen technique.

1. **Sensitivity ($\Delta T_{min}$):** This is defined as the smallest change in the antenna temperature $\Delta T_{min}$ that the instrument can detect successfully. Even for a

perfectly stable receiver, a higher $T_{\text{Sys}}$ can reduce the sensitivity. The equation for $\Delta T_{\min}$ is shown in Eq. (6.5), where $\Delta v_{\text{HF}}$ is the equivalent rectangular RF passband (same as IF), $t_{\text{LF}}$ is the integration time, and $K_s$ is the factor of sensitivity for the radiometer.

$$\Delta T_{\min} = K_s \frac{T_{\text{Sys}}}{\sqrt{t_{\text{LF}}\,\Delta v_{\text{HF}}}} \tag{6.5}$$

To achieve higher sensitivity, $\Delta T_{\min}$ must be reduced by increasing $\Delta v_{\text{HF}}$ and $t_{\text{LF}}$. Presently, very low-noise LNA is available that can also reduce $T_{\text{Sys}}$ and $\Delta T_{\min}$.

2. **Stability:** Over the period of observation, variations in gain, bandwidth, and receiver noise temperature must be minimized, although it may not be possible to make the system 100% stable. However, good LNA designs and the use of switched radiometers may solve many of these problems. We shall discuss this in the following sections.

## 6.4  TYPES OF RADIOMETER

The basic principles of different radiometers used in radio astronomy are discussed here. Some of these radiometers are of historical importance. The objective for these designs was to improve the sensitivity and stability.

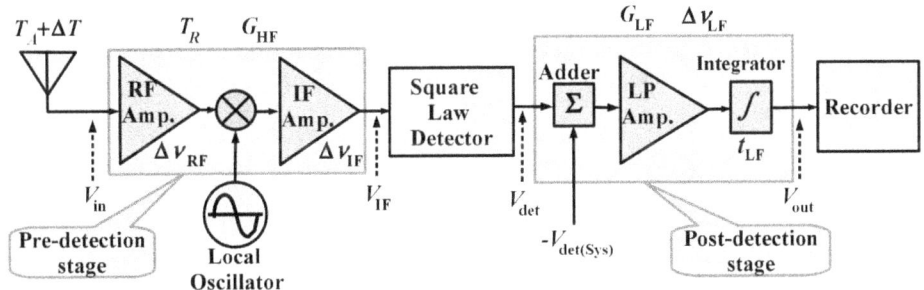

**FIGURE 6.3:** Functional blocks of a total power receiver. The RF and IF filters are not shown but are assumed to exist and to have identical bandwidths $\Delta v_{\text{RF}}$ and $\Delta v_{\text{IF}}$, respectively.

### 6.4.1 Total Power Receiver (Direct Radiometer)

The total power receiver measures the total noise power from the antenna. The block diagram is shown in Fig. 6.3. The rectangular passbands of the RF and IF amplifiers are $\Delta v_{\text{RF}}$ and $\Delta v_{\text{IF}}$, respectively. They are the same other than having different center frequencies. Let $G_{HF}(v)$ be the power gain of the predetection stage across the rectangular passband $\Delta v_{RF}$. The predetection stage generates a temperature $T_R$ near the antenna terminal. Let $T_A$ be the antenna temperature of the cold sky, and let $\Delta T$ be the incremental temperature when the antenna is moved from cold sky to a radio source. The IF power output $W_{HF}$ is expressed in Eq. (6.6), where $T_{Sys}$ is the system temperature with the antenna pointed to cold sky, and $k_B$ is Boltzmann constant.

$$W_{HF} = G_{HF}\left(T_{Sys} + \Delta T\right) k_B \, \Delta v_{RF} \tag{6.6}$$

Being a square-law device, the detector output voltage $V_{det}$ is proportional to the input RF power. This is shown in Eq. (6.7), where $K_{det}$ is the proportionality constant of the detector.

$$\left.\begin{aligned} V_{det} &= K_{det} \, G_{HF}\left(T_{Sys} + \Delta T\right) k_B \, \Delta v_{RF} = V_{det(Sys)} + V_{det(Src)} \\ \text{where,} \quad V_{det(Sys)} &= T_{Sys}\left(K_{det} \, G_{HF} \, k_B \, \Delta v_{RF}\right) \\ \text{and} \quad V_{det(Src)} &= \Delta T\left(K_{det} \, G_{HF} \, k_B \, \Delta v_{RF}\right) \end{aligned}\right\} \tag{6.7}$$

Note that $V_{det(Src)}$ is due to the source, and $V_{det(Sys)}$ is due to the cold sky. On most occasions, $V_{det(Sys)} \times V_{det(Src)}$. Hence, amplification of $V_{det(Src)}$ alone is required. Thus we have to get rid of $V_{det(Sys)}$ and amplify the rest. This is done by adding a negative voltage $-V_{det(Sys)}$ in the postdetection stage. Hence, when pointed to cold sky, $V_{out} = 0$. It becomes nonzero when the antenna is moved towards a radio source for which the emissions are stronger than the sky noise. Hence, the output voltage $V_{out}$ to the recorder is proportional to the input power. This is expressed in Eq. (6.8), where $K_{LF}$ is the voltage gain of the postdetection stage.

$$V_{out} = K_{LF} \, \Delta T\left(K_{det} \, G_{HF} \, k_B \, \Delta v_{RF}\right) \tag{6.8}$$

The pseudo voltage waveforms at different stages of the receiving system are shown in Fig. 6.4. The IF output voltage $V_{IF}$ is like a carrier frequency $V_{IF}$ in that its amplitude is randomly modulated, as shown in Fig. 6.4a. The detector output voltage $V_{det}$ is shown in Fig. 6.4b. The output voltage $V_{out}$ to the recorder is shown in Fig. 6.4c.

The compensating dc voltage in the post detection stage is kept at nearly 1 Volt. This requires a very large gain of the RF predetection stage. Hence, the amplifiers must be highly stable. The gain stability determines how small a signal can be detected.

The sensitivity, or minimum detectable temperature $\Delta T_{min}$, is shown in Eq. (6.9), where $T_{Sys}$ is the system temperature (cold sky), and $\Delta v_{HF}$ and $\Delta v_{LF}$ are given in Eqs. (6.10) and (6.11), respectively.

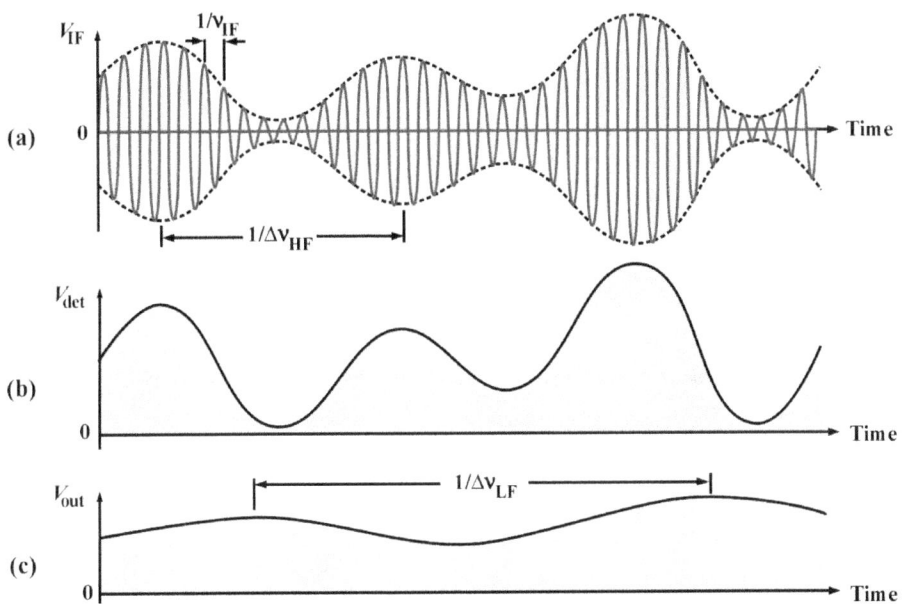

**FIGURE 6.4:** Pseudo voltage waveforms at different stages of the receiving system. (a) IF output voltage. (b) Detector output voltage. (c) Output voltage to the recorder.

$$\Delta T_{\min} = T_{\text{Sys}} \sqrt{\frac{2\,\Delta\nu_{\text{LF}}}{\Delta\nu_{\text{HF}}}} = \left(T_A + T_R\right)\sqrt{\frac{2\,\Delta\nu_{\text{LF}}}{\Delta\nu_{\text{HF}}}}$$

$$(6.9)$$

$$\Delta\nu_{\text{HF}} = \frac{\left[\int_0^\infty G_{\text{HF}}(\nu)\,d\nu\right]^2}{\int_0^\infty \left[G_{\text{HF}}(\nu)\right]^2 d\nu}$$

$$(6.10)$$

$$\Delta\nu_{\text{LF}} = \frac{\int_0^\infty G_{\text{LF}}(\nu)\,d\nu}{G_{\text{LF}}(0)}$$

$$(6.11)$$

Note that $G_{\text{LF}}(0)$ represents the postdetection dc power gain. If the integration time is $t_{\text{LF}}$, then $\Delta\nu_{\text{LF}}$ can be also be written as Eq. (6.12). Substituting this in Eq. (6.9), we obtain $\Delta T_{\min}$, as shown in Eq. (6.13).

$$\Delta\nu_{\text{LF}} = \frac{1}{2t_{\text{LF}}}$$

$$(6.12)$$

$$\Delta T_{\min} = \frac{T_{\text{Sys}}}{\sqrt{\Delta v_{\text{HF}}\, t_{\text{LF}}}} \tag{6.13}$$

The effective integration time $t_{\text{LF}}$ for a smoothing filter can be obtained using Eqs. (6.11) and (6.12), as shown in Eq. (6.14).

$$t_{\text{LF}} = \frac{G_{\text{LF}}(0)}{2\int_0^\infty G_{\text{LF}}(v)\,dv} \tag{6.14}$$

Fluctuations in $V_{\text{out}}$ due to amplifier gain variations are independent of the system noise fluctuations. If $\Delta G$ is the effective variation in $G_{\text{HF}}$, then the actual sensitivity can be expressed as in Eq. (6.15).

$$\Delta T_{\min} = T_{\text{Sys}} \sqrt{\frac{1}{\Delta v_{\text{HF}}\, t_{\text{LF}}} + \left(\frac{\Delta G}{G_{\text{HF}}}\right)^2} \tag{6.15}$$

Postdetection gain variations do not affect the sensitivity if dc compensation is used. Bandwidth and noise temperature variations of the receiver reduce the sensitivity during reception with gain instability. An automatic gain control or AGC is not used here, for it can vary the receiver bandwidth and noise figure, thereby introducing spurious noise fluctuations.

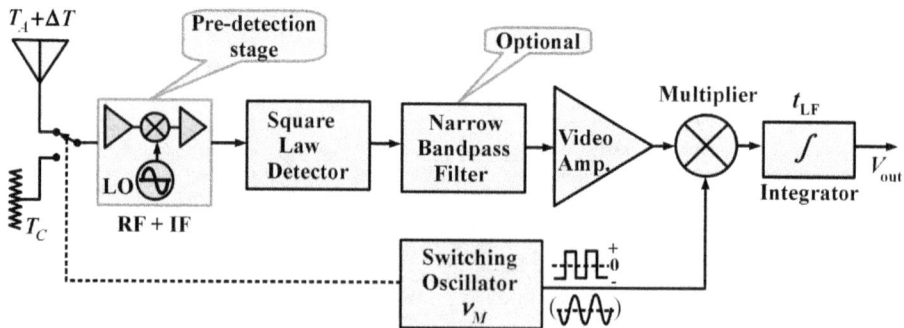

**FIGURE 6.5:** Functional blocks of a Dicke receiver.

### 6.4.2 Dicke Receiver

Fig. 6.5 shows the block diagram of Dicke receiver. The predetection stage is similar to that of a total power receiver. A square-law detector is used. The input to the predetection stage is switched between a resistive load and the antenna at a frequency $v_M$. In general, $v_M$ varies between 10 to 10,000 Hz, and the duty cycle is maintained at 50%. The postdetector stage consists of a bandpass filter (optional) and a video amplifier, followed by a multiplier and an integrator. The multiplier is used to change the sign of the signals at the switching frequency $v_M$. The output from the multiplier the results from the antenna and the resistor develop opposite signs. After integration only the difference signal exists. If the noise temperature $T_C$ of the resistor is adjusted so that it becomes the same as the antenna temperature of cold sky, the integrator output will be a function of the incremental temperature $\Delta T$ alone. Thus when the antenna is pointed to cold sky, the integrator output becomes zero, and when pointed to the source, it becomes positive.

Let $G_{HF}$ be the gain of the predetection stage, and let $\Delta G$ be the gain variations in $G_{HF}$. When pointed to a radio source, the incremental temperature $\Delta T$ is given as in Eq. (6.16), where $T_A$ is the antenna temperature of cold sky.

$$\Delta T = \left(T_A - T_C\right)\frac{\Delta G}{G_{HF}} \tag{6.16}$$

Because the switching duty cycle is 50%, the signal arrives for half of the observation time. Hence the sensitivity reduces by a factor of $\sqrt{2}$. Due to subtraction of the noise signals, it is further degraded by a factor of $\sqrt{2}$. The sensitivity of a Dicke receiver is given by Eq. (6.17), where $T_{Sys}$ is the system temperature (cold sky), $\Delta v_{HF}$ is the predetection bandwidth, and $t_{LF}$ is the effective integration time.

$$\Delta T_{min} = 2\frac{T_{Sys}}{\sqrt{\Delta v_{HF} t_{LF}}} \tag{6.17}$$

The switching waveform fed to the multiplier can be a square wave (preferable) or a sine wave. If it is a square wave, the video amplifier bandwidth should be about $10v_M$ to accommodate all the important harmonics. If the video bandwidth is restricted to a few Hz centered at $v_M$, with the aid of the optional narrow-bandpass filter shown, some reduction in sensitivity takes place. The first harmonic amplitude of a square wave is $4/\pi$ times its amplitude. Hence its effective value is $4/\sqrt{2}\pi$. Thus the sensitivity in this case is given in Eq. (6.18), which is obtained by multiplying Eq. (6.17) by $\sqrt{2}\pi/4$.

$$\Delta T_{min} = \sqrt{\frac{\pi}{2}}\frac{T_{Sys}}{\sqrt{\Delta v_{HF} t_{LF}}} \tag{6.18}$$

The sensitivity reduction is about 10. If a sinusoidal modulation (shown within brackets in Fig. 6.5) is used, the sensitivity is further reduced by a factor of $4/\pi$.

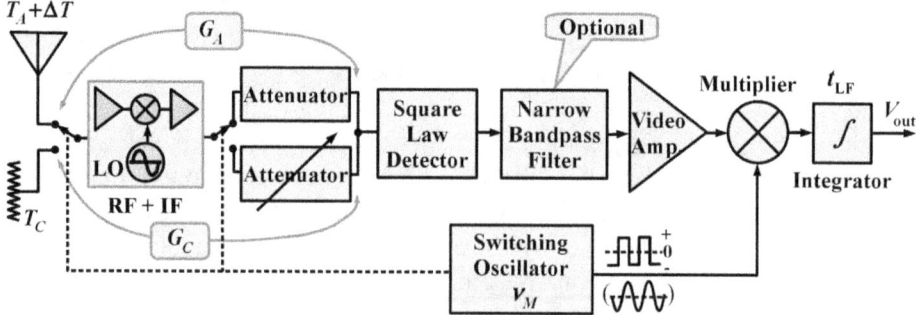

**FIGURE 6.6:** Functional blocks of a gain-modulated Dicke receiver.

### 6.4.3 Gain-Modulated Dicke Receiver

One of the modified Dicke receivers, known as the *gain-modulated Dicke receiver*, is shown in Fig. 6.6. Two passive attenuators are used to control the gains of the paths to the square-law detector (*i*) from the antenna, and (*ii*) from the noise generator. One of the attenuators, preferably in the path of the resistive noise source, is variable. The attenuators are selected alternately by switching in synchronization with the input to the RF stage together with the modulating signal to the multiplier. The balance condition between cold sky and the noise source is given by Eq. (6.19), where $T_A$, $T_R$, and $T_C$ are, respectively, the antenna temperature (cold sky), receiver temperature, and noise resistor temperature, and $G_A$ and $G_C$ are, respectively, the effective gains of the signal paths from the antenna and the resistor to the square-law detector.

$$G_A\left(T_A + T_R\right) = G_C\left(T_C + T_R\right) \tag{6.19}$$

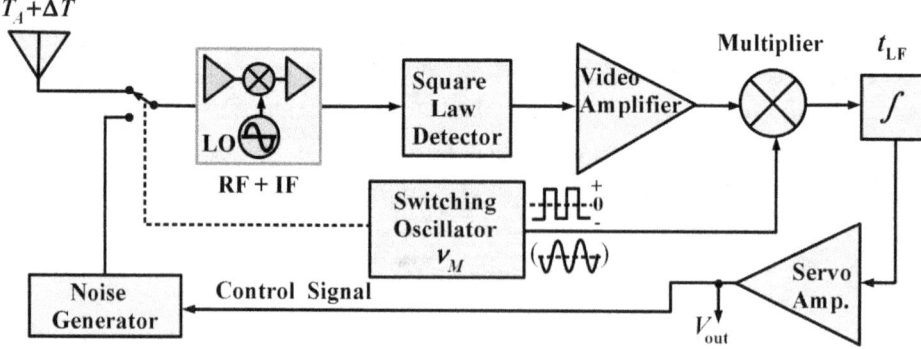

**FIGURE 6.7:** Functional blocks of a null-balancing Dicke receiver made by Machin, Ryle, and Vonberg.

### 6.4.4 Null-Balancing Dicke Receiver

The two Dicke receivers discussed above suffer from gain instability, especially for stronger signals. As a result, the shape of the signal curve is distorted, which reduces the accuracy. Further modifications by Machin, Ryle, and Vonberg to the Dicke receiver allow it to remain in a balanced condition. This is shown in Fig. 6.7. This modified receiver makes use of an adjustable noise source through a servo loop control mechanism. The power from this adjustable noise source is feedback controlled by the integrator output such that it is always zero. The control signal voltage $V_{out}$ to the noise generator is proportional to the incremental temperature $\Delta T$, which can be recorded. The noise generator may be constructed using a noise diode.

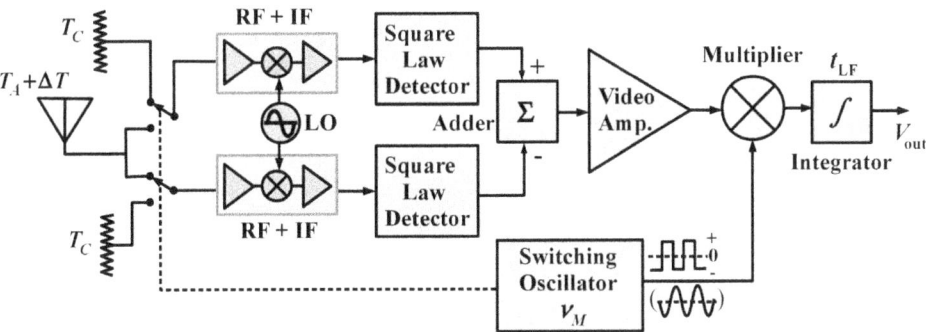

**FIGURE 6.8:** Functional blocks of a Graham's receiver.

### 6.4.5 Graham's Receiver

The signal power available in a simple Dicke receiver is only for half of the observation time. Hence, a large amount of signal goes unused. A Graham's receiver avoids this loss. As shown in Fig. 6.8, there are two receivers that are time multiplexed to process the incoming signals. Hence, the full observing efficiency is obtained. If the two receivers are alike, the sensitivity increases by a factor of $\sqrt{2}$ .

### 6.4.6 Correlation Receiver

Fig. 6.9 shows the block diagram of a correlation receiver. Signals from the antenna are split equally into two halves and fed to two identical heterodyne receivers (RF + IF) that are operated using a common local oscillator. Hence the output signals from these receivers are in phase. However, the noise generated from the two receivers is uncorrelated. When the signals are multiplied, the product of the actual signal parts are boosted, while the noise parts are attenuated, thereby enhancing the

SNR. The low-pass filter amplifier boosts the lower frequencies and feeds them to the integrator. The SNR further improves after integration.

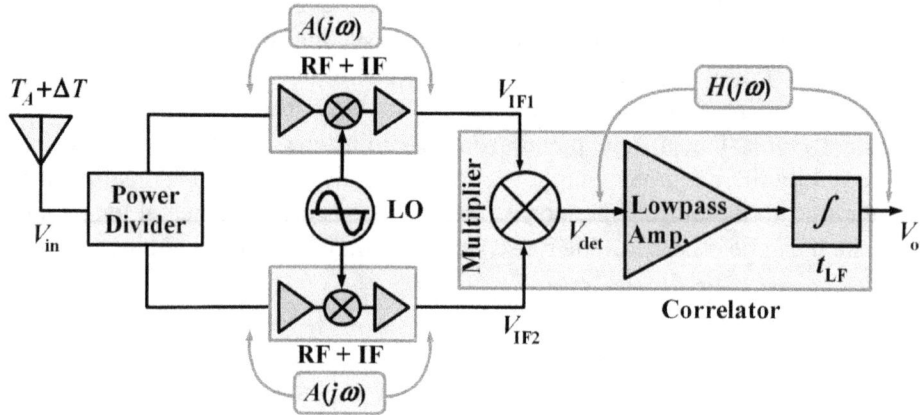

**FIGURE 6.9:** Functional blocks of a correlation receiver.

Because the two RF-plus-IF sections are identical, they have the same receiver temperatures $T_R$. The system temperature $T_{Sys}$ for each RF-plus-IF section is shown in Eq. (6.20).

$$T_{Sys} = \frac{1}{2}T_A + T_R \qquad (6.20)$$

The expected value $E[V_{out}]$ of the product $V_{det}$ from the $V_{IF1}$ and $V_{IF2}$ that result from a Gaussian input $V_{in}$ can be shown as Eq. (6.21), where $A(j\omega)$ represents the frequency response of each of the RF-plus-IF sections, $T_A$ is the temperature of cold sky, $\Delta T$ is the incremental temperature when moved from cold sky to a source, $k_B$ is the Boltzmann constant and $\omega = 2\pi v$.

$$E[V_{out}] = \frac{1}{4}k_B(T_A + \Delta T)\int_{-\infty}^{\infty}|A(j\omega)|^2 dv \qquad (6.21)$$

Let $\sigma_{out}$ be the standard deviation of $V_{out}$. If $\Delta T \ll T_A$, the relative deviation $\sigma_{out}$ can be expressed as in Eq. (6.22), where $\Delta v_{LF}$ is the bandwidth of the low-pass amplifier and $\Delta v_{HF}$ is the bandwidth of RF-plus-IF section. The sensitivity $\Delta T_{min}$ is shown in Eq. (6.23), where $t_{LF}$ is the effective integration time.

$$\sigma_{out}^* = \frac{\sigma_{out}}{E[V_{out}]} = \frac{2T_{Sys}}{\Delta T}\sqrt{\frac{\Delta v_{LF}}{\Delta v_{HF}}}\sqrt{1+\left(\frac{T_A}{2T_{Sys}}\right)^2} \qquad (6.22)$$

$$\Delta T_{min} = \sqrt{2}\,\frac{T_{Sys}}{\sqrt{\Delta \nu_{HF} t_{LF}}}\sqrt{1+\left(\frac{T_A}{2T_{Sys}}\right)^2} \qquad (6.23)$$

### 6.4.7 Additive Interferometer Receiver

We have already discussed the basic additive radio interferometer in Chapter 5 (Section 5.3). Fig. 6.10 shows the block diagram of an additive interferometer receiver. It is also known as a *simple interferometer receiver*. It consists of a pair of identical antennas connected to two identical preadder stages (RF + IF) having a common LO (local oscillator). The outputs are summed as shown in Eq. (6.24) and sent to a square-law detector. This is followed by postdetection (video amplifier and integrator).

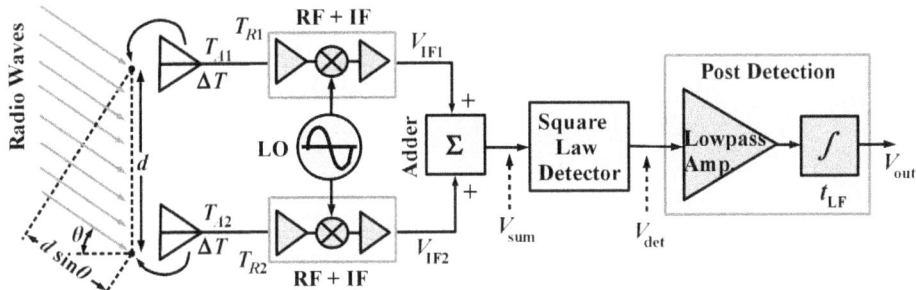

FIGURE 6.10: Functional blocks of an additive interferometer (simple interferometer) receiver.

A phase difference may exist between the signals arriving from the two antennas depending on the source direction, but the signal magnitudes will be identical. The voltage $V_{sum}$ will be a maximum if Eq. (6.25) is satisfied. It is a minimum when Eq. (6.26) is satisfied. Here, $d$ is the separation between the antennas, $\theta$ is the inclination angle (see Fig. 6.10), $n$ is an integer, and $\lambda$ is the wavelength.

$$V_{sum} = V_{IF1} + V_{IF2} \qquad (6.24)$$

$$d\sin\theta = n\lambda \qquad (6.25)$$

$$d\sin\theta = \frac{(2n-1)\lambda}{2} \qquad (6.26)$$

While observing a discrete radio source, the signal (incremental) and noise (cold sky) powers delivered by antenna 1 are, respectively, $k_B \Delta T \Delta \nu_{HF}$ and $k_B T_{A1} \Delta \nu_{HF}$, where $k_B$ is the Boltzmann constant and $\Delta \nu_{HF}$ is the available preadder bandwidth.

For antenna 2, these are $k_B \Delta T \Delta \nu_{HF}$ and $k_B T_{A2} \Delta \nu_{HF}$. The receiver noise powers are $k_B T_{R1} \Delta \nu_{HF}$ and $k_B T_{R2} \Delta \nu_{HF}$. The noise power quantities $k_B T_{A1} \Delta \nu_{HF}$, $k_B T_{A2} \Delta \nu_{HF}$, $k_B T_{R1} \Delta \nu_{HF}$, and $k_B T_{R2} \Delta \nu_{HF}$ are assumed to be independent. Hence, in absence of a discrete radio source, the square-law detector receives a noise power $W_{det}$ given by Eq. (6.27), where $G_{HF}$ is the gain of the preadder stages. $T_{Sys1}$ and $T_{Sys2}$ are, respectively, the system temperatures of the two preadder sections given by Eqs. (6.28) and (6.29).

$$W_{det} = k_B \, G_{HF} \left(T_{Sys1} + T_{Sys2}\right) \Delta \nu_{HF} \tag{6.27}$$

$$T_{Sys1} = T_{A1} + T_{R1} \tag{6.28}$$

$$T_{Sys2} = T_{A2} + T_{R2} \tag{6.29}$$

Eq. (6.30) expresses the fluctuating noise power output $W_{LF}$ from the integrator, where $\Delta \nu_{LF}$ is the postdetection bandwidth and $C'$ is a constant.

$$W_{LF} = G_{LF} \, 2C' k_B{}^2 \left(T_{Sys1} + T_{Sys2}\right)^2 \Delta \nu_{HF} \, \Delta \nu_{LF} \tag{6.30}$$

For a weak discrete source, the noise voltages from each IF amplifier are proportional to $\sqrt{k_B \Delta T \Delta \nu_{HF}}$ . If these two are in phase, the integrator output voltage $V_{out}$ is proportional to $4 k_B \Delta T \Delta \nu_{HF}$. If there exists a phase difference $\phi$ between the two, then $V_{out}$ becomes proportional to $4 k_B \Delta T \Delta \nu_{HF} (1 + \cos\phi)$. The integrator output power $W_{out}$ is shown in Eq. (6.31).

$$W_{out} = G_{LF} \, C' \left[2 k_B \, \Delta T \, \Delta \nu_{HF} \left(1 + \cos\phi\right)\right]^2 \tag{6.31}$$

The sensitivity $DT_{min}$ of the additive interferometer receiver is given in Eq. (6.32), where $t_{LF} = 1(2 \Delta \nu_{LF})$ is the effective integration time. If $T_{Sys1} = T_{Sys2} = T_{Sys}$, we obtain a simplified expression as shown in Eq. (6.33).

$$\Delta T_{min} = \frac{T_{Sys1} + T_{Sys2}}{2\left(1 + \cos\phi\right)\sqrt{\Delta \nu_{HF} \, t_{LF}}} \tag{6.32}$$

$$\Delta T_{min} = \frac{T_{Sys}}{\left(1 + \cos\phi\right)\sqrt{\Delta \nu_{HF} \, t_{LF}}} \tag{6.33}$$

The simple interferometer has drawbacks similar to those of a total power receiver.

### 6.4.8 Multiplicative Interferometer Receiver

A multiplicative interferometer receiver is also known as an *interferometer using a correlator receiver*. The block diagram of the this receiver is shown in Fig. 6.11. Here, the IF output voltages ($V_{IF1}$ and $V_{IF2}$) are multiplied. Because, noise from the premultiplier stage (RF plus IF) are uncorrelated, they contribute very little to the output voltage $V_{out}$. The source signal contributions on the other hand are correlated and contribute to $V_{out}$. The noise power output from the integrator $W_{LF}$ is shown in Eq. (6.34), where $\Delta v_{LF}$ is the postdetection bandwidth and $C'$ is a constant.

$$W_{LF} = G_{LF}\, 2C' k_B^{\,2}\, T_{Sys1}\, T_{Sys2}\, \Delta v_{HF}\, \Delta v_{LF} \tag{6.34}$$

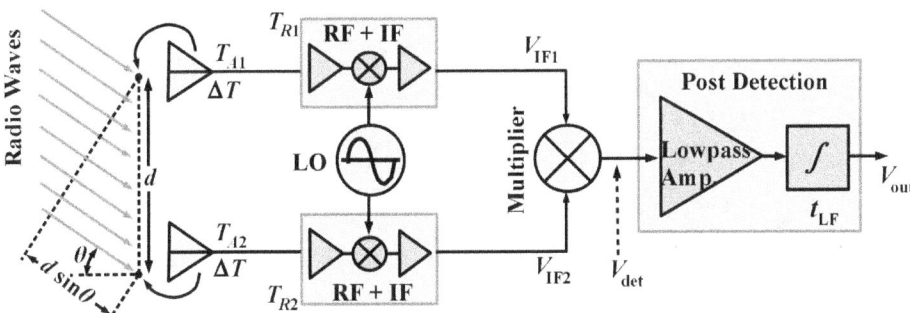

**FIGURE 6.11:** Functional blocks of a multiplicative interferometer (correlator) receiver.

For a weak discrete source, the noise voltages from each of the IF amplifiers are proportional to $\sqrt{k_B\, \Delta T\, \Delta v_{HF}}$, where $\Delta v_{HF}$ is the premultiplier bandwidth. The output voltage $V_{det}$ from the multiplier detector is proportional to $k_B \Delta T \Delta v_{HF} \cos\phi$, where $\phi$ is the phase difference between $V_{IF1}$ and $V_{IF2}$. The signal power output $W_{out}$ from the integrator is given by Eq. (6.35). The sensitivity $\Delta T_{min}$ is given by Eq. (6.36). If $T_{Sys1} = T_{Sys2} = T_{Sys}$, we obtain the simplified expression shown in Eq. (6.37), where $t_{LF} = 1(2\Delta v_{LF})$ is the effective integration time.

$$W_{out} = G_{LF}\, C' \left[ k_B\, \Delta T\, \Delta v_{HF} \right]^2 \cos^2\phi \tag{6.35}$$

$$\Delta T_{min} = \frac{1}{\cos\phi} \sqrt{\frac{T_{Sys1}\, T_{Sys2}\, \Delta v_{LF}}{\Delta v_{HF}}} \tag{6.36}$$

$$\Delta T_{min} = \frac{1}{\cos\phi} \frac{T_{Sys}}{\sqrt{2\, \Delta v_{HF}\, t_{LF}}} \tag{6.37}$$

The sensitivity is $2\sqrt{2}$ times than that of a Dicke receiver. Because the cor-related noise alone can contribute to $V_{out}$, the gain instability seldom affects the sensitivity. However, gain variations can affect the receiver calibrations. Random phase variations in the premultiplier stage can also affect $V_{out}$. The scintillations in the ionosphere also produce phase variations that are similar to the phase-variation problem and can reduce the sensitivity in a similar fashion. Because the RF switch is not used for the input, it avoids additional loss of sensitivity due to an increase in the system temperature.

### 6.4.9 Phase-Switched Receiver

Fig. 6.12 shows the blocks of a phase-switched receiver built by Ryle. It also uses the correlation principle. Two identical heterodyne systems (RF plus IF) fed from a common LO are connected to the two antennas of an interferometer. The IF sig-nal $V_{IF2}$ goes through a phase-reversal switch driven at a frequency $v_M$. If $V_{IF1}$ and $V_{IF2}$ are uncorrelated, the switching will have no effect on the square-law detector output voltage $V_{det}$. When $V_{IF1}$ and $V_{IF2}$ contain some correlated components, $V_{det}$ is different for $V_{IF1} + V_{IF2}$ than it is for $V_{IF1} - V_{IF2}$. This implies that $V_{det}$ changes at a frequency $v_M$ because of the correlated signal. If we assume that the desired signals alone are correlated, then the sensitivity of the phase-switched receiver be-comes the same as that of the Dicke receiver, provided the latter also uses a similar low-frequency section.

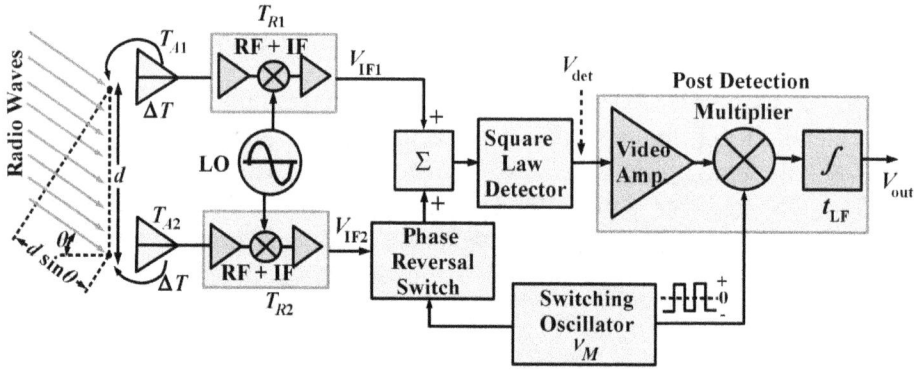

**FIGURE 6.12:** Functional blocks of a phase-switched receiver.

### 6.4.10 Comparison of Sensitivities of Radiometers

We may express the temperature sensitivity $\Delta T_{min}$ (minimum detectable temperature) of any receiver as shown in Eq. (6.38), where $K_s$ is the sensitivity constant, $t_{Sys}$ is the system temperature, $\Delta v_{HF}$ is the equivalent rectangular predetection bandwidth, $t_{LF}$ is the effective integration time (postdetection). We compare the sensitivities by substituting various values obtained from different receivers in Eq. (6.38). Table 6.1 lists the different values of $K_s$ obtained from different radiometers.

$$\Delta T_{\min} = K_s \frac{T_{\text{Sys}}}{\sqrt{\Delta \nu_{\text{HF}} \, t_{\text{LF}}}} \qquad (6.38)$$

**TABLE 6.1:** Relative sensitivities obtained using different radiometers.

| Receiver Type | $K_s$ |
|---|---|
| Total power receiver (Fig. 6.3). | 1 |
| Dicke receiver using square-wave modulation, broadband video amplifier followed by square-wave multiplication (Fig. 6.5). | 2 |
| Dicke receiver using square-wave modulation, narrowband video amplifier followed by sine-wave multiplication (Fig. 6.5). | $\frac{\pi}{\sqrt{2}} \approx 2.22$ |
| Dicke receiver using sine-wave modulation, narrowband video amplifier followed by sine wave multiplication (Fig. 6.5). | $2\sqrt{2} \approx 2.83$ |
| Graham's receiver (Fig. 6.8). | $\sqrt{2} \approx 1.41$ |
| Correlation receiver having a small antenna noise in comparison to the receiver noise (Fig. 6.9). | $\sqrt{2} \approx 1.41$ |
| Additive interferometer with identical antennas (Fig. 6.10). | $\frac{1}{2} = 0.5$ |
| Multiplicative interferometer with identical antennas (Fig. 6.11). | $\frac{1}{\sqrt{2}} \approx 0.71$ |
| Phase-switched interferometer with identical antennas, square-wave switching, and square-wave multiplication (Fig. 6.12). | 2 |

## 6.5   CALIBRATION OF RECEIVERS AND RADIOMETERS

There are several noise sources available for calibration, but the most commonly used are (*i*) thermally controlled resistors, and (*ii*) current-controlled noise diodes. The resistor generates a noise power $W_{\text{Res}}$ given by Eq. (6.39), where $k_B$ is Boltzmann constant, $T$ is the physical temperature of the resistor, and $\Delta \nu$ is the bandwidth across which the noise is measured. Thus by changing the physical temperature of the resistor, an equivalent amount of noise can be generated.

$$W_{\text{Res}} = k_B T \Delta v \qquad (6.39)$$

However, when large temperatures of the order of 1000 K and above are required, noise diodes are preferable. These diodes are operated in the avalanche breakdown region and can work at frequencies up to 20 GHz. If the diode-biasing current is $I_0$, then the mean square noise current $\overline{I^2_{\text{noise}}}$ is given by Eq. (6.40), where $e^-$ is the charge of an electron.

$$\overline{I^2_{\text{noise}}} = 2I_0 e^- \Delta v \qquad (6.40)$$

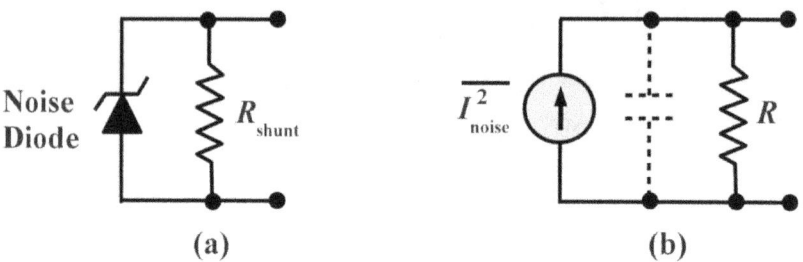

**(a)**      **(b)**

**FIGURE 6.13:** (a) Circuit diagram of an avalanche diode noise generator. (b) Equivalent RF circuit.

The circuit diagram of a avalanche diode noise generator is shown in Fig. 6.13a. The shunt resistor $R_{\text{shunt}}$ is used to match the transmission line. Fig. 6.13b shows the RF equivalent circuit where the effective resistance is $R$, which is the same as the characteristic impedance of the transmission line. The capacitance shown by dotted lines is the junction capacitance of the diode that decides the maximum frequency of usage. Hence, the deliverable noise power $W_G$ by the generator to the receiver under calibration is shown in Eq. (6.41), where $k_B T_0 \Delta v$ is generated by the matching resistor due to its physical temperature $T_0$. The equivalent noise temperature $T_G$ of the noise generator is given by Eq. (6.42)

$$W_G = \frac{\overline{I^2_{\text{noise}}}}{4} R + k_B T_0 \Delta v = \frac{I_0 e^- R \Delta v}{2} + k_B T_0 \Delta v \qquad (6.41)$$

$$T_G = \frac{W_G}{k_B \Delta v} = \frac{I_0 e^- R}{2k_B} + T_0 = (20 I_0 R + 1) T_0 \qquad (6.42)$$

### 6.5.1 Calibration of Receiver Temperature

The receiver noise temperature can be measured by connecting its input to a noise generator and measuring the output using a power meter as shown in Fig. 6.14. The internal impedance of the noise generator must be equal to that of the antenna

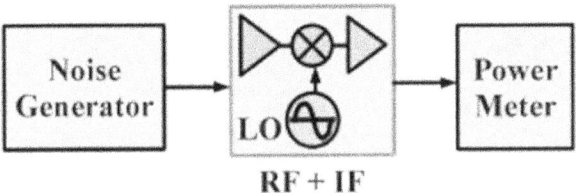

**FIGURE 6.14:** Setup for measurement of the noise temperature of a receiver.

and should not fluctuate with time. Let the physical temperature of the noise generator be $T_0$. Under the unfired condition, the noise temperature provided by the generator will be $T_G = T_0$. Let the power meter reading corresponding to this be $P_{off}$. Under the fired condition, let the noise temperature of the generator be $T_G$ and the corresponding power meter reading be $P_{on}$. The ratio $K_1$ of these two readings is shown in Eq. (6.43).

$$\frac{P_{on}}{P_{off}} = \frac{T_G + T_R}{T_0 + T_R} = K_1 \tag{6.43}$$

The receiver temperature is given by Eq. (6.44). If the noise generator is adjustable, $T_G$ can be adjusted so that, $P_{on} = 2\, P_{off}$ and $T_R$ may be expressed as in Eq. (6.45).

$$T_R = \frac{P_{off} T_G - P_{on} T_0}{P_{off} - P_{on}} \tag{6.44}$$

$$T_R = T_G - 2T_0 \tag{6.45}$$

### 6.5.2 Calibration of Radiometers

Standard noise sources are used with receiving systems for calibrations. Calibrations can be of two types: (*i*) internal, and (*ii*) external. In the former, a noise generator is connected permanently to the receiving system using a directional coupler, as shown in Fig. 6.15a. The noise temperature of the generator is $T_0$ in the unfired condition and $T_G$ in the fired condition. Generally, $T_0$ is close to the ambient temperature (300 K). A variable attenuator is used to control the noise flow to the receiver. If the attenuation is $L$ and the coupling loss of the directional coupler is $L_D$, the excess noise temperature $T_{excess}$ added to system can be estimated by Eq. (6.46).

$$T_{excess} = \frac{1}{L_D}\left[\frac{T_G}{L} + \left(1 - \frac{1}{L}\right)T_0\right] - \frac{T_0}{L_D} = \frac{T_G - T_0}{L\, L_D} \tag{6.46}$$

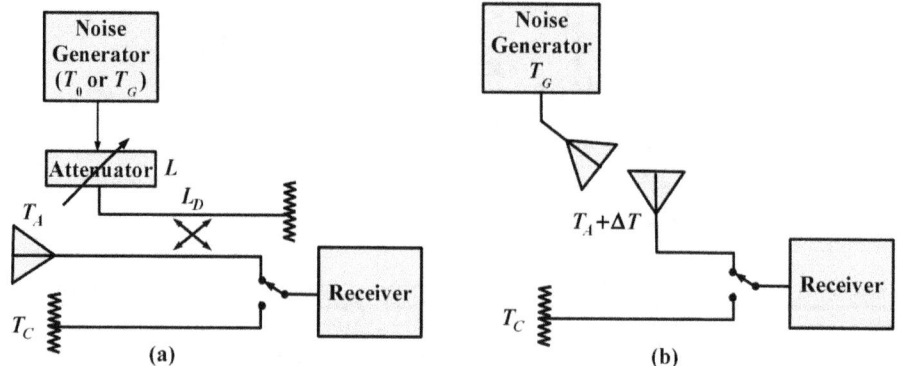

**FIGURE 6.15:** Calibration configurations. (a) Noise generator connected directly to the receiver through a directional coupler. (b) Noise generator is coupled through antennas.

Fig. 6.15b shows the arrangement for calibrating the radio telescope as a whole. Here, the noise generator is connected to an auxiliary antenna that radiates. This is received at a distance by the radio telescope antenna. The attenuation between the telescope antenna and the auxiliary antenna must be accurately known for absolute calibration. A radio source whose flux density is known accurately can also be used for calibration, provided the effective aperture area $A_e$ of the telescope antenna is accurately known. In this case, the incremental temperature $\Delta T$ is given by Eq. (6.47), where $S$ is the spectral flux density near the antenna aperture due to a point source, and $k_B$ is the Boltzmann constant.

$$\Delta T = \frac{S\,A_e}{2k_B} \tag{6.47}$$

Receiver calibrations are essential for providing an absolute scale of the antenna temperature. Because the gain and noise temperature of the receiver may vary during observation, calibration checks must be frequently made before, between, and after observations.

## 6.6   SPECTRAL-LINE RADIOMETERS

Special radiometers were developed for correctly measuring the 1420 MHz line radiation from neutral hydrogen in the Milky Way and other galaxies. Later, frequency resolution was increased by adding more channels with reduced channel bandwidths. This enabled viewing details of the spectra within a narrow bandwidth covering the spectral line. Although modern radio telescopes use interferometers for spectral-line observations, we will here describe some earlier versions that are of historical interest.

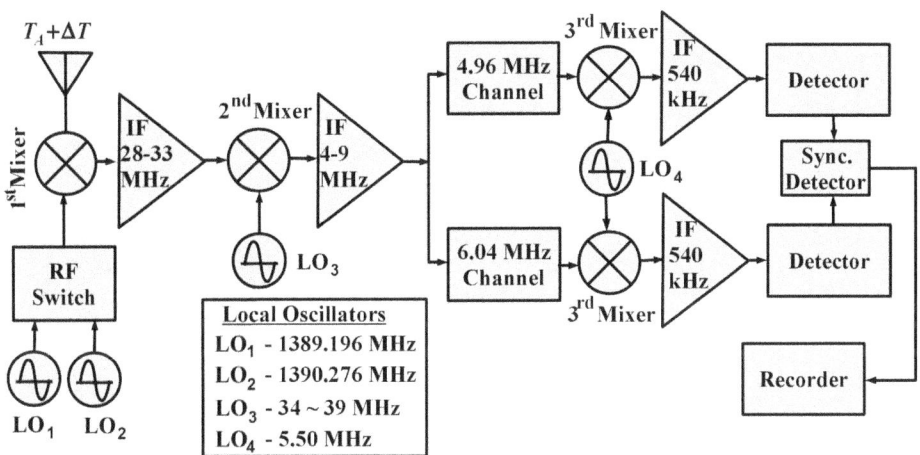

**FIGURE 6.16:** Functional blocks of a frequency-switched hydrogen-line radiometer.

## 6.6.1 Frequency-Switched Radiometer

A simplified functional block diagram of the frequency-switched radiometer is shown in Fig. 6.16. The antenna signals are alternatively mixed with two crystal local oscillators $LO_1$ and $LO_2$ at 400 Hz using an RF switch (not shown). The frequencies of $LO_1$ and $LO_2$ are, respectively, 1389.196 MHz and 1390.276 MHz. The difference between the two is 1.08 MHz. The mixer output is fed to an IF filter amplifier having a frequency band of 28 to 33 MHz. After amplification it goes into the second mixer, whose local oscillator $LO_3$ (set at 35.764 MHz) can be swept between 34 to 39 MHz. The second IF output is amplified within a range of 4 to 9 MHz using another filter amplifier. It is then fed to two separate channel filters centered at 4.96 MHz and at 6.04 MHz. The same frequency difference of 1.08 MHz is maintained between the channels. These two channel outputs are mixed with $LO_4$ signal (5.50 MHz) over a bandwidth of 35 kHz and processed through two IF filter amplifiers at 540 kHz. These are next sent to two detectors. The detected signals are sent to a synchronous detector that is also supplied with the 400 Hz switching signal. If a 1420 MHz signal is found, it will emerge through the 4.96 MHz channel when $LO_1$ operates and through the 6.04 MHz channel when $LO_2$ operates.

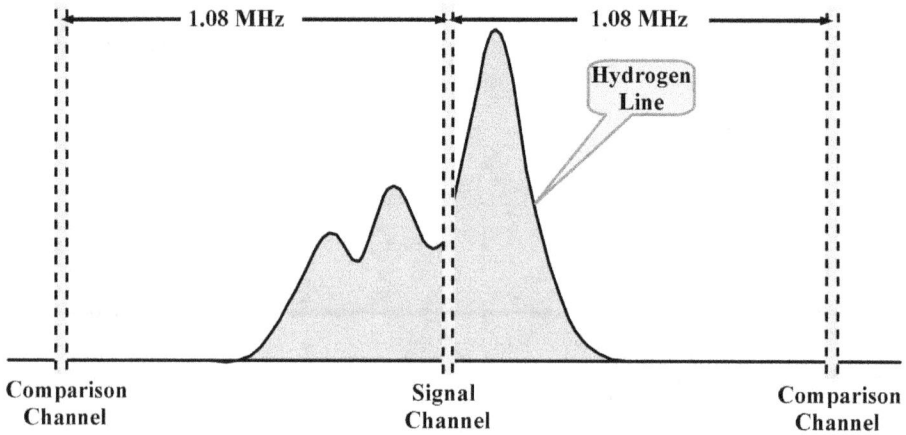

**FIGURE 6.17:** The location of channels in a frequency-switched radiometer with respect to a hydrogen-line profile.

The required information at 1420 MHz will be found at one of these two channels, while the other channel, which has a 35 kHz bandwidth, lies 1.08 MHz below or above 1420 MHz. The synchronous detector output measures (*i*) the difference between the noise power at 1420 MHz, and (*ii*) the mean of the noise power at frequencies equally spaced below and above 1420 MHz. Fig. 6.17 shows these channels along with a hydrogen-line profile under observation. By suitably tuning $LO_3$, the sensitive band of the receiver may be swept across the hydrogen-line profile. Switching can improve the receiver's overall stability because the effects of the receiver noise are minimized. The line is observed in both switching positions to avoid loss of sensitivity.

### 6.6.2 Multichannel Radiometers

Fig. 6.18 shows the blocks of a multichannel hydrogen-line radiometer built at NRAO. It is also based upon the principle of frequency switching along with sampling of the whole frequency range. A good stability is obtained by frequency switching. The 5 MHz IF is sent into the bank of twenty filters. Each of these has a separate synchronous detector. The filter channel outputs are sampled and recorded digitally, which produces the hydrogen-line profile. This receiver was designed for studying the hydrogen content of other galaxies. It was expected that the hydrogen profile might extend over 2 MHz. Twenty channels of 95 kHz individual widths have been used. The observation of galactic hydrogen requires more channels having narrower bandwidths.

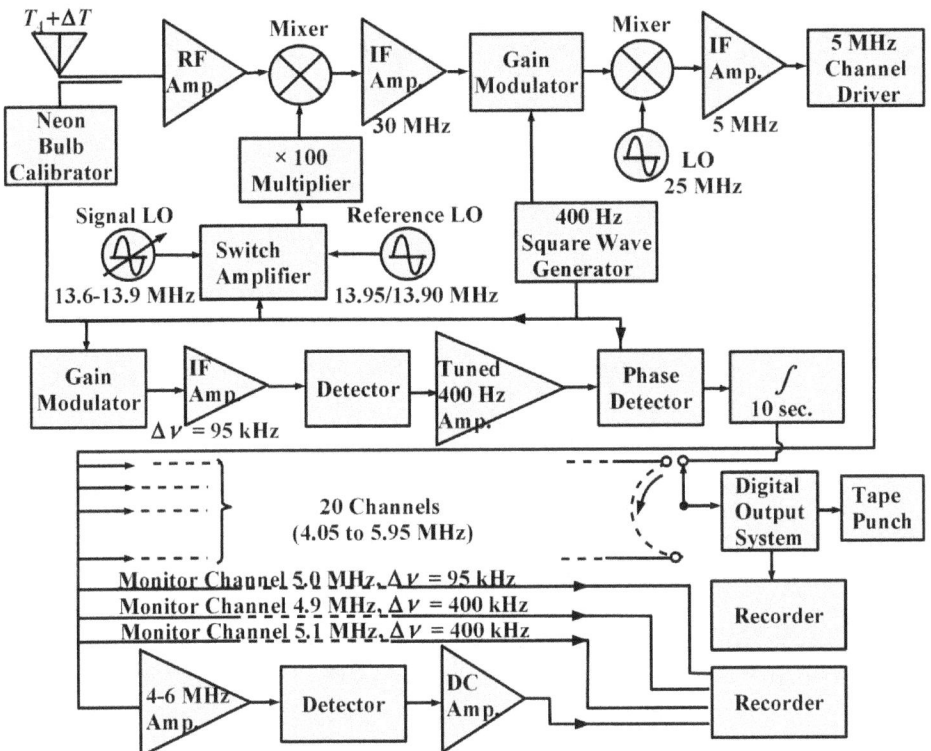

**FIGURE 6.18:** Functional blocks of a multichannel hydrogen-line radiometer.

## 6.7 PULSAR RECEIVERS

Pulsars may be broadly grouped into two types: (*i*) those whose pulses remain constant over many pulse periods, and (*ii*) those whose pulses vary from pulse to pulse. The same receiver hardware can be used for both; however, data handling of these can be completely different. To obtain integrated pulse characteristics, the receiver output must be averaged synchronously with the pulsar period and then recorded over relatively large time intervals on the order of several minutes or more. This may be done on a computer or by using special hardware, provided the pulse period is accurately known before hand.

Pulsar receivers differ from conventional radio astronomy receivers in three ways: (*i*) synchronous detection (comparison switching) is avoided because the pulses themselves provide switching information; (*ii*) four identical signal channels are used to instantaneously determine the polarization states of the rapidly

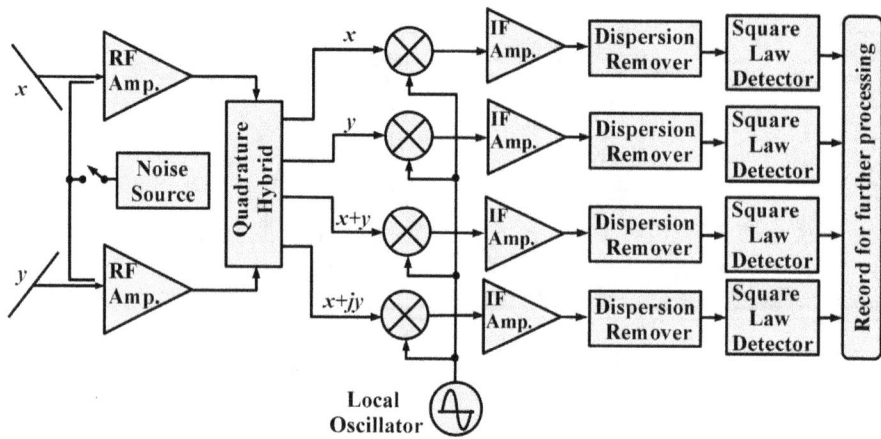

**FIGURE 6.19:** System blocks of a pulsar receiver.

varying signals without feed rotation; (*iii*) to compensate for the dispersion effects from the interstellar medium, dispersion removers are used in all four signal paths.

Fig. 6.19 shows the basic system blocks of a pulsar receiver. Here *x* and *y* represent the two orthogonal linear polarizations of the antenna. The noise source is for simultaneous calibration of both polarizations. After amplification, the signals are added in a quadrature hybrid to generate $x + y$ and $x + jy$ components. All of these ($x, y, x + y$, and $x + jy$) are individually mixed with a common LO signal, and the respective IF signals are generated. These are amplified and passed through dispersion removers and then fed to square-law detectors. The detector outputs are recorded (kept for further processing, e.g., pulse averaging).

### 6.7.1 Dispersion Measure (DM)

Pulsar signals get dispersed in the interstellar medium, and the dispersion there is frequency dependent. Hence the use of large bandwidths for increasing sensitivity is restricted. For a desired time resolution of $\Delta t$ (sec), the maximum usable bandwidth $\Delta v_{max}$(MHz) is given by Eq. (6.48), where $dv/dt$ (MHz/sec) is the rate of change of the frequency of the dispersed signal, $v$ is the observing frequency, and DM ($cm^{-3}$ pc) is the dispersion measure.

$$\Delta v_{max} = \Delta t \frac{dv}{dt} \approx 1.2 \times 10^{-4} \Delta t \, v^2 \, DM^{-1} \qquad (6.48)$$

DM represents the total column density of free electrons existing between the observer and the pulsar. It is the integral of the electron density along the line of sight, as shown in Eq. (6.49), where $D$ represents the distance between the pulsar and the observer and $n_e$ is the electron density of the ISM (interstellar medium).

$$\text{DM} = \int_0^D n_e(s)\,ds \qquad (6.49)$$

### 6.7.2 Correction of Dispersion

To remove the dispersion effects, a complex Fourier analysis of the IF signal is performed. This complex spectrum is multiplied with a frequency-dependent phase factor that is precomputed from the known DM of the pulsar. The inverse Fourier transform of this gives the desired signal (free from the dispersion effects). The dispersion correction methods may be broadly categorized into two types: (*i*) incoherent dedispersion, and (*ii*) coherent dedispersion.

#### 6.7.2.1 Incoherent Dedispersion

Incoherent dedispersion separates the frequency channels of the incoming signal using a number of narrowband filters, applies different time delays based on their respective channel frequencies, and finally recombines them as output. These delays compensate for the frequency-dependent delays applied by the interstellar medium to the signal. Hence the corrected signal channels are in phase and are added to obtain the actual signal. The difference in arrival times $\Delta t$(ms) between a pulse received by a channel of frequency $v_{\text{chan}}$ relative to the center frequency $v_0$ of the observed band is approximately given by Eq. (6.50), where DM is in (cm$^{-3}$ pc).

$$\Delta t = 4.15 \times 10^6 \left( \frac{1}{v_0^2} - \frac{1}{v_{\text{chan}}^2} \right) \times \text{DM} \qquad (6.50)$$

For a finite bandwidth near a center frequency $v_0$, the dispersive delay $t_{\text{DM}}$ (ms) across a frequency channel of bandwidth $\Delta v$ with condition that $v \gg \Delta v$ is given by Eq. (6.51).

$$t_{\text{DM}} = 8.3 \times 10^6 \times \text{DM} \times \Delta v \times v^{-3} \qquad (6.51)$$

Hence, channel bandwidths must be carefully chosen to prevent $t_{\text{DM}}$ from becoming a significant fraction of the pulse period.

**Analog Filter-bank Spectrometers:** The system blocks of an analog filter-bank spectrometer are shown in Fig. 6.20. It is the most popular and simplest data-acquisition device for incoherent dedispersion. Using narrow bandpass filters, the broad-

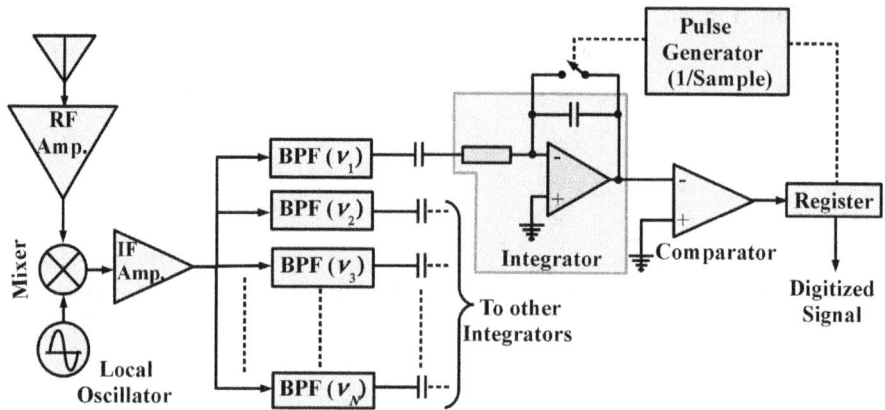

FIGURE 6.20: An analog filter-bank spectrometer using a single-bit digitization scheme.

band IF signal is split into $N$ adjacent channels having frequencies $\nu_1$, $\nu_2$, . . ., $\nu_N$, followed by integration and digitization. The digitizing scheme shown here is a one-bit sampler. The integrator output is the running mean of the signal. It is compared with zero voltage inside the comparator. The comparator output is digital: either 0 or 1. Sampling is achieved by reading the comparator output in a register at a rate of $1/t_{\text{samp}}$, where $t_{\text{samp}}$ is the sampling period. The pulses are also sent to the integrator switch for discharging the capacitor in each cycle. The integrator acts as a low-pass filter and rejects nearly all frequency components above $1/t_{\text{samp}}$. This technique avoids the need for any anti-aliasing filters. All the frequency channels are digitized separately.

This simple single-bit digitizing scheme is very useful in large projects that search for pulsars. The loss is about 20 from ideal sampling. Almost two thirds of today's known pulsars have been discovered using this technique.

**Correlation Spectrometers:** An autocorrelation spectrometer can also provide filter-bank style output. Fig. 6.21a shows the functional blocks of correlation spectrometers. Any single-polarized channel is split and multiplied with its delayed version, followed by integration. The time delays used are integer multiples of $\Delta t$, which is the minimum time delay adopted. Cross-correlation products between the two orthogonal linear polarizations A and B are also generated, as shown in Fig. 6.21b.

Let $R(\tau)$ represent the autocorrelation and $H(\nu)$ be its power spectrum. Let $r(\tau)$ be the cross-correlation. From the *Wiener-Khinchin* theorem (Section 1.4.3.2), it may be said that $R(\tau)$ obtained from an interferometer is the Fourier transform of the cosmic signal's power spectrum $H(\nu)$ with a bandwidth equal to the system

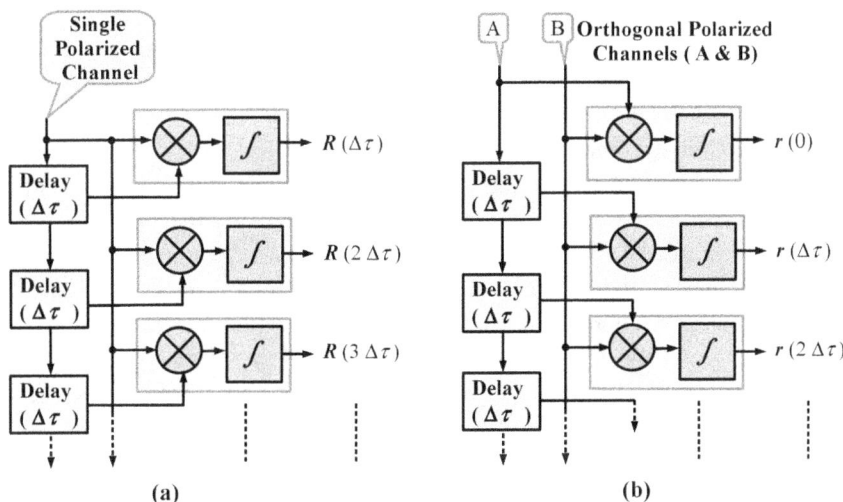

**FIGURE 6.21:** Functional blocks used in a correlation spectrometer. (a) Autocorrelation. (b) Cross-correlation

passband. If $H_{AA}$ and $H_{BB}$ are, respectively, the power spectra for the channels A and B resulting from autocorrelations, then the total intensity I = $H_{AA}$ + $H_{BB}$. Let $H_{AB}$ and $H_{BA}$ be the cross-correlation power spectra between the channels A and B. Using these, the Stokes parameters $I$, $Q$, $U$, and $V$ (see Section 3.9) from polarimetric observations using a dual-polarized linear antenna feed can be shown as in Eq. (6.52). If the antenna feed is dual circularly polarized, the Stokes parameters are given by Eq. (6.53).

$$\begin{bmatrix} I \\ Q \\ U \\ V \end{bmatrix} = \begin{bmatrix} H_{AA} + H_{BB} \\ H_{AA} - H_{BB} \\ 2\,\Re e\left[H_{AB}\right] \\ 2\,\Im m\left[H_{BA}\right] \end{bmatrix} \tag{6.52}$$

$$\begin{bmatrix} I \\ Q \\ U \\ V \end{bmatrix} = \begin{bmatrix} H_{AA} + H_{BB} \\ 2\,\Im m\left[H_{BA}\right] \\ 2\,\Re e\left[H_{AB}\right] \\ H_{BB} - H_{AA} \end{bmatrix} \tag{6.53}$$

The Stokes parameters obtained using this method are simpler than those obtained using analog filter banks. As a precaution, the auto- and cross-correlations must be calibrated before analysis and interpretation.

### 6.7.2.2 Coherent Dedispersion

Here, the phase information obtained from the telescope output voltage can be used to completely remove the phase-change effects caused by the ISM. The propagating signal through the ISM is treated like a phase change caused by a phase filter with a transfer function $H(\nu)$.

Let a signal possess a bandwidth $\Delta\nu$ centered at a frequency $\nu_0$. Let $\nu_{in}(t)$ and $\nu_{out}(t)$ be, respectively, the input and output voltages of the phase filter. Let their Fourier transforms be $V_{in}(\nu)$ and $V_{out}(\nu)$, respectively. These quantities can be related by using $H(\nu)$, as shown in Eq. (6.54). The objective is to regain the actual pulsar signal $\nu_{in}(t)$ from $\nu_{out}(t)$ with the ISM acting as the phase filter.

$$V_{out}(\nu_0 + \nu) = V_{in}(\nu_0 + \nu)H(\nu_0 + \nu) \quad \text{where,} \quad |\nu| \le \Delta\nu/2 \qquad (6.54)$$

The delay caused by the ISM may be represented by phase rotations that depend on frequency and path length of travel. Hence, the phase change $\Delta\phi$ can be shown as in Eq. (6.55), where $k(\nu)$ is the wave number at a frequency $\nu$, and $d$ is the distance of travel. We now express $H(\nu)$ as shown in Eq. (6.56). The wave number in terms of the plasma frequency $\nu_p$ and cyclotron frequency $\nu_B$ is given by Eq. (6.57).

$$\Delta\phi = -k(\nu_0 + \nu)d \qquad (6.55)$$

$$H(\nu_0 + \nu) = e^{-jk(\nu_0+\nu)d} \qquad (6.56)$$

$$k(\nu_0 + \nu) = \frac{2\pi}{c}(\nu_0 + \nu)\sqrt{1 - \frac{\nu_p^2}{(\nu_0 + \nu)^2} \mp \frac{\nu_p^2\,\nu_B}{(\nu_0 + \nu)^3}} \qquad (6.57)$$

In ISM, $\nu_p \sim$ 2kHz and $\nu_B \sim$ 3Hz. For observations above 100 MHz, the last two terms of the above equation are on the orders of $10^{-10}$ and $10^{-18}$. Hence, the wave number is approximately given as in Eq. (6.58). Substituting Eq. (6.58) into Eq. (6.56) and simplifying, we obtain $H(\nu)$ as shown in Eq. (6.59).

$$k(\nu_0 + \nu) \approx \frac{2\pi}{c}(\nu_0 + \nu)\left[1 - \frac{\nu_p^2}{2(\nu_0 + \nu)^2}\right] \qquad (6.58)$$

$$H(v_0 + v) = e^{-j\frac{2\pi}{c}d\left[\left(v_0 - \frac{v_p}{2v_0^2}\right) + \left(1 + \frac{v_p^2}{v_0^2}\right)v - \frac{v_p^2}{2(v_0+v)^2}v^2\right]} \tag{6.59}$$

The first term in the exponent gets eliminated by the square-law detector and cannot be determined. The second term links to a time delay. It may be nullified by suitably shifting the arrival time. The final term is responsible for the phase rotations within those bands that are quadratic in frequency. This term is important for recovering the original pulsar signal. Eq. (6.59) can be simplified further and expressed as Eq. (6.60).

$$H(v_0 + v) = e^{j\frac{2\pi}{c}d\left[\frac{v_p^2}{2(v_0+v)^2}\right]v^2} \tag{6.60}$$

A dispersion constant $\mathcal{D}$ is defined in relation to the plasma frequency $v_p$, as in Eq. (6.61), where $c$ is the speed of light and $n_e$ is the average electron density along the line of sight. The dispersion constant $\mathcal{D}$ may be related to the DM as shown in Eq. (6.62). Relating these two equations with Eq. (6.60), the transfer function $H(v)$ is finally expressed in Eq. (6.63).

$$\mathcal{D} = \frac{v_p^2}{2c\,n_e} \tag{6.61}$$

$$\frac{v_p^2\,d}{2c} = \text{DM} \times \mathcal{D} \tag{6.62}$$

$$H(v_0 + v) = \exp\left[\frac{j2\pi\mathcal{D}}{(v_0 + v)^2\,v_0^2}\,\text{DM}\,v^2\right] \tag{6.63}$$

$$V_{\text{in}}(v) = V_{\text{out}}(v)\,H^{-1}(v_0 + v) \tag{6.64}$$

The inverse transform $H^{-1}$ is now applied to the voltage data $V_{\text{out}}(v)$ to obtain the voltage $V_{\text{in}}(v)$ corresponding to the pulsar signal, as shown above in Eq. (6.64). This process, however, is slightly complicated for practical implementation because digitization is involved. Generally, $H^{-1}$ is multiplied with a tapering function $T(v)$, which takes care of the practical problems. The resulting product $C(v_0, v)$ is known as the chirp function. It is shown in Eq. (6.65).

$$C(v_0, v) = T(v)\,H^{-1}(v_0 + v) \tag{6.65}$$

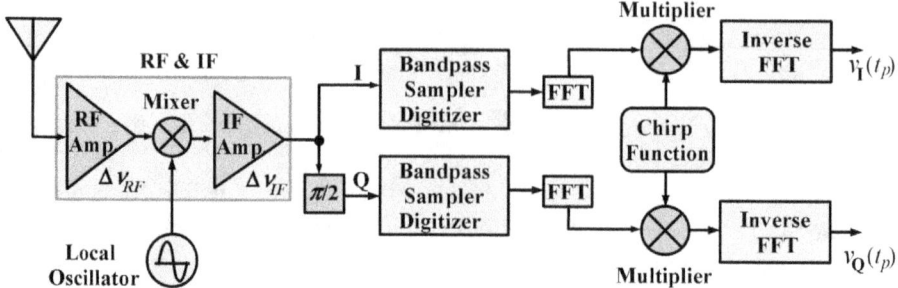

**FIGURE 6.22:** Functional blocks of coherent dedispersion receiver.

Fig. 6.22 shows the functional blocks of a coherent dedispersion receiver and a digital data-preprocessing scheme. After downconversion, the IF output generates two orthogonal functions $I$ and $Q$ using a 90° phase shifter. These are bandpass sampled and digitized. Next, FFT is performed over a fixed number of samples. These are then multiplied with a digitized chirp function. Finally, an inverse FFT is performed and the time-domain digitized values of the signal are obtained, which are free from dispersion effects.

By combining the two orthogonal linear polarizations ($X$ and $Y$) of the antenna feed, the Stokes parameters $I$, $Q$, $U$, and $V$ are obtained as shown in Eq. (6.66), where $X^*$ represents the complex conjugate of $X$.

$$\begin{bmatrix} I \\ Q \\ U \\ V \end{bmatrix} = \begin{bmatrix} |X|^2 + |Y|^2 \\ |X|^2 - |Y|^2 \\ 2\,\mathrm{Re}\left[ X^* Y \right] \\ 2\,\mathrm{Im}\left[ X^* Y \right] \end{bmatrix} \tag{6.66}$$

Similarly, for a dual circularly polarized antenna feed with $L$ and $R$ as the left and right circularly polarized components, the Stokes parameters can be obtained using Eq. (6.67).

$$\begin{bmatrix} I \\ Q \\ U \\ V \end{bmatrix} = \begin{bmatrix} |L|^2 + |R|^2 \\ 2\,\mathrm{Re}\left[ L^* R \right] \\ 2\,\mathrm{Im}\left[ L^* R \right] \\ |R|^2 + |L|^2 \end{bmatrix} \tag{6.67}$$

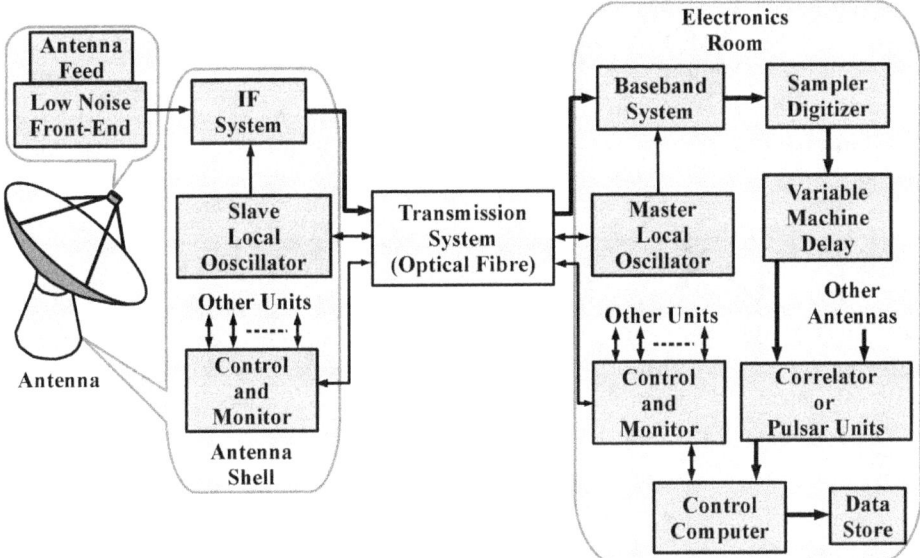

**FIGURE 6.23:** A generalized layout of the complete receiving and control chain from the observatory building to the antennas of a radio array.

## 6.8   A GENERALIZED RECEIVING-CHAIN LAYOUT

A generalized layout of the receiving chain followed in radio observatories with arrays is shown in Fig. 6.23. The low-noise front-end system is positioned near the antenna feed. It contains RF amplifiers. The output is brought through a RF cable inside the antenna shell where the remaining front-end electronics exist. IF signals are obtained after mixing with the LO (local oscillator) signal. The LO is locked to a master oscillator, usually kept in the observatory building. Other control units like servo for antenna azimuth and elevation rotations and their driving units, feed positioning systems, protective sentinel systems, etc., are also kept inside the shell. They receive commands from a computer located in the observatory building. The transmission of commands and LO signals and the reception of IF signals from the antennas is usually done using one or two optical fibers. Optical modulators and demodulators are used for conversion from RF to optical and optical to RF, respectively.

These units are located at the antenna shell and the observatory building. The distance from the observatory to an antenna may vary between a few hundred meters to several km. Once the IF signals from the antennas are brought to the observatory building, they are converted into several basebands. This is followed by sampling and digitizing. Variable instrumental delay is added for compensating for the geometric delay in order to achieve the fringe-stopping condition. Thereafter the signal goes into a correlator or pulsar backend and is finally saved in a data storage system. Signals from all antennas are managed in a similar manner.

| BOX 6.1 | A Pulsar Dispersion and Distance |
|---------|----------------------------------|

Pulsar signals experience dispersion traveling through the ISM, where lower frequency pulse components experience larger time delays. Here we calculate the time delay between pulsar signals measured at two different frequencies, $v_2$ and $v_1$. The group velocity as a function of frequency $v$ is

$$v_{gr} = c\sqrt{1 - \frac{v_p^2}{v^2}},$$  (6.68)

where the plasma frequency $v_p$ in MKS units is

$$v_p = \frac{1}{2\pi}\sqrt{\frac{n_e e^2}{\varepsilon_0 m_e}},$$  (6.69)

and where $n_e$ is the density of electrons. The time delay $t$ as a function of frequency is

$$t = \int_0^D \frac{ds}{v_{gr}},$$  (6.70)

where $D$ is the distance to the pulsar. Typical ISM values of $v_p$ are in the kHz range, compared to the observing frequencies $v \sim 10$ MHz. We can therefore make use of the binomial theorem for $v \gg v_p$ to evaluate Eq. (6.70).

$$\frac{1}{v_{gr}} = \frac{1}{c}\left(1 - \frac{v_p^2}{v^2}\right)^{-1/2} \approx \frac{1}{c}\left(1 + \frac{v_p^2}{2v^2}\right).$$  (6.71)

The time delay is thus

$$t = \frac{1}{c}\int_0^D \left(1 + \frac{v_p^2}{2v^2}\right) ds.$$  (6.72)

Substituting the plasma frequency, we have

$$t = \frac{1}{c}\int_0^D \left(1 + \frac{1}{v^2}\frac{e^2 n_e(s)}{2\pi m_e}\right) ds = \frac{D}{c} + \frac{e^2}{2\pi c m_e}\frac{1}{v^2}\int_0^D n_e(s)\, ds. \qquad (6.73)$$

The time delay $\Delta t$ between two measurements at frequencies $v_2$ and $v_1$, with $v_1 > v_2$ is

$$\Delta t = t_2 - t_1 = \frac{e^2}{2\pi c m_e}\left(\frac{1}{v_2^2} - \frac{1}{v_1^2}\right)\int_0^D n_e(s)\, ds, \qquad (6.74)$$

which can be expressed in terms of the dispersion measure DM,

$$\Delta t = t_2 - t_1 = \frac{e^2}{2\pi c m_e}\left(\frac{1}{v_2^2} - \frac{1}{v_1^2}\right) \times DM. \qquad (6.75)$$

This is of the form of Eq. (6.50) where numerically

$$\Delta t = 4.15 \times 10^6 \left(\frac{1}{v_2^2} - \frac{1}{v_1^2}\right) \times DM \qquad (6.76)$$

with $\Delta t$ in ms, DM in $cm^{-3}$ pc. The distance to the pulsar may be estimated if we approximate $DM = n_e D$, where $n_e$ is the number of electrons per cubic centimeter and $D$ is the pulsar distance in parsecs.

## REVIEW QUESTIONS

1. Astronomical radio signals may be affected by noise due to atmospheric absorption. Would you prefer to observe a source at 1.4 GHz at the time it rises or when it is at the zenith? Justify your answer.   [*Hint:* See Fig. 6.1]

2. Using a diagram, describe the functional blocks of a radiometer.
   [*Hint:* See Fig. 6.2]

3. Explain the following terms as applicable to a radiometer receiver: (*i*) center frequency, (*ii*) gain, (*iii*) bandwidth, (*iv*) integration time, and (*v*) receiver temperature.   [*Hint:* See Section 6.2]

4. For a radiometer, explain the terms: (*i*) sensitivity, and (*ii*) stability.
   [*Hint:* See Section 6.3]

5. A receiver consists of an LNA followed by an RF amplifier. The LNA gain is 20 dB and its noise figure as 1.1 dB. The RF amplifier has a gain of

60 dB and a noise figure of 3.5 dB. If the efficiency of the transmission line is unity, find the receiver noise temperature given its physical temperature is 300 K.

[**Hint:** Convert the noise figures into noise factors, and use Eqs. (6.3) and (6.4).]

6. With the help of a diagram, briefly explain the total power receiver. Why is it called a direct radiometer? [**Hint:** See Fig. 6.3]

7. Using a diagram, briefly explain the Dicke receiver. [**Hint:** See Fig. 6.5]

8. With the help of a diagram, briefly explain the gain-modulated Dicke receiver. [**Hint:** See Fig. 6.6]

9. Draw a block diagram of a null-balancing Dicke receiver and explain its functionality. [**Hint:** See Fig. 6.7]

10. With the help of a diagram, briefly explain Graham's receiver.

[**Hint:** See Fig. 6.8]

11. Draw a block diagram of a correlation receiver and explain its functionality.

[**Hint:** See Fig. 6.9]

12. Describe the functionality of an additive interferometer receiver using a block diagram. [**Hint:** See Fig. 6.10]

13. Draw the block diagram of a multiplicative interferometer receiver, and explain it in detail. [See Fig. 6.11]

14. Explain the functional blocks of a phase-switched receiver using a diagram. [**Hint:** See Fig. 6.12]

15. An LNA is kept at room temperature (300 K) with its input connected to a matched load. The output power of the LNA is measured using a power meter. When the load is at room temperature, the power meter reading is 1 unit. When it is dipped in liquid nitrogen (77 K) the power meter reads 0.40 unit. Find the receiver temperature. Also, calculate the noise figure of the LNA. [**Hint:** Use Eqs. (6.44) and (6.3)]

16. A noise generator is used in receiver calibration with a directional coupler and an attenuator (see Fig. 6.15a). It has a fired temperature of 9900 K. If the directional coupler has 20 dB coupling, find the required attenuation to obtain a calibration signal temperature of $1 \pm 0.05$ K.

[**Hint:** Use Eq. (6.46)]

17. A point source produces a flux density of 100 Jansky near the aperture of a single-polarized radio telescope antenna. If the aperture area is $3m^2$, find the incremental temperature. [*Hint:* Use Eq. (6.47)]

18. What is a frequency-switched radiometer, and where it is used? Explain briefly. [*Hint:* See Section 6.6.1 and Fig. 6.16]

19. Briefly describe the operation of a multichannel hydrogen-line radiometer using a block diagram. [*Hint:* See Fig. 6.18]

20. Using a block diagram, explain the operation of a basic pulsar receiver. [*Hint:* See Fig. 6.19]

21. What is an analog bank spectrometer? Explain with a block diagram.

    [*Hint:* See Fig. 6.20]

22. Explain the functionality of a correlation spectrometer using a diagram. [*Hint:* See Fig. 6.21]

23. What is the difference between incoherent and coherent dedispersion in pulsar receivers? [*Hint:* See Sections 6.7.2.1 and 6.7.2.2]

24. Draw a generalized block layout of the complete receiving and control chain from the observatory building to the antennas of a radio array.

    [*Hint:* See Fig. 6.23]

# *INTERFEROMETER APERTURE SYNTHESIS AND RADIO MAPPING*

## 7.1 INTRODUCTION

In earlier chapters, we have seen some techniques for data receiving and data processing. Which of these techniques we wish to use depends on our final goal. For example, spectral-line observations are made to find elements, velocity of the object, etc. Pulsar data, on the other hand, may be used for analyzing such things as the timing, magnetic field, and intensity of the pulsar. Interferometer continuum data can be used for creating an image map of the source; this map is a two-dimensional (right ascension and declination) image of the source intensity distribution. Here we introduce some basic techniques for gathering interferometer array data in order to produce radio maps.

An optical paraboloid mirror produces an image from the parallel rays arriving at its focus. Theoretically, this is true at all wavelengths. However, this process of making an image is practically impossible in the radio part of the spectrum due to limitations on the size of the dish. Even if we do so, the resolution of the image will be poor (see the Rayleigh criterion in Chapter 4). To construct a radio image at a wavelength $\lambda = 50$ cm having a resolution visible to an unaided human eye, the required diameter of the dish would be about 1 km. Other conditions to be satisfied are (*i*) the surface accuracy of the dish must be within $\lambda/20$, and (*ii*) the SNR should be greater than unity. All these point towards increasing the aperture size.

In the earlier days of radio astronomy, a single antenna was multiplexed in time and space to synthesize a large aperture. The process of synthesizing a large antenna aperture using small antennas is known as *aperture synthesis*. In later stages, the natural rotation of the Earth was used to move the antennas in time and space with respect to the observed radio source. Interferometer arrays improved the resolution and reduced observing time. In 1955, the first synthesis array designed to use the Earth's rotation was commissioned. In 1946, Ryle and Vonberg published the first interferometric astronomical measurements at radio wavelengths. However, it is also said that Joseph Pawsey was the first to make interferometric measurements.

## 7.2 SYNTHESIZING A LARGE ANTENNA APERTURE WITH SMALL ANTENNAS

Fig. 7.1a shows a large aperture of size $L \times L$ to be synthesized by using multiple small apertures of size $l \times l$. The main lobes of the beam patterns are shown perpendicular to the aperture planes. The large aperture has been broken into nine equal parts of size $l \times l$, as shown by the dotted lines. Note that the main beam of the smaller aperture is broader and has reduced gain. Instead of nine, it is possible to use only two fractional apertures $X$ and $Y$ for the synthesis. This is done by keeping $X$ at a fixed location, and multiplexing $Y$ in time and space. For a distant radio source, eight paired data sets are obtained. Finally, the data is properly arranged for the nine fractions and then added. This gives equivalent data to that of the synthesized aperture.

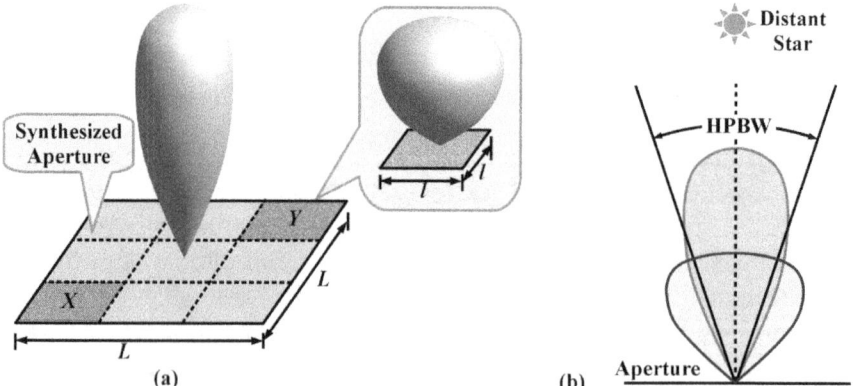

**FIGURE 7.1:** Synthesis of a large aperture using small apertures. (a) A large aperture of size $L \times L$ is synthesized using many small apertures of size $l \times l$. (b) The observed radio source must be within the HBBW (half-power beam width) of the large aperture for a successful synthesis.

Instead of nine, let the synthesized aperture be divided into $N$ equal parts. Let $I_n e^{j\phi_n}$ be the current delivered by the $n^{\text{th}}$ element, where $I_n$ and $\phi_n$ represent the magnitude and phase, respectively. Because the current magnitudes of all the elements are identical, the resulting current $I_{\text{total}}$ from the contribution of all elements is expressed in Eq. (7.1), where $I_n = I_0$ is the magnitude of the current of an element. The power $P$ delivered to the receiver is given by Eq. (7.2), where $k_1$ is a constant.

$$I_{\text{total}} = \sum_{n=1}^{N} I_n e^{j\phi_n} = I_0 \sum_{n=1}^{N} e^{j\phi_n} \qquad (7.1)$$

$$P = k_1 I_0^2 \sum_{n=1}^{N} e^{j2\phi_n} \qquad (7.2)$$

Before data addition, phase corrections are made using the data from the fixed aperture, which is the reference aperture. This is necessary because the wave front varies with time. It is also necessary for the radio source under observation to lie within the HPBW of the synthesized aperture, as shown in Fig. 7.1b. For nontracking antennas, the observation time is very much limited.

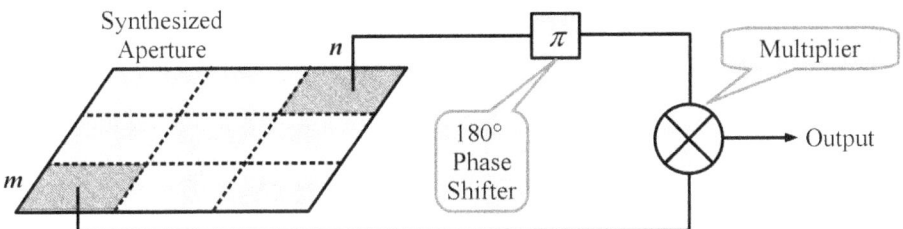

**FIGURE 7.2:** Phase correction before combining the data from all the elements. The output of element $m$ is multiplied with the output of another element $n$, which has been phase-shifted by 180°.

The arrangement shown in Fig. 7.2 can be used to find the phases of other elements with reference to a fixed element. Here, the output of the fixed element $m$ is multiplied with the 180°-phase-shifted output of any other element $n$. The path lengths from any element to the multiplier are considered identical. Let $Ve^{j\phi_m}$ and $Ve^{j\phi_n}$ represent the voltages of the $m$ and $n$ elements, respectively. The multiplier output $F$ is given by Eq. (7.3), which is a measure of phase difference between the two.

$$F = V^2 e^{j(\phi_m - \phi_n)} \qquad (7.3)$$

This function $F$ can be suitably combined with data obtained from each location $n$ to correct the respective phases before addition. In this way, the entire aperture is synthesized.

These methods were used in the early days of radio astronomy. Angular resolutions of 45' at 7.9 m and about 25' at 1.7 m wavelengths were achieved, and nearly 5,000 radio sources were resolved. The movable antenna was set on railway tracks.

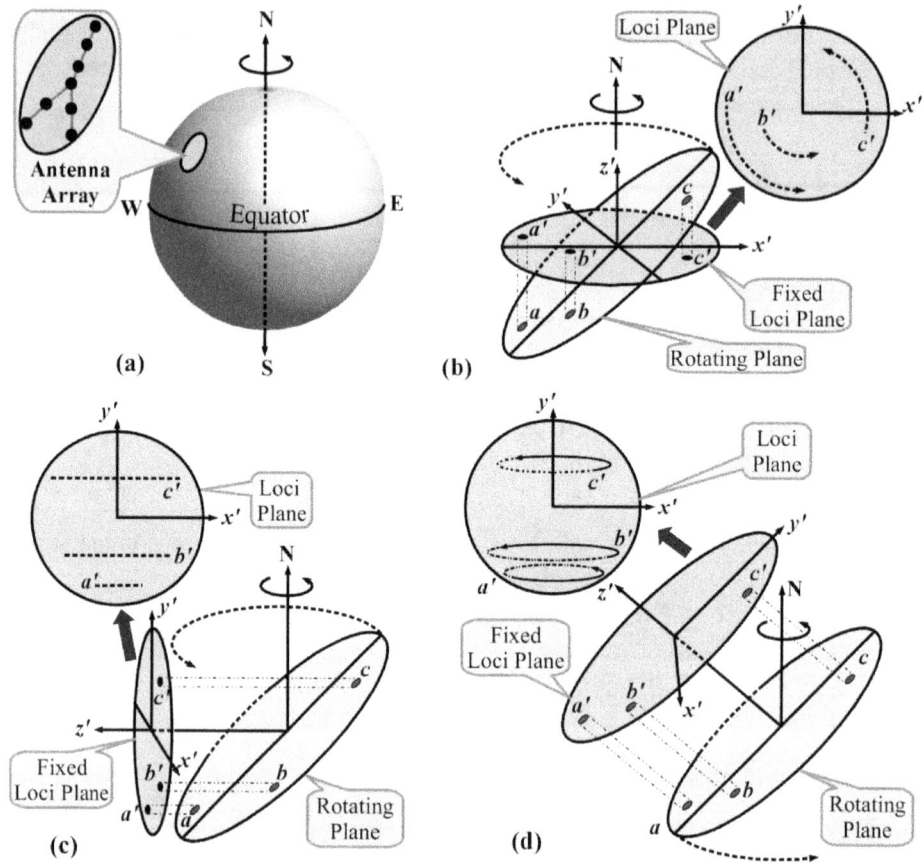

**FIGURE 7.3:** Principles of super-synthesis. (a) An array of antennas rotates with the Earth. (b) Observing a radio source towards the CNP. (c) Observing a source on the celestial equator. (d) Observing a source at a celestial latitude above the equator but below the celestial pole.

## 7.3   BASIS OF SUPER-SYNTHESIS USING EARTH'S ROTATION

We already have a taste of the super-synthesis technique from Chapter 1 (Section 5.3). We now develop this idea a bit more. Consider an antenna array spread over a plane area at some latitude between 20° and 70° on the Earth as shown in Fig. 7.3a. These antennas are tracking a distant radio source on the celestial sphere. As seen from the radio source, the entire array appears to rotate with the Earth. Consider a rectangular coordinate system $(x', y', z')$ with its $z'$-axis pointed towards the source, as shown in Fig. 7.3b. Let the center of the antenna plane be located at its origin. Let the $x$-' and $y'$-coordinates be stationary in the source frame. As seen from the source, the position of the antennas move over the $x'$-$y'$ plane due to the Earth's rotation. Consider three antennas $a$, $b$, and $c$, whose projections on the $x'$-$y'$ plane will be investigated for the following three cases:

**Case (*i*):**   Let the radio source be located at the CNP (celestial north pole). As shown in Fig. 7.3b, the loci $a'$, $b'$, and $c'$, respectively, for the antennas $a$, $b$, and $c$, will trace circles on the $x'$-$y'$ plane over a period of twenty-four hours.

**Case (*ii*):**   Let the radio source be directed towards a point on the celestial equator. As shown in Fig. 7.3c, the loci $a'$, $b'$, and $c'$ trace straight lines on the $x'$-$y'$ plane. As a special case, if all the antennas are positioned on a single east-west line, all the loci will trace a single straight line and the population on the $x'$-$y'$ plane will be one dimensional. This is not suitable for making a two-dimensional map and must be avoided. Hence, for observing on the celestial equator, some antennas must have separations along the north-south axis.

**Case (*iii*):**   As shown in Fig. 7.3d, if the radio source is directed at a celestial latitude less than or greater than 0°, the loci will trace ellipses.

### 7.3.1 Synthesizing the Aperture

The voltage outputs from individual antennas can be recorded at short intervals of time. These values can be placed on the $x'$-$y'$ plane on the loci of the antenna. At very short time intervals, we may record the voltage outputs of individual antennas and place these values on the $x'$-$y'$ plane. The greater the observation time and the number of antennas in the array, the more the $x'$-$y'$ plane is populated. After 24h the $x'$-$y'$ plane will be highly populated with the voltage values resulting from the radio source. This populated area on the $x'$-$y'$ plane represents the aperture of a large synthesized antenna.

## 7.4   INTERFEROMETER APERTURE SYNTHESIS

A single interferometer with adjustable baselines can be used for aperture synthesis, aided by the Earth's rotation. One of the antenna elements is used for phase reference. A clear understanding of equatorial systems (see Appendix A.3 and Appendix A.4), the $u$, $v$, $w$ coordinate system (see Appendix A.9), and the $X$, $Y$, $Z$ coordinate system (see Appendix A.8) are required. We begin by visualizing these coordinates together in relation to a radio source and their interrelationships.

### 7.4.1 Various Coordinate Systems in Relation to a Radio Source

Astronomical coordinates measure only two angles, such as $\alpha$, $\delta$. Hence, distance information is not directly available. However, in interferometry, a three-dimensional coordinate system like $X$, $Y$, $Z$ is necessary to measure the baseline distance and relate it to the astronomical coordinates.

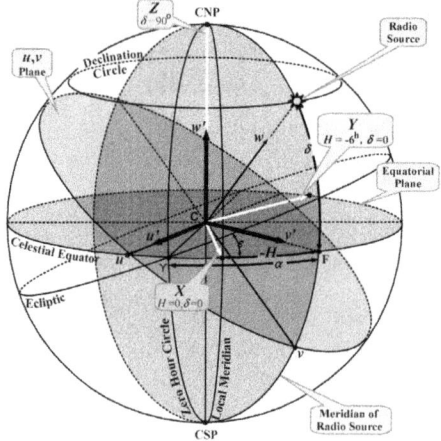

**FIGURE 7.4:** Different coordinate systems in relation to a radio source under observation.

Consider a radio source under observation and the right-handed $u$, $v$, $w$ coordinate system as shown in Fig. 7.4. The CNP and CSP represent the celestial north and south poles, respectively, and $\Upsilon$ represents the vernal equinox. The coordinates $u$, $v$, and $w$ are measured by the number of wavelengths (dimensionless). The $w$-axis is always kept directed towards the source. Hence, the $u$-$v$ plane always remains perpendicular to the source direction. The $v$-axis is kept in a longitudinal plane touching the three points: (*i*) the two celestial poles, and (*ii*) the direction of observation. Using the universal equatorial system, on the figure are marked the right ascension $\alpha$ and the declination $\delta$ with reference to the celestial equator. Using the local equatorial system, the hour angle $H$ with respect to the local meridian is shown. The angle $\varepsilon$ between the ecliptic and the equatorial planes is the obliquity. Note that as $\alpha$ and $\delta$ of the source change with time, the $u$ and $v$ also change their orientations accordingly. The $Z$-axis of the right-handed $X$, $Y$, $Z$ coordinate system points to the CNP. The $Y$-xis faces east and the $X$-axis lies in the local meridian plane. The $u'$, $v'$, $w'$ coordinate system is a special case of the $u$, $v$, $w$ coordinates when the source is at the CNP. It is obtained when the $u$-$v$ plane aligns with the equatorial plane by rotating the former around the $u$-axis.

### 7.4.2 Single Interferometer with Source at a Pole

Consider a single interferometer with two antennas of equal diameter $D$. These are positioned on an east-west line having the same latitude and separated by a distance $d$, as shown in Fig. 7.5a. Let the source be towards the CNP. The projection $d\cos\theta$ of the baseline $d$ on the $u$-$v$ plane rotates with the Earth. The antenna positions for three time instants ($t_1$, $t_2$, and $t_3$) are shown in Fig. 7.5a. The $u$-$v$ plane is parallel to the equatorial plane because the former is always perpendicular to the source direction. If we center the $u$-$v$ plane on antenna A, the position of antenna B for $t_1$, $t_2$, and $t_3$ can be seen as illustrated in Fig. 7.5b. For a twelve-hour observation, the trace of antenna B produces a half-circular ring of thickness $D/\lambda$ on the $u$-$v$ plane. If we now center the $u$-$v$ plane on antenna B, the trace of antenna A produces another half-circular ring of thickness $D/\lambda$ on the other side of the $u$-$v$ plane. We can join the two halves to form a complete ring of thickness $D/\lambda$ in the $u$-$v$ plane. This is shown in Fig. 7.5c. For this particular case, the projected baseline is identical to the actual baseline $d$ because the source is towards the CNP. On the $u$-$v$ plane, the central radius of the ring is projected as $u = d/\lambda$ and $v = d/\lambda$ , where $\lambda$ is the wavelength of observation.

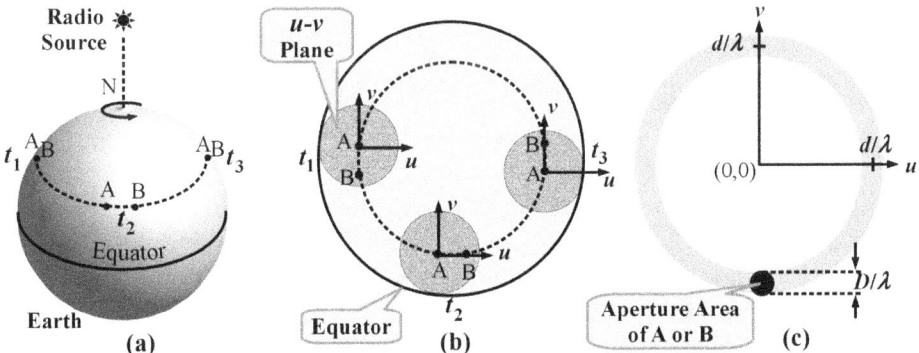

**FIGURE 7.5:** An interferometer with an east-west baseline is observing a source at the CNP. (a) Antenna positions change with time $t_1$, $t_2$, and $t_3$. (b) The $u$-$v$ plane is centered on one antenna. The position of the other antenna changes on the $u$-$v$ plane. (c) After twelve hours, the traces of A and B produce a ring on the $u$-$v$ plane.

### 7.4.3 Populating the *u-v* Plane by Repositioning One Antenna

The synthesized aperture should lie on the *u-v* plane for making a radio map of the source. Reducing the distance between the two antennas reduces the baseline, resulting in smaller circles on the *u-v* plane. Hence, after several twelve-hour observations with different baselines $d_{\lambda n} = d_n/\lambda$, the *u-v* plane can be well populated. Here, $d_n$ represents the distance between the two antennas, $n = 1, 2, 3,...$ represents the baselines, and $\lambda$ represents the wavelength of observation.

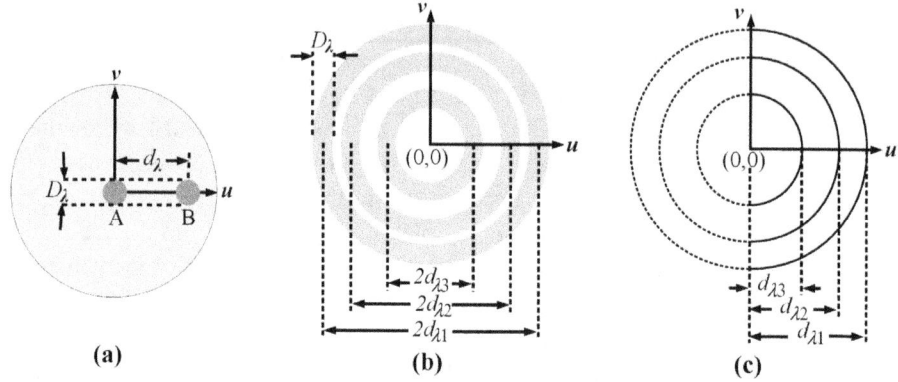

**(a)**  **(b)**  **(c)**

**FIGURE 7.6:** (a) Instantaneous *u-v* plane coverage using an interferometer formed from two small equal-aperture antennas A and B. (b) Theoretical *u-v* plane coverage in twelve hours using three different baselines. (c) Actual *u-v* coverage in twelve hours using three different baselines.

Consider an interferometer formed from two equal-diameter ($D$) antennas A and B. As shown in Fig. 7.6a, let the *u, v* coordinates be centered on A. Hence, antenna A is called the *reference* antenna. With Earth's rotation, one half circle is traced on the *u-v* plane every twelve hours. The semicircular ring has a radius equal to the baseline $d_\lambda$ and thickness $D_\lambda = D/\lambda$. The other half of the ring is constructed using the relation $\mathcal{V}(-u, -v) = \mathcal{V}^*(u, v)$, where $\mathcal{V}^*(u, v)$ is the complex conjugate of the visibility $\mathcal{V}(u, v)$. A complete circular ring is obtained by joining the two halves. The *u-v* plane coverage for three different baselines $d_{\lambda 1}$, $d_{\lambda 2}$, and $d_{\lambda 3}$ is shown in Fig. 7.6b. Thus by changing baselines, it is theoretically possible to populate the *u, v* plane. Practically, the thickness of the rings are zero, as shown in Fig. 7.6c. The fault lies with the antenna output that converts the entire dish response to a single output. The visibility data contains only resultant information produced from (*i*) a band of spatial frequencies determined by the aperture diameter of individual antennas, and (*ii*) its distance from the center of the *u-v* plane (projection of the baseline). Hence the synthesized aperture in the *u-v* plane is only an approximation. Furthermore, the maximum diameter of the rings is restricted by the maximum possible physical distance between the two antennas. Hence, above a certain diameter in the *u-v* plane, there is no data. Similarly, the central portion of the *u-v* plane also remains empty, which is determined from the shortest baseline. The shortest baseline possible is *D* because two antennas cannot physically accommodate themselves to less than this dimension.

### 7.4.4 Interferometer With Source Away From Poles

We now consider an interferometer observing a source away from the celestial poles. The antenna alignment can be generalized under two groups: (*i*) the baseline parallel to the celestial equator (lying east-west), and (*ii*) the baseline with a component perpendicular to the celestial equator (lying north-south).

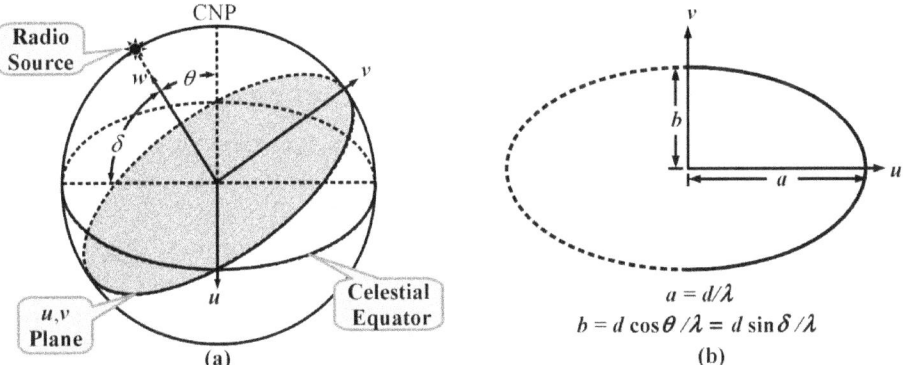

**FIGURE 7.7:** (a) A radio source making an angle $\theta$ with the CNP. The geometry of the observation, declination angle $\Delta$, and the *u-v* plane, are shown. (b) Due to the inclination (source) with the CNP, the locus of antenna B traces an ellipse in the *u-v* plane.

#### 7.4.4.1 A Baseline Parallel to Celestial Equator

Consider an interferometer having a east-west baseline formed by antennas A and B, with A as the reference antenna. The baseline will be parallel to the celestial equator. Let the observed source be inclined at an angle $\theta$ from the CNP, as shown in Fig. 7.7a. As shown in Fig. 7.7b, antenna B traces an ellipse on the *u-v* plane with Earth's rotation. In twelve hours, half of the ellipse (shown by a thick line) is formed. Using the visibility relation $\mathcal{V}(-u, -v) = \mathcal{V}^*(u, v)$, the other half (shown by dotted lines) is generated from the data. The major and minor axes of the ellipse represented by *a* and *b*, respectively, are given by Eqs. (7.4) and (7.5), where *d* is the baseline (distance between the antennas), $\theta$ is the inclination of the source from CNP, $\delta$ is the declination of the radio source, and $\lambda$ is the wavelength.

$$a = \frac{d}{\lambda} \tag{7.4}$$

$$b = \frac{d \cos \theta}{\lambda} = \frac{d \sin \delta}{\lambda} \qquad (7.5)$$

As a special case, when the source is on the celestial equator ($\theta = 0$), Eq. (7.5) gives $b = 0$. Hence, antenna B traces a straight line on the $u$-$v$ plane. For any source below and above the celestial equator, antenna B traces an ellipse on the $u$-$v$ plane. Because the baseline used here is parallel to the celestial equator, the center of the ellipse coincides with that of the $u$-$v$ plane. This is true for any source declination.

### 7.4.4.2 A Baseline Component Orthogonal to The Celestial Equator

If the interferometer antennas are not exactly on the east-west line, then the baseline has a component along the north-south line. The ellipse splits, and their centers keep changing with the declination angle. To understand this, we go back to the $X$, $Y$, $Z$ coordinate system, as shown in Fig. 7.4. The local hour angle $H$ and the declination $\delta$ are marked. They are used for finding all the baselines when the number of antennas is greater than two. Later, these are transformed into $u$, $v$, $w$ coordinates.

Let the components of a baseline $d_\lambda$, when expressed in the $X$, $Y$, $Z$ coordinate system, be $X_\lambda = X/\lambda$, $Y_\lambda = Y/\lambda$, and $Z_\lambda = Z/\lambda$, where $\lambda$ is the observing wavelength. These are the coordinate differences for any two antennas in the antenna array. The conversion of $X$, $Y$, $Z$ to $u$, $v$, $w$ coordinates is given by Eq. (7.14) of Appendix A, which is reproduced as Eq. (7.6).

$$\begin{bmatrix} u \\ v \\ w \end{bmatrix} = \begin{bmatrix} \sin H & \cos H & 0 \\ -\sin\delta\cos H & \sin\delta\sin H & \cos\delta \\ \cos\delta\cos H & -\cos\delta\sin H & \sin\delta \end{bmatrix} \begin{bmatrix} X_\lambda \\ Y_\lambda \\ Z_\lambda \end{bmatrix} \qquad (7.6)$$

The quantities $H$ and $\delta$ are fixed on the phase reference position[1]. The phase reference position is a point on the source, usually at its center, such that the visibility has maximum amplitude (the fringe-stopping condition). The $w$-axis is directed at this point, and all measurements are done with this as the reference. Let us consider one antenna as a reference. With respect to this antenna, let $h$ and $d$, respectively, represent the hour angle and the declination of the other antenna forming the baseline (see Fig. A.4a of Appendix A). This is like positioning the reference antenna at the origin of an $X$, $Y$, $Z$ coordinate system and measuring $h$ and $d$ (declination) of the other antenna. From the reference antenna, the $X$, $Y$, $Z$

---

[1]In VLBI observations, the $X$-axis is set on the Greenwich Meridian, and $H$ is measured from there instead of locallly.

coordinates of the other antenna are obtained by Eq. (7.7), where $d_\lambda$ is the baseline expressed in number of wavelengths. By substituting Eq. (7.7) into Eq. (7.6) and simplifying, we obtain Eq. (7.8), which expresses the baseline $d_\lambda$ in terms of $u$, $v$, and $w$, whose dimensions are in spatial frequency (number of wavelengths).

$$\begin{bmatrix} X_\lambda \\ Y_\lambda \\ Z_\lambda \end{bmatrix} = d_\lambda \begin{bmatrix} \cos d \cos h \\ -\cos d \sin h \\ \sin d \end{bmatrix} \tag{7.7}$$

$$\begin{bmatrix} u \\ v \\ w \end{bmatrix} = d_\lambda \begin{bmatrix} \cos d \sin(H - h) \\ \sin d \cos \delta - \cos d \sin \delta \cos(H - h) \\ \sin d \sin \delta + \cos d \cos \delta \cos(H - h) \end{bmatrix} \tag{7.8}$$

All the baselines of an interferometer array are brought into the $u$, $v$, $w$ domain with the help of Eq. (7.8). If the telescope antennas use an altazimuth system, it would be preferable to use Eq. (7.9) for determining the components $X_\lambda$, $Y_\lambda$, and $Z_\lambda$, where $\mathcal{A}$ and $\mathcal{E}$ represent, respectively, the altitude and elevation differences between any two antennas, and $\mathcal{L}$ is the latitude of the reference point of the coordinate system (approximately the center of the baseline).

$$\begin{bmatrix} X_\lambda \\ Y_\lambda \\ Z_\lambda \end{bmatrix} = d_\lambda \begin{bmatrix} \cos \mathcal{L} \sin \mathcal{E} - \sin \mathcal{L} \cos \mathcal{E} \cos \mathcal{A} \\ \cos \mathcal{E} \sin \mathcal{A} \\ \sin \mathcal{L} \sin \mathcal{E} + \cos \mathcal{L} \cos \mathcal{E} \cos \mathcal{A} \end{bmatrix} \tag{7.9}$$

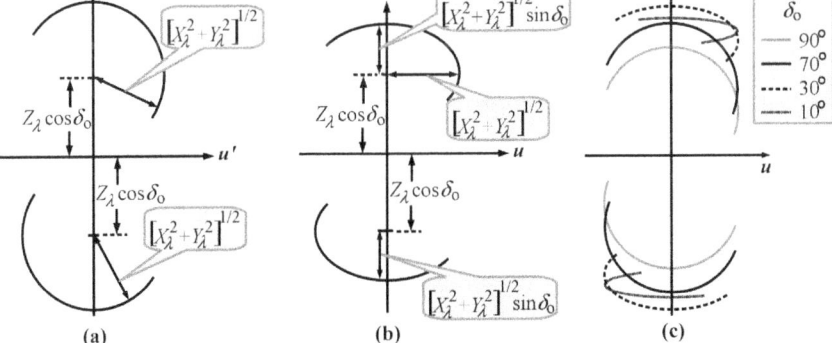

**FIGURE 7.8:** (a) Locus of the ellipse on the $u'$-$v'$ plane for a baseline having a nonzero $Z_\lambda$ observing a radio source at the CNP. (b) Locus of the ellipse for a baseline having a nonzero $Z_\lambda$ observing a radio source at declination $\delta_0$. (c) Tentative plots for different declinations for a baseline having a nonzero $Z_\lambda$.

If the phase reference position of the radio source is $(H_0, \delta_0)$, then using the first and second rows of the matrix shown in Eq. (7.6), one can obtain Eq. (7.10), which represents an ellipse. Note that if the component $Z_\lambda$ is not zero (a baseline or its component exists along the north-south direction), the ellipse splits into two halves in the $u$-$v$ plane.

$$u^2 + \left[ \frac{v - Z_\lambda \cos \delta_0}{\sin \delta_0} \right]^2 = X_\lambda^2 + Y_\lambda^2 \qquad (7.10)$$

The traces while observing a radio source at the CNP are shown in Fig. 7.8a. The coordinates used here are $u'$,$v'$, which are a special case of the $u$, $v$ coordinates. The radius of the split ellipse is a constant that is a split circle. When the source is away from the CNP, we get split ellipses as shown in Fig. 7.8b. The major axis of the ellipse stays along $u$, and its length is same as the radius of the circle of the previous case. The length of the minor axis along $v$ varies with the declination $\delta_0$ of the phase reference point. However, the distance between the major axes of the partial ellipses remain unchanged. Tentative plots for different values of $\delta_0$ are shown in Fig. 7.8c. In any case, if the baseline component $Z_\lambda$ becomes zero, the split ellipses join together.

### 7.4.5 Populating the *u-v* Plane Using Interferometer Arrays

We have seen how an interferometer fills the $u$-$v$ plane with visibility $\mathcal{V}(u, v)$. We also know that by employing more antennas, the number of baselines grows to $N(N-1)/2$. A small increase in $N$ results in a large increase of baselines, thereby filling the $u$-$v$ faster, provided each baseline size is unique. A redundancy in baseline size produces duplicate data sets and so is of no use. However, baseline redundancy is sometimes useful for phase corrections at the cost of less sensitivity. Some antennas must be aligned along the north-south direction for observing radio sources close to the celestial equator, with the result that the $u$-$v$ plane is filled in two dimensions.

The VLA (very large array, New Mexico) observatory consists of twenty-seven antennas in a Y-configuration, as shown in Fig. 7.9a. We have already seen a star-like data set on the $u$-$v$ plane which is obtained at any instant of time when the VLA antennas observe a source at the zenith (see Fig. 7.21 of Chapter 1). With the rotation of the Earth, the star-like structure also changes its position and shape across the $u$-$v$ plane, depending on the declination of the source. Fig. 7.9b illustrates a typical $u$-$v$ plane coverage obtained in a few hours by tracking a source with changing declination. The VLA antennas are set on rail tracks. Hence after a twelve-hour period, the antennas can be repositioned for new unique baselines and

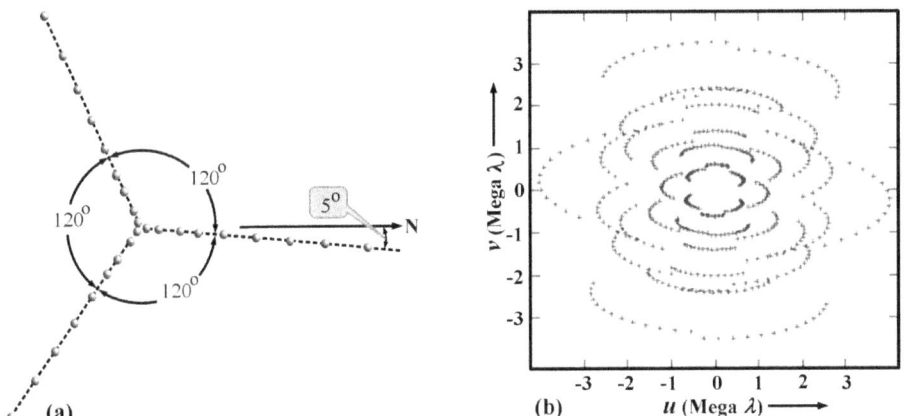

**FIGURE 7.9:** (a) The VLA consists of twenty-seven antennas in a Y-configuration. (b) A typical u-v plane coverage obtained from tracking a source with changing declination.

the observation can be repeated. In this way the u-v plane may be more and more densely populated.

## 7.5 INTERFEROMETRY RADIO MAPPING

The next step after filling the synthesized aperture with visibility data is to generate an image of the radio source. The procedure involves a mapping of the visibilities in the u-v plane with the source intensity distribution $I(l, m)$ on the sky (l-m plane). The $(l, m)$ coordinates (see Appendix A.10) are in the spatial domain, whereas the $u, v$ coordinates are in the spatial frequency domain. Intuitively we find a Fourier-transform-like relationship between the two. We have already talked about the basis of radio mapping between the source intensity and the aperture electric field of a single-dish radio telescope in Chapter 1 (Section 1.5.2) and in Chapter 4 (Section 4.6). The mapping between the source intensity distribution $I(l, m)$ and the visibilities $\mathcal{V}(u, v, w)$ occurs through the van Cittert-Zernike equation, which we have seen in Section 1.5.4.2. We now derive this equation to obtain a detailed understanding of the mapping procedure. For this, we first establish a relationship between the visibility and the correlation, which was mentioned in Section 1.5.4.1.

### 7.5.1 Relationship Between Visibility and Correlation

The visibility $\mathcal{V}(u, v, w)$ or the spatial coherence function is computed from the values of cross-correlations $r(t_g)$ obtained from the correlator. Visibilities are actually scaled values of the cross-correlations that are functions of the $u, v, w$ coordinates.

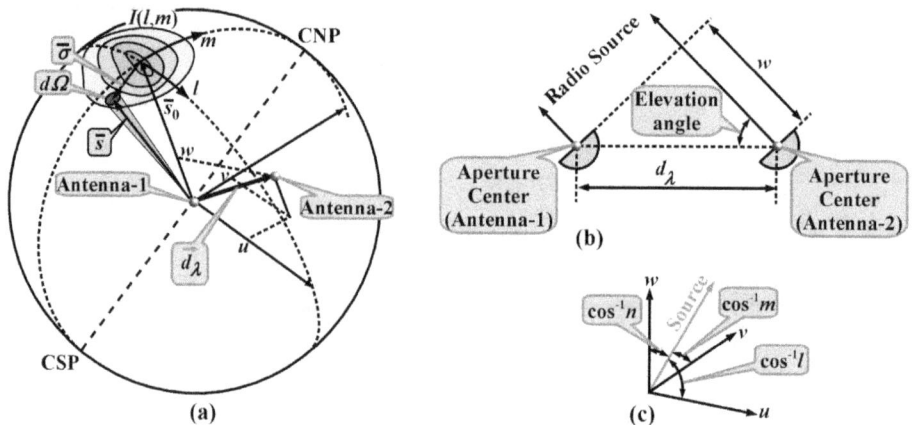

**FIGURE 7.10:** (a) A radio source with an intensity distribution $I(l, m)$ is observed using an interferometer. (b) A lower elevation angle increases $w$. (c) The relationship between the $u$, $v$, $w$ coordinates and the $l$, $m$, $n$ coordinates (see Appendix A.10).

Consider an interferometer having two antennas that is observing a spatially incoherent radio source on the celestial sphere, as shown in Fig. 7.10a. Antenna 1 is positioned at the origin of the $u$, $v$, $w$ coordinate system. The intensity distribution of the source is represented as $I(l, m)$. The origin of the $l$, $m$ coordinate system is located on the phase reference position, usually at the center of the source, which is also known as the *phase tracking center*. Let $\bar{s}_0$ be a space vector pointed to the phase tracking center from antenna 1. With reference to the phase tracking center, any nearby position in the sky is represented by another space vector $\bar{\sigma}$. This position in the sky can also be assessed from antenna 1 with a space vector $\bar{s}$ pointed to that position in the sky, such that $\bar{s} = \bar{s}_0 + \bar{\sigma}$. For an observing frequency $v$, let $I(\bar{s})$ be the intensity (brightness) of the sky along the direction $\bar{s}$, and let $A_e(\bar{s})$ be the effective aperture area of an antenna along that direction. Over a solid angular element $d\Omega$ along $\bar{s}$, the signal power received over a bandwidth $\Delta v$ is given by Eq. (7.11). Hence the correlated signal power $dr$ per solid angle $d\Omega$ is given by Eq. (7.12).

$$A_e(\bar{s})I(\bar{s})\Delta v d\Omega \tag{7.11}$$

$$dr = A_e(\bar{s})I(\bar{s})\Delta v d\Omega \cos(2\pi v \tau_g) \tag{7.12}$$

Because the antennas have three-dimensional patterns, the correlator power $r$ can be obtained by integrating Eq. (7.12) over the celestial sphere. This is shown in Eq. (7.13), where $\vec{d}_\lambda$ represents the baseline space vector (in number of wavelengths) pointing from antenna 1 to antenna 2 (see Fig. 7.10a).

$$r(\vec{d}_\lambda, \overline{s}) = \Delta v \int_{-\infty}^{\infty} A_e(\overline{s}) I(\overline{s}) \cos\left[2\pi(\vec{d}_\lambda \cdot \overline{s})\right] d\Omega \tag{7.13}$$

Because the antenna beam widths are extremely small, only regions very close to the phase reference position $\overline{s}_0$ are relevant. Hence we make measurements with respect to $\overline{s}_0$. Thus we express $r$ in terms of $\overline{\sigma}$ and $\overline{s}_0$, as shown in Eq. (7.14). Because the source is spatially incoherent, which is applicable to all cosmic radio sources, the radiated waveforms from different elements of the source within $d\Omega$ are uncorrelated.

$$
\left.
\begin{aligned}
r(\vec{d}_\lambda, \overline{s}_0) = &\Delta v \cos\left[2\pi(\vec{d}_\lambda \cdot \overline{s}_0)\right] \\
&\times \int_{-\infty}^{\infty} A_e(\overline{\sigma}) I(\overline{\sigma}) \cos\left[2\pi(\vec{d}_\lambda \cdot \overline{\sigma})\right] d\Omega \\
&- \Delta v \sin\left[2\pi(\vec{d}_\lambda \cdot \overline{s}_0)\right] \\
&\times \int_{-\infty}^{\infty} A_e(\overline{\sigma}) I(\overline{\sigma}) \sin\left[2\pi(\vec{d}_\lambda \cdot \overline{\sigma})\right] d\Omega
\end{aligned}
\right\} \tag{7.14}
$$

A normalized antenna beam pattern $P_n(\overline{\sigma})$ is obtained as $P_n(\overline{\sigma}) = A_e(\overline{\sigma})/A_{e0}$, where $A_{e0}$ is the maximum effective aperture area of the antenna that is along the direction of the radio source. Because the visibility $\mathcal{V}$ is a complex function, it can be expressed in relation to the antenna pattern as shown in Eq. (7.15), where $\phi_v$ is the phase of the visibility.

$$\mathcal{V} = |\mathcal{V}| e^{j\phi_v} = \int_{-\infty}^{\infty} P_n(\overline{\sigma}) I(\overline{\sigma}) e^{-j2\pi\vec{d}_\lambda \cdot \overline{\sigma}} \, d\Omega \tag{7.15}$$

The real and imaginary parts of the visibility can now be separately shown using Eqs. (7.16) and (7.17).

$$|\mathcal{V}| \cos\phi_v = \int_{-\infty}^{\infty} P_n(\overline{\sigma}) I(\overline{\sigma}) \cos(2\pi\vec{d}_\lambda \cdot \overline{\sigma}) \, d\Omega \tag{7.16}$$

$$|\mathcal{V}| \sin\phi_v = -\int_{-\infty}^{\infty} P_n(\overline{\sigma}) I(\overline{\sigma}) \sin(2\pi\vec{d}_\lambda \cdot \overline{\sigma}) \, d\Omega \tag{7.17}$$

Relating the above two equations with Eq. (7.14), we finally establish a relationship between visibility and correlation, as shown in Eq. (7.18). This indicates that the interferometer is an instrument that measures the visibility, which is the spatial coherence function having a different normalization.

$$r(\vec{d}_\lambda, \overline{s}_0) = A_{e0} \, \Delta v |\mathcal{V}| \cos\left[2\pi(\vec{d}_\lambda \cdot \overline{s}_0 - \phi_v)\right] \tag{7.18}$$

### 7.5.2 The van Cittert-Zernike Equation

It is necessary to know the values of the visibilities and their positions on the $u$, $v$, $w$ coordinate system for making a radio image in the $l$, $m$ coordinate system. Hence we need to relate Eq. (7.18) to both the $u$, $v$, $w$ and the $l$, $m$, $n$ coordinate systems. As shown in Fig. 7.10b, with decreasing elevation angle, the value of $w$ increases. A point on the celestial sphere is defined using the $l$-, $m$-, and $n$-axes, which are the direction cosines related to the $u$-, $v$-, and $w$-axes, respectively, as shown in Fig. 7.10c (see Appendix A.10). A synthesized image on the $l - m$ surface is therefore a projection of that region of the celestial sphere onto a plane that is tangent to the origin of the $l$, $m$ coordinates. Various relations among the baseline vector $\vec{d}_\lambda$ (in number of wavelengths), the point of observation, the $u$, $v$, $w$ coordinates, and the $l$, $m$, $n$ coordinates, are shown in Eqs. (7.19) through (7.21).

$$\vec{d}_\lambda \cdot \overline{s} = ul + vm + wn \tag{7.19}$$

$$\vec{d}_\lambda \cdot \overline{s_0} = w \tag{7.20}$$

$$d\Omega = \frac{dl\,dm}{n} = \frac{dl\,dm}{\sqrt{1 - l^2 - m^2}} \tag{7.21}$$

We can now express the visibility $\mathcal{V}$ obtained from Eq. (7.18) as a function of $u$, $v$, $w$, as shown below in Eq. (7.22). This is known as the *van Cittert-Zernike equation*.

$$\mathcal{V}(u, v, w) = \int_{-\infty}^{\infty} \int_{-\infty}^{\infty} \frac{P_n(l, m)\,I(l, m)}{\sqrt{1 - l^2 - m^2}} e^{-j2\pi\left[ul + vm + w\left(\sqrt{1 - l^2 - m^2} - 1\right)\right]} dl\,dm \tag{7.22}$$

The van Cittert-Zernike equation proves that a complex visibility function $\mathcal{V}(u, v, w)$ is a Fourier-like integral of the sky brightness $I(l, m)$, multiplied by the primary beam response of the interferometer, $P_n(l, m)$, and a factor of $1/\sqrt{1 - l^2 - m^2}$. The $u$, $v$, $w$ coordinate system has been defined in such a way that the $w$-axis always points towards the radio source.

When the source extent is small, the quantities $|l|$ and $|m|$ are also small. Hence the term $w\left(\sqrt{1 - l^2 - m^2} - 1\right)$ in the exponent becomes $-0.5(l^2 + m^2)w$, which is very small and therefore can be neglected. With this simplification, Eq. (7.22) can be re-expressed as Eq.(7.23).

$$\mathcal{V}(u, v, w) = \mathcal{V}(u, v, 0) = \int_{-l_1}^{l_2} \int_{-m_1}^{m_2} \frac{P_n(l, m)\,I(l, m)}{\sqrt{1 - l^2 - m^2}} e^{-j2\pi(ul + vm)} dl\,dm \tag{7.23}$$

Hence, within a small range of $l(l_1, l_2)$ and $m$ $(m_1, m_2)$, the visibility $\mathcal{V}(u, v, w)$ can be thought to be independent of $w$. Taking the inverse Fourier transform of Eq. (7.23), and after some rearrangement, we get an expression for $I(l, m)$ in terms of visibility, as shown in Eq. (7.24).

$$I(l,m) = \frac{\sqrt{1-l^2-m^2}}{P_n(l,m)} \int_{-l_1}^{l_2} \int_{-m_1}^{m_2} \mathcal{V}(u,v) e^{j2\pi(ul+vm)} du\,dv \qquad (7.24)$$

Assuming the antenna beam widths are limited to $(l_1, l_2)$ and $(m_1, m_2)$, we may extend the limits of integration to $\pm\infty$ because the results are unchanged. Hence Eq. (7.24) can be rewritten as Eq. (7.25), as shown below.

$$I(l,m) = \frac{\sqrt{1-l^2-m^2}}{P_n(l,m)} \int_{-\infty}^{\infty} \int_{-\infty}^{\infty} \mathcal{V}(u,v) e^{j2\pi(ul+vm)} du\,dv \qquad (7.25)$$

This equation is useful for constructing radio maps of the intensity distribution of the source, provided the source is small enough to be accommodated within the primary beam pattern.

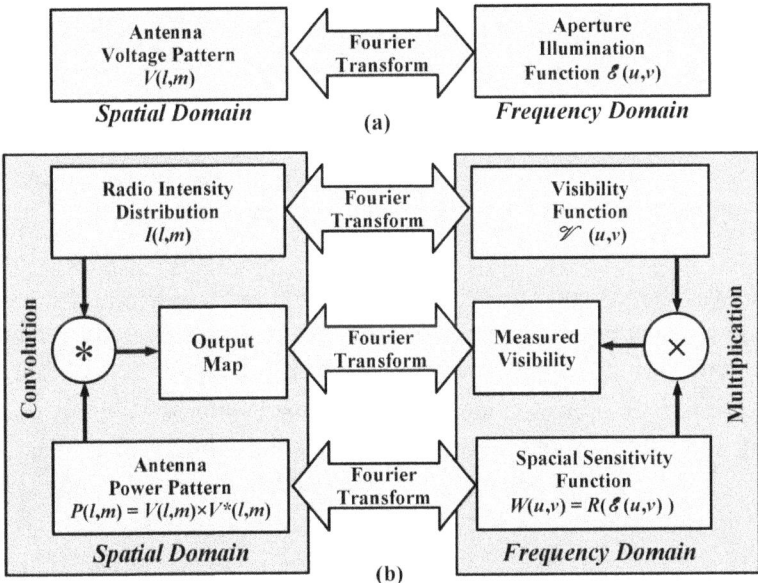

FIGURE 7.11: (a) Relation between the antenna voltage pattern $V(l, m)$ and the aperture illumination function $\mathcal{E}(u, v)$ (see Section 1.5.2). (b) Relationships between the intensity distribution $I(l, m)$, the antenna power pattern $P(l, m)$, and the output map, respectively, with the true visibility function $\mathcal{V}(u, v)$, the spatial sensitivity function $W(u, v)$ and the measured visibility.

### 7.5.3 Summary of Mapping Functions Between *u,v* and *l,m*

The various block relations existing between the functions of the spatial domain $(l, m)$ and the spatial frequency domain $(u, v)$ are summarized in Fig. 7.11. The two domains are connected by a two-dimensional Fourier transform.

Fig. 7.11a shows the relationship between the antenna voltage pattern $V(l, m)$ and the aperture illumination function $\mathcal{E}(u, v)$, as explained in Section 5.2 of Chapter 1.

Fig. 7.11b shows the relationships of the intensity distribution $I(l, m)$, the antenna power pattern $P(l, m)$, and the output map, respectively, with the true visibility function $V(u, v)$, the spatial sensitivity function $W(u, v)$, and the measured visibility. The big asterisk symbol has been used to denote convolution. The small superscript asterisk has been used to denote the complex conjugate. Note that the measured visibility is different from the true visibility due to the convolution of the intensity distribution with the beam pattern (see Section 2.10). The power pattern $P(l, m)$ of the antenna is the product of antenna voltage pattern $V(l, m)$ with its conjugate $V^*(l, m)$ in the spatial domain. The spatial sensitivity function $W(u, v)$ is obtained from the autocorrelation of the aperture illumination function $\mathcal{E}(u, v)$ in the frequency domain.

## REVIEW QUESTIONS

1. Describe the basic technique for synthesizing a large aperture using two small aperture antennas when the source is at the zenith.

   [***Hint:*** See Section 7.2]

2. If the synthesized aperture required is $100\text{m}^2$ for a source at the zenith, how many locations must the second antenna be positioned, given that the aperture size of the antenna elements is $2\text{m}^2$?     [***Hint:*** See Section 7.2]

3. Do we really require two antennas for aperture synthesis? What is the difficulty in using only one antenna? Justify your answer using equations.

   [***Hint:*** See Section 7.2 and Eq. 7.3]

4. What do you understand by "Earth rotation super-synthesis?"

   [***Hint:*** See Section 7.3]

5. Two antennas are positioned at a latitude of 45° north on Earth and observe a source in the sky. Let a rectangular coordinate system $(x, y, z)$ have its origin at the center of the Earth with its $z$-axis always pointed to the source.

Using diagrams, illustrate the projections of the antennas in the *x-y* plane as seen from the source for the cases of the source located at (*i*) the CNP, (*ii*) the celestial equator, and (*iii*) a celestial latitude of 45° north.

[*Hint:* See Section 7.3 and Fig. 7.3]

6. Illustrate with the help of a diagram the geometrical relations between (*i*) the *u, v, w* and *u', v', w'* coordinate systems, (*ii*) the ecliptic and the celestial equator. [*Hint:* See Fig. 7.4]

7. For a single dish having a diameter *D* operated at a radio wavelength $\lambda$, what is the range of spatial frequencies this single dish covers ?

[Ans: 0 to $D/2\lambda$ ]

8. While observing a source above the celestial equator for twelve hours, an interferometer generates visibilities $\mathcal{V}(u, v)$ whose locus on the *u-v* plane is half an ellipse. Using the same data, how can you create the other half of the ellipse? [Ans: $\mathcal{V}(-u, -v) = \mathcal{V}^*(u, v)$]

9. An interferometer situated along the east-west line observes a radio source. The visibilities obtained are spread across the *u-v* plane. Let the wavelength be $\lambda$, *d* the distance between the antennas, and $\theta$ the angle the source makes with respect to the CNP. At any instant of time, what is distance from the center of the *u-v* coordinates at which the visibility data will lie?

[*Hint:* Use Eqs. 7.4 and 7.5]

10. An east-west interferometer tracks a radio source. Unless the source is on the celestial equator, the traces of visibilities on the *u-v* plane vary from an ellipse to a circle (at the celestial pole). What happens if the antennas are along the north-south direction? Explain in detail.

[*Hint:* see Fig. 7.8 and Eq. 7.10]

11. Explain the advantages of building an interferometer array having baseline components along the north-south direction. Give an example of such an array.

[*Hint:* Radio images can be made even for sources located on the celestial equator.]

12. Explain the relation between visibility and correlation, using an equation.

[*Hint:* Use Eq. 7.18]

13. State and explain the significance of the van Cittert-Zernike equation.

[*Hint:* See Eq. 7.22]

14. With the help of a block diagram, illustrate the relationships existing between (*i*) the radio intensity distribution, (*ii*) the output map, (*iii*) the antenna power pattern, (*iv*) the visibility function, (*v*) the measured visibility, and (*vi*) the spatial sensitivity function. [***Hint:*** Use Fig. 7.11]

# CHAPTER 8

# INTERFEROMETER DATA CALIBRATION AND IMAGE PROCESSING

## 8.1 INTRODUCTION

The ultimate goal of radio interferometry is to construct the images of astronomical radio sources. The process broadly involves three steps: (*i*) data sorting, (*ii*) data calibration, and (*iii*) construction of an image. Data sorting involves removal of those data that are affected by instrumental problems, RFI, etc. Data calibration is necessary to free it from instrumental gain and phase contributions. Image construction involves a transformation and a lot of postprocessing to improve the image quality. Here we discuss calibration and image construction.

## 8.2 BASIC IDEAS OF IMAGE CONSTRUCTION

In a previous chapter (Section 7.5.2), we studied the van Cittert-Zernike equation in detail, and it forms the base of image construction. We now investigate how to apply this in image construction.

Consider an extended quasi-monochromatic source being observed by an interferometer, as shown in Fig. 8.1. The spacing between the antenna pair is measured in wavelengths using the $u$, $v$, $w$ coordinates. Note that the $w$-axis always points towards the source. The brightness distribution $I(l, m)$ on the celestial sphere is a function of the direction cosines $l$, $m$, and $n$. The interferometer response

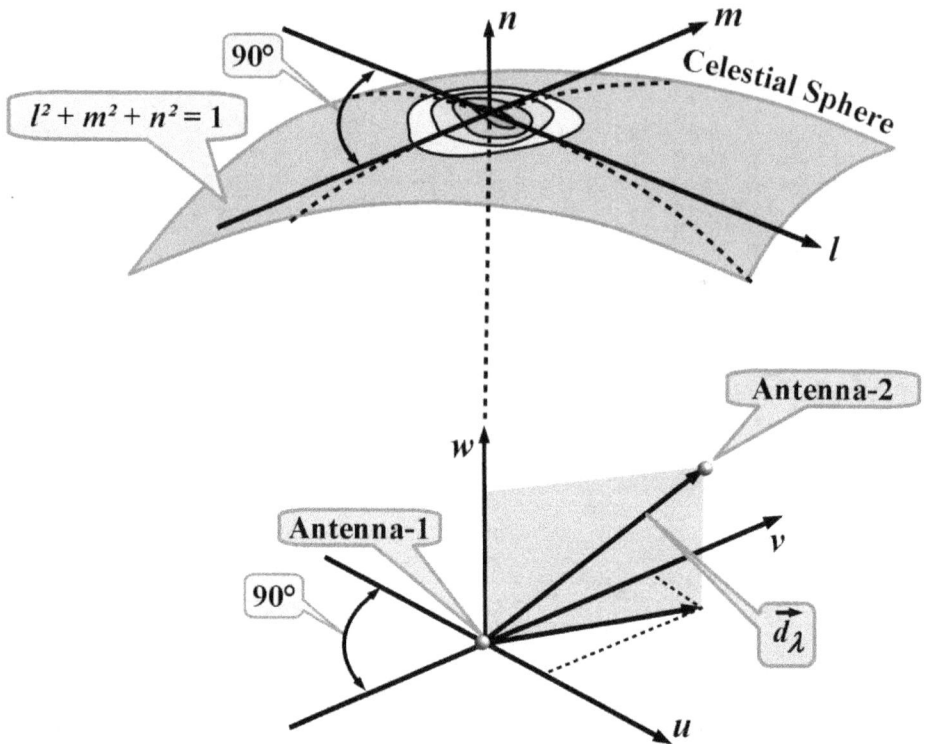

**FIGURE 8.1:** Two interferometer antennas are separated by a wavelength distance $d_\lambda$. The $u$, $v$, $w$ coordinates measure the projections of $d_\lambda$. A point on the celestial sphere is defined by the direction cosines $l$, $m$, and $n$, such that $l^2 + m^2 + n^2 = 1$.

(spatial coherence function $\mathcal{V}(u, v, w)$) can be expressed using the van Cittert-Zernike equation, which is shown in Eq. (8.1), where $P_n(l, m)$ is the normalized primary beam pattern of an antenna and $n = \sqrt{1 - l^2 - m^2}$ .

$$\mathcal{V}(u,v,w) = \int_{-\infty}^{\infty}\int_{-\infty}^{\infty} \frac{P_n(l,m)\,I(l,m)}{n} e^{-j2\pi[ul+vm+w(n-1)]} dl\,dm \qquad (8.1)$$

The above equation shows that $\mathcal{V}(u, v, w)$ is a Fourier-like integral of the product of (*i*) $I(l, m)$, (*ii*) the normalized primary beam pattern $P_n(l, m)$, and (*iii*) $1/n$. It can be used directly for imaging, especially when the source extents are large. This kind of imaging is generally called *wide-field* imaging.

For very small sources that are observed with interferometers and that have narrow primary beam widths, it is convenient to assume (*i*) $n \approx 1$, and (*ii*) $P_n(l, m) \approx$

1. This reduces the complexity of the computation because the van Cittert-Zernike equation becomes independent of $n$ and $w,$ as shown in Eq. (8.2).

$$\mathcal{V}(u,v) = \int_{-\infty}^{\infty}\int_{-\infty}^{\infty} I(l,m)\exp\left[-j2\pi(lu+mv)\right]dl\,dm \qquad (8.2)$$

This approximation $P_n(l,\,m) \approx 1$ makes it impossible to image large fields. Some effects of approximating $P_n(l,\,m) \approx 1$ may be corrected in the final stages of image analysis. This two-dimensional relationship has been used over a long period of time for nearly all radio imaging, so we shall call it *conventional* imaging. Currently, researchers are devising ways to avoid these approximations so that imaging larger fields is allowed.

Interferometer arrays contain more than two antennas and have several different baselines. The number of baselines is $^{N}C_2$, where $N$ is the number of antennas. An interferometer array aided by super-synthesis gives a good amount of data in the $(u,\,v,\,w)$ domain.

We begin with the study of general calibration techniques, and then we move on to the conventional imaging techniques (for small fields of view). Towards the end, we will show some techniques of wide-field imaging. Finally, we will discuss spectral- line observations.

## 8.3  GENERAL CALIBRATION TECHNIQUES

The interferometer data contain system-dependent terms, which are different for different antennas. The receiving systems may introduce extra gain and or change the phase. These variations can be long or short. Hence the calibration processes are broadly of two types: (*i*) long-term, and (*ii*) short-term calibrations.

### 8.3.1 Long-Term Calibrations

Due to the self-weights of the antenna structures, mechanical deformations may take place. These are elastic in nature and are position dependent. They include (*i*) antenna coordinates and baselines, and (*ii*) antenna pointing corrections caused from misalignments. Calibrations for these problems are generally done every week and also before observing the target source. A radio source of known flux density is generally used for this purpose. The antennas are pointed to this source and the zero points are set in such a way that delays from the antennas to the correlator are the same for all. These settings are expected to function throughout the actual observation period.

### 8.3.2 Short-Term Calibrations

Short-term calibrations are done intermittently within the observation period. A point radio source with known flux density that is located in a direction close to the target source is generally used. The calibration data are recorded along with the target source data. The calibration data are applied during data analysis (after observation). However, the visibility data must be carefully edited before applying these, so that the data is free from any evidence of equipment malfunctioning or traces of radio interference. The data is examined for any unexpected level of gain or phase variation. These are eliminated by a process called *flagging*.

Short-term calibrations are made to correct the gain and phase of the telescope, which includes the antenna and the correlator. Phase corrections are important for eliminating the phase fluctuations caused by changing ionospheric behavior, especially at frequencies below 1 GHz. This increases with the wavelength size. These are also important for correcting the constant component of atmospheric attenuation that changes as we move away from the zenith.

Other problems like shadowing of one antenna by any other antenna when the projected baselines become small at lower elevation angles can be overcome by short-term calibrations. Due to ground temperature pickup by antenna feed spillover, mesh leakage, and side lobes, the system noise temperature changes with antenna elevation. This may cause the ALC (automatic level control) to change the system gain, which is used in many telescopes. The receiving system can be checked intermittently by injecting a known quantity of noise into the receiver (using a directional coupler near the LNA) and observing its effect at the correlator output.

Instead of a calibrated radio astronomical source whose position and power are known, a more recent technique called *self-calibration* may be used. For VLBI observations, additional steps are required for matching time, phase, and frequency of different observatory data.

#### 8.3.2.1 Radio Sources as Calibrators

Let $G_{pq}(t)$ be the factor by which the system gain of an interferometer (consisting of antennas $p$ and $q$) changes as a function of observation time. Eq. (8.2) can be combined with $G_{pq}(t)$ and the uncalibrated interferometer response can be written as Eq. (8.3), where $V_{uc}(u, v)$ represents the uncalibrated visibility.

$$V_{uc}(u,v) = G_{pq}(t)\int_{-\infty}^{\infty}\int_{-\infty}^{\infty} I(l,m)\, e^{-j2\pi[ul+vm]} dl\, dm \qquad (8.3)$$

The complex gain factor $G_{pq}(t)$ is a function of the gain parameters of the antennas $p$ and $q$ that form the interferometer. This varies with time. Comparing Eq. (8.3) with Eq. (8.2), we express $V_{uc}(u, v)$ as shown in Eq. (8.4).

$$\mathcal{V}_{uc}(u,v) = G_{pq}(t)\mathcal{V}(u,v) \tag{8.4}$$

When the interferometer is pointed to an unresolved calibrator source of known flux density $S_c$ (Watt/m$^2$/Hz), such that the source is located at the phase center of its field, the measured visibility $\mathcal{V}_c(u, v)$ is given by Eq. (8.5).

$$\mathcal{V}_c(u,v) = G_{pq} S_c \tag{8.5}$$

Comparing Eq. (8.4) with (8.5), we obtain Eq. (8.6). This gives the amplitude $|G_{pq}|$ and phase $\angle G_{pq}$ of $G_{pq}(t)$, respectively, as shown by Eqs. (8.7) and (8.8).

$$\mathcal{V}(u,v) = \mathcal{V}_{uc}(u,v)\left(\frac{S_c}{\mathcal{V}_c(u,v)}\right) \tag{8.6}$$

$$|G_{pq}| = \left|\frac{\mathcal{V}_c}{S_c}\right| \tag{8.7}$$

$$\angle G_{pq} = \angle\left(\mathcal{V}_c / S_c\right) \tag{8.8}$$

We may express the gain factor $G_{pq}$ using Eqs. (8.7) and (8.8) in terms of amplitude and phase, as shown in Eq. (8.9). The instrumental gain factor $G_{pq}$, visibility $\mathcal{V}$, and the uncalibrated visibility $\mathcal{V}_{uc}$, after determinng their magnitude and phase form, can be related as in Eq. (8.10).

$$G_{pq} = |G_{pq}|e^{j\angle G_{pq}} = \left|\frac{\mathcal{V}_c}{S_c}\right|e^{j\angle(\mathcal{V}_c/S_c)} \tag{8.9}$$

$$\mathcal{V} = |\mathcal{V}|e^{j\angle\mathcal{V}} = \frac{|\mathcal{V}_{uc}|e^{j\angle\mathcal{V}_{uc}}}{|G_{pq}|e^{j\angle G_{pq}}} = \left|\frac{\mathcal{V}_{uc}}{G_{pq}}\right|e^{j\left(\angle\mathcal{V}_{uc}-\angle G_{pq}\right)} \tag{8.10}$$

It is seen from Eq. (8.10) that the magnitude of the visibility $\mathcal{V}$ is the ratio of the magnitudes of the uncalibrated visibility $\mathcal{V}_{uc}$ to that of the instrumental gain factor $G_{pq}$. Again, the phase of the visibility $\mathcal{V}$ is the phase of the uncalibrated visibility $\mathcal{V}_{uc}$ minus the phase of $G_{pq}$. These are shown in Eqs. (8.11) and (8.12), respectively. Calibrations for both polarizations must be separately performed.

$$|\mathcal{V}| = \frac{|\mathcal{V}_{uc}|}{|G_{pq}|} = \left|\frac{\mathcal{V}_{uc} S_c}{\mathcal{V}_c}\right| \tag{8.11}$$

$$\angle\mathcal{V} = \angle\mathcal{V}_{uc} - \angle G_{pq} = \angle\mathcal{V}_{uc} - \angle\left(\mathcal{V}_c / S_c\right) \tag{8.12}$$

It is preferable to perform gain calibrations immediately before the actual observation by means of a high-power unresolved calibrator source. Phase calibrations are intermittently performed throughout the observation period by interrupting the target observation process and moving to an unresolved phase calibrator source lying in a direction very close to the target. This allows the telescope to make use of the almost identical ionosphere medium present between the target source and the antennas so as to detect the correct phase variations. As mentioned before, the instrumental phase obtained using the calibrator must be subtracted from the phase of the target visibility. Gain calibrators must be strong for obtaining good SNR within a short time. Strong sources are few in number, and one may not always be available close to the target source. Hence, separate calibrators for gain and phase are used. Commonly used gain or flux calibrators are 3C48, 3C147, 3C286, and 3C295. Thermal sources like the compact planetary nebula NGC7027 can be very useful for short baselines.

### 8.3.2.2 Self-Calibration

Self-calibration is based on (*i*) phase closure, and (*ii*) gain closure obtained from a few antennas. The target source being observed is used as a calibrator. Though there are different gain pairs $G_{pq}$ formed from different antenna pairs, still the data contain good observables. For an array of $N$ antennas, $N(N-1)/2$ baselines are formed that are equal to the number of visibilities obtained at any instant. We express $G_{pq} = g_p g_q^*$, where $g_p$ and $g_q$ are, respectively, the complex gain factors of the signal paths associated with antennas $p$ and $q$. Hence there are $N$ unknown complex antenna gain factors $g_p$, which means $N$ unknown amplitude factors and $N$ unknown phases. Thus there are $N(N-1)/2 - N$ uncorrupted complex quantities due to antenna effects that are called *closure quantities,* as we see shall see next.

(*i*) **Phase closure:** This is the sum of the visibility phases around a triangle formed using three antennas. Let the antenna pairs $pq$, $qr$, and $rp$, respectively, give the calibrated visibilities $\mathcal{V}^{pq}$, $\mathcal{V}^{qr}$, and $\mathcal{V}^{rp}$. Let the corresponding uncalibrated visibilities be $\mathcal{V}_{uc}^{pq}$, $\mathcal{V}_{uc}^{qr}$, and $\mathcal{V}_{uc}^{rp}$, respectively. Then at any instant of time, the sum of the phases of the calibrated visibilities is equal to the sum of the phases of the uncalibrated visibilities, as shown in Eq. (8.13).

$$\angle \mathcal{V}^{pq} + \angle \mathcal{V}^{qr} + \angle \mathcal{V}^{rp} = \angle \mathcal{V}_{uc}^{pq} + \angle \mathcal{V}_{uc}^{qr} + \angle \mathcal{V}_{uc}^{rp} \qquad (8.13)$$

(*ii*) **Gain closure:** Gain closure requires a minimum of four antennas. Let these four antennas be $p$, $q$, $r$, and $s$. It can be shown that the gain magnitudes of the uncalibrated and calibrated visibilities are given as in Eq. (8.14).

$$A_{pqrs} = \frac{\left|\mathcal{V}_{uc}^{pq}\right|\left|\mathcal{V}_{uc}^{rs}\right|}{\left|\mathcal{V}_{uc}^{pr}\right|\left|\mathcal{V}_{uc}^{qs}\right|} = \frac{\left|\mathcal{V}^{pq}\right|\left|\mathcal{V}^{rs}\right|}{\left|\mathcal{V}^{pr}\right|\left|\mathcal{V}^{qs}\right|} \qquad (8.14)$$

Three different gain closures ($A_{pqrs}$, $A_{pqsr}$, $A_{qrsp}$) can be calculated if all six interferometers formed are correlated, but only two of these are independent.

Before starting the actual observation, a set of data is taken. This is calibrated from the current best estimate of the antenna gains, and a model of the sky is generated using imaging techniques. Considering this sky model as correct, visibilities are obtained by which new antenna gains are found that reduce the differences between the model and the measured visibilities. Let $\mathcal{V}_{mod}^{pq}$ and $\mathcal{V}_{uc}^{pq}$ represent the model and measured (uncalibrated) visibilities, respectively. Assuming $\mathcal{V}_{uc}^{pq} = G_{pq} \mathcal{V}_{mod}^{pq}$ from Eq. (8.4), the quantity $\epsilon^2$ shown in Eq. (8.15) has to be minimized to obtain the values of $g_p$, $g_q$, and $G_{pq}$.

$$\epsilon^2 = \sum_{p,q} \left| \mathcal{V}_{mod}^{pq} - \frac{\mathcal{V}_{uc}^{pq}}{g_p \, g_q^{\star}} \right|^2 \tag{8.15}$$

Actual observations are made with these calibrated values. It is assumed that these gain solutions remain valid over the entire observation.

**Additional techniques of self-calibration:** Certain techniques can be used to aid self-calibrations. Two are of them are commonly used and are described next.

(*i*) **Self-calibration with a redundant array:** An array is designed to measure a common Fourier component of a brightness distribution using more than one baseline. It is called a *redundant array* because some of the baselines are identical. An example of a redundant array of five antennas is shown in Fig. 8.2a. The antennas are each separated by a distance $d$. The total number of baselines is 10. The various baselines and their redundancies are listed in Table 8.1.

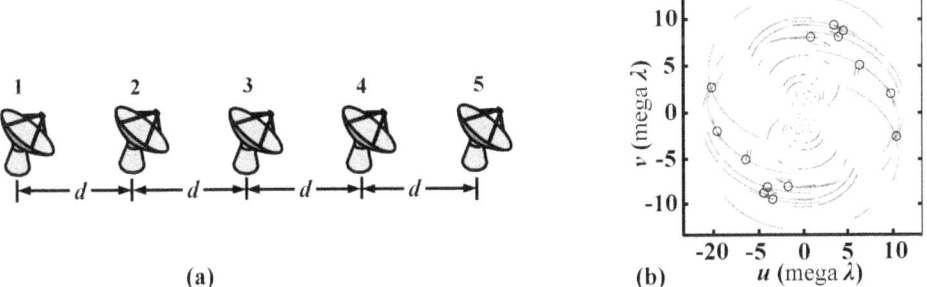

**FIGURE 8.2:** Additional calibration methods. (a) An example of a redundant array of five antennas. (b) Crossing of visibility data on the *u-v* plane if north-south baselines exist. The crossing points are encircled.

**TABLE 8.1:** Baselines formed by a five-element redundant array.

| Antenna pairs | Baseline distances | No. of baselines |
|---|:---:|---|
| 1-2, 2-3, 3-4, 4-5 | $d$ | 4 (redundant) |
| 1-3, 2-4, 3-5 | $2d$ | 3 (redundant) |
| 1-4, 2-5 | $3d$ | 2 (redundant) |
| 1-5 | $4d$ | 1 |

By substituting $G_{pq} = g_p g_q^*$ into Eq. (8.4), the uncalibrated visibility $\mathcal{V}_{uc}^{pq}$ can be shown in terms of the true visibility $\mathcal{V}^{pq}$, as in Eq. (8.16).

$$\mathcal{V}_{uc}^{pq}(u,v) = \mathcal{V}^{pq}(u,v)\, g_p g_q^*, \text{ or } \mathcal{V}^{pq}(u,v)\, g_p^* g_q \tag{8.16}$$

The equations obtained using baselines of length $d$ from the redundant array (Fig. 8.2a) are shown in Eq. (8.17). Similarly, for the baselines having $2d$ spacings, we obtain Eq. (8.18), and from $3d$ spacings we obtain Eq. (8.19).

$$\begin{bmatrix} \mathcal{V}_{uc}^{1,2} \\ \mathcal{V}_{uc}^{2,3} \\ \mathcal{V}_{uc}^{3,4} \\ \mathcal{V}_{uc}^{4,5} \end{bmatrix} = \begin{bmatrix} \mathcal{V}^{1,2}\, g_1 g_2^* \\ \mathcal{V}^{2,3}\, g_2 g_3^* \\ \mathcal{V}^{3,4}\, g_3 g_4^* \\ \mathcal{V}^{4,5}\, g_4 g_5^* \end{bmatrix} \tag{8.17}$$

$$\begin{bmatrix} \mathcal{V}_{uc}^{1,3} \\ \mathcal{V}_{uc}^{2,4} \\ \mathcal{V}_{uc}^{3,5} \end{bmatrix} = \begin{bmatrix} \mathcal{V}^{1,3}\, g_1 g_3^* \\ \mathcal{V}^{2,4}\, g_2 g_4^* \\ \mathcal{V}^{3,5}\, g_3 g_5^* \end{bmatrix} \tag{8.18}$$

$$\begin{bmatrix} \mathcal{V}_{uc}^{1,4} \\ \mathcal{V}_{uc}^{2,5} \end{bmatrix} = \begin{bmatrix} \mathcal{V}^{1,4}\, g_1 g_4^* \\ \mathcal{V}^{2,5}\, g_2 g_5^* \end{bmatrix} \tag{8.19}$$

Solving the above equations, we obtain the complex gain ratios shown in Eq. (8.20). It may be possible to estimate the individual antenna gains from a standard map or if one of the antenna gains is known. The main drawback of using a redundant array comes from the reduced $u$-$v$ coverage, which occurs because common data sets are generated.

$$\begin{bmatrix} g_1/g_3 \\ g_2/g_4 \\ g_3/g_5 \\ g_5/g_1 \end{bmatrix} = \begin{bmatrix} \mathcal{V}_{uc}^{1,2}/\mathcal{V}_{uc}^{2,3} \\ \left(\mathcal{V}_{uc}^{2,3}/\mathcal{V}_{uc}^{3,4}\right)^* \\ \mathcal{V}_{uc}^{3,4}/\mathcal{V}_{uc}^{4,5} \\ \mathcal{V}_{uc}^{3,5}/\mathcal{V}_{uc}^{1,3} \end{bmatrix} \tag{8.20}$$

(*ii*) **Self-calibration using redundant data (cross-overs in the *u-v* plane):** If an interferometer array has north-south baseline components, the observed data may contain repetitions (redundancy) of the same spatial frequency components produced by different interferometers. This is shown in Fig. 8.2b. The redundant data encircled in the picture may be used for calibration.

Self-calibration works well for strong sources, or when the SNR for the visibility data is reasonably good. Though both phase and gain calibrations are required, the former is more effective and can be accurately done by self-calibration. As a precautionary note, self-calibration using low-SNR visibility data can result in artifacts in the image.

## 8.4    CONVENTIONAL IMAGING METHOD

The van Cittert-Zernike equation (Eq. (8.2)) forms the basis of conventional imaging. Images are made only after calibrating the good visibility data. There are two steps involved: (*i*) creating the best possible image (called a *dirty image*) from the data by applying weighting techniques; and (*ii*) improving the quality of the dirty image by certain image processing functions like *deconvolution*.

We begin the construction of a dirty image with the assumption that the visibility data is already calibrated for gain and phase. As shown in Fig. 7.11 of Chapter 7, the measured and calibrated visibility $V_{mc}(u, v)$ is a product of the true visibility $V(u, v)$ and the spatial sensitivity function $w(u, v)$. This is given by Eq. (8.21).

$$V_{mc}(uv) = V(u,v)W(u,v) \tag{8.21}$$

The scaled data is now called *weighted* for reasons that will be clear later. The measured, calibrated, and weighted visibilities $V_{mcw}(uv)$ are expressed in Eq. (8.22), where $V(u, v)$ is the true visibility, $W(u, v)$ is the spatial sensitivity function, and $W(u, v)$ is the weighting function.

$$V_{mcw}(u,v) = V(u,v)W(u,v)w(u,v) \tag{8.22}$$

An inverse Fourier transform recovers the sky brightness distribution $I(l, m)$, as shown in Eq. (8.2). Applying this to Eq. (8.22) on both sides, we obtain Eq. (8.23), where $I_{mcw}(l, m)$ represents a distorted source intensity distribution resulting from the synthesized beam pattern, where $P_{syn}(l, m)$ represents the synthesized beam pattern. The symbol "*" represents convolution.

$$I_{mcw}(l,m) = I(l,m) \star P_{syn}(l,m) \tag{8.23}$$

We may expand these terms as shown below in Eqs. (8.24) through (8.26). Note that we can modify the synthesized beam $P_{syn}(l, m)$ by changing the weights $w(u, v)$.

$$I_{mcw}(l,m) = \int_{-\infty}^{\infty} \int_{-\infty}^{\infty} \mathcal{V}_{mcw}(u,v) e^{-j2\pi(lu+mv)} \, dl \, dm \tag{8.24}$$

$$I(l,m) = \int_{-\infty}^{\infty} \int_{-\infty}^{\infty} \mathcal{V}(u,v) e^{-j2\pi(lu+mv)} \, dl \, dm \tag{8.25}$$

$$P_{syn}(l,m) = \int_{-\infty}^{\infty} \int_{-\infty}^{\infty} W(u,v) w(u,v) e^{-j2\pi(lu+mv)} \, dl \, dm \tag{8.26}$$

The interferometer does not give a continuous distribution of visibility across the $u$, $v$ plane. Instead, it appears at a discrete set of points $u_i$, $v_i$, where $i$ is an integer. It contains an ensemble of $N_{vis}$ pairs of points that appear symmetrically on both sides of the origin of the $u$, $v$ axes. Eq. (8.27) represents them in discrete form, where $I_D(l, m)$ is known as the *dirty image*.

$$I_D(l,m) = \sum_{i=1}^{N_{vis}} w_i \left[ \mathcal{V}_{mc}(u_i,v_i) e^{j2\pi(lu_i+mv_i)} + \mathcal{V}_{mc}(-u_i,-v_i) e^{-j2\pi(lu_i+mv_i)} \right] \tag{8.27}$$

Let both antennas of the interferometer have identical polarizations. The measured and calibrated visibilities on both sides of the origin of $u$, $v$ coordinates form a complex-conjugate pair, as shown in Eq. (8.28). Hence the dirty image $I_D(l, m)$ obtained using Eq. (8.27) is real.

$$\mathcal{V}_{mc}(-u_i,-v_i) = \mathcal{V}_{mc}^{\star}(u_i,v_i), \qquad i = 1,2,...,N_{vis} \tag{8.28}$$

### 8.4.1 Noise in Visibility

The measured visibilities from the correlator also contain noise for which the percentage does not change after calibration. The measured and calibrated visibility $\mathcal{V}_{mc}$ is the sum of noiseless visibility $\mathcal{V}(u, v)$ and the noise $\epsilon(u, v)$, as shown in Eq. (8.29).

$$\mathcal{V}_{mc}(uv) = \mathcal{V}(u,v) + \epsilon(u,v) \tag{8.29}$$

Let $\epsilon_{eff}$ represent the effective (rms) noise in the visibility. A fictitious unresolved radio source having a flux density $S_{noise}$ (in W m$^{-2}$ Hz$^{-1}$) can be thought to be located at the phase reference point of a noise-free interferometer system that produces this system-effective noise output $\epsilon_{eff}$ as shown in Eq. (8.60) of Chapter 5. It is reproduced here in Eq. (8.30).

$$\epsilon_{eff} = \sqrt{2} \, S_{noise} \tag{8.30}$$

If $\epsilon_i$ is the complex noise in the $i^{th}$ visibility sample, then it can be expressed as Eq. (8.31), where $\epsilon_i^{Re}$ and $\epsilon_i^{Im}$, respectively, represent its real and imaginary parts.

$$\epsilon_i = \epsilon_i^{Re} + j\epsilon_i^{Im} \tag{8.31}$$

Let $\tau_0$ represent the total observation time, and let $\tau_d$ be the integration time of each visibility data point. The number of data points $n_d$ on the $u$-$v$ plane is given by Eq. (8.32), where $n_p$ represents the number of interferometers (antenna pairs) formed by the array. The intensity $I_0$ at the center of the map is given by Eq. (8.33), where $w_i$ is the weight factor and $\mathcal{V}$ is the noise-free visibility.

$$n_d = n_p \frac{\tau_0}{\tau_a} \tag{8.32}$$

$$I_0 = \frac{\sum_{i=1}^{n_d} w_i \left( \mathcal{V} + \epsilon_i^{Re} \right)}{\sum_{i=1}^{n_d} w_i} \tag{8.33}$$

Note that $\varepsilon_i^{Im} = 0$ is at the origin of the $u$-$v$ plane. The neighboring points possess identical rms noise levels in their real and imaginary parts. For any two different locations on the $u$-$v$ plane, the noise terms are uncorrelated. Mathematically, $\left\langle \epsilon_i^{Re} \epsilon_j^{Re} \right\rangle = 0$ for $i \neq j$. Eq. (8.34) expresses the variance $\sigma_m^2$ of the estimated intensity. The bracket symbols $<>$ denote the expected value.

$$\sigma_m^2 = \left\langle I_0^2 \right\rangle - \left\langle I_0 \right\rangle^2 = \frac{\sum_{i=1}^{n_d} w_i^2 \left\langle \left( \epsilon_i^{Re} \right)^2 \right\rangle}{\left( \sum_{i=1}^{n_d} w_i \right)^2} \tag{8.34}$$

Let $w_{mean}$ and $w_{rms}$ represent the mean and rms weighting factors, respectively. These are related to $w_i$, as shown in Eqs. (8.35) and (8.36).

$$w_{mean} = \frac{1}{n_d} \sum_{i=1}^{n_d} w_i \tag{8.35}$$

$$w_{rms} = \sqrt{\frac{1}{n_d} \sum_{i=1}^{n_d} w_i^2} \tag{8.36}$$

### 8.4.2 Putting Weight on Visibilities

Practical observations can only partially fill the $u$, $v$ plane because (*i*) visibilities are sampled at discrete points, and (*ii*) visibility samples are fewer in number at large projected baselines on the $u$, $v$ plane. Also the *u-v* plane is never populated as a square or a rectangle due to Earth's rotation. An example of *u-v* coverage is shown in Fig. 8.3. Consider a small rectangular area ($du \times dv$) on the data. Let us define the *area density* $\rho_\sigma$, which is the number of data points available within the rectangle. As shown in the figure, the area densities $k_1$, $k_2$, and $k_3$ (at three different locations) are not the same. Hence, $\rho_\sigma$ varies from place to place and can be zero at some locations. High data concentrations are seen for shorter $u$, $v$ spacings. The synthesized beam $P_{\mathrm{syn}}(l, m)$ produces high side lobe levels (see Fig. 8.4) due to empty areas that result in artifacts in the image.

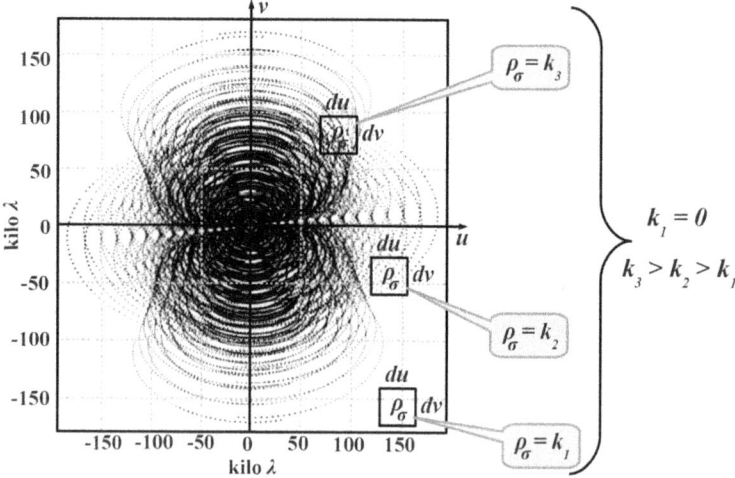

**FIGURE 8.3:** An example of a visibility distribution across the *u-v* plane obtained over 8 hours using VLA for a source at $\delta = 45°$. The area density $\rho_\sigma$ is shown at three locations.

The reciprocal of the area density $\rho_\sigma$ is sometimes used for weighting the visibilities. Care must be taken that $\rho_\sigma$ is nonzero, however, which may be achieved by carefully selecting the area of the rectangle.

#### 8.4.2.1 Imaging Using Natural Weights

In natural weighting, the values of $w_i$ are kept the same (usually unity) for all $i$, which means $w_{\mathrm{mean}} = w_{\mathrm{rms}}$. Natural weights give maximum sensitivity at the cost of high side lobes in the synthesized beam $P_{\mathrm{syn}}(l, m)$, which results in large artifacts in the dirty image. There exists a relationship between the natural weights $w_i$ and the effective noise of the visibilities, as shown in Eq. (8.37).

$$w_i = \frac{1}{\sigma_i^2}, \quad \text{where } \sigma = \epsilon_{eff} \tag{8.37}$$

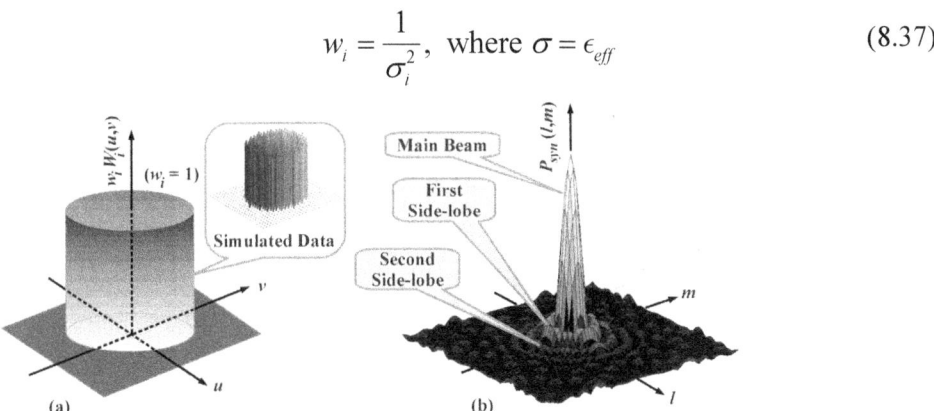

**FIGURE 8.4:** High side lobes occur in the synthesized beam using natural weights. (a) A naturally weighted spatial sensitivity function. (b) A synthesized beam pattern. (Simulated by the author.)

A naturally weighted spatial sensitivity function and the resulting synthesized beam are shown in Fig. 8.4. Note that the side lobes are strong enough to produce artifacts in the image. The main reason for side lobes is the stepped fall of the spatial sensitivity function beyond a certain radius. In actual data, there exists a hole in the middle portion of the cylinder due to the absence of data below minimum baseline distances. Also, the data density falls with increasing baseline length and abruptly stops above the maximum baseline length. The shape of the data area need not be circular as shown. All of these contribute to form different side-lobe structures.

### 8.4.2.2 Imaging Using Uniform Weights
As observed in Fig. 8.3 the visibility population on the u-v plane is not uniform. Because the visibilities are discrete, many places on the u-v plane are empty. As the projected baselines become smaller, the discrete data becomes denser on the u-v plane. Hence the SNR is not uniform across the u-v plane. The first step towards obtaining a uniform SNR is to evaluate the area density $\rho_\sigma$ as a function of $u$, $v$. A least-squares fitting may be used for the general case. After evaluating the function $\rho_{\sigma i} = \rho_\sigma(u_i, v_i)$, weights are applied as shown in Eq. (8.38), where $k_0$ is a constant that is usually unity.

$$w_i = \frac{k_0}{\rho_{\sigma i}}, \quad 1 \le i \le n_d \tag{8.38}$$

When the area density function $\rho_{\sigma i}$ is uniform across the u-v plane, the weighting turns out to be natural.

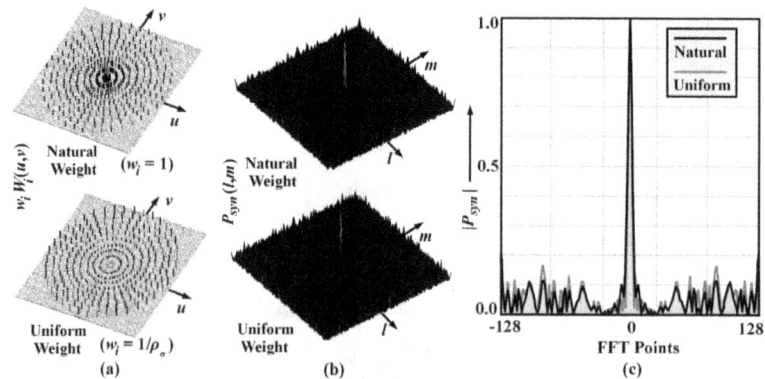

**FIGURE 8.5:** Simulated beam shapes of a nonredundant east-west array using natural and uniform weights. A source at the CNP is observed for 12 hours (a) Spatial sensitivity functions. (b) Beam shapes. (c) Beam shapes in one dimension. (Simulated by the author.)

For an east-west array observing a source at the CNP for 12 hours, the spatial sensitivity functions for the natural and uniform weights are shown in Fig. 8.5a. The simulated beam shapes obtained using the natural and uniform weights are shown in both Fig. 8.5b and Fig. 8.5c. There are only ten unique baselines that are multiples of the smallest baseline. Hence, the visibilities discretely appear along circles over the $u$, $v$ plane. The radii of these circles are multiples of the smallest baseline. The upper diagram of Fig. 8.5a shows a naturally weighted spatial sensitivity function that appears as cylindrical fences of different radii but equal heights. Observe that $\rho_\sigma$ is lowest on the largest circle and maximum on the smallest circle because the number of data points on each circle are the same. Hence, the $\rho_\sigma$ on the circles are inversely proportional to their radii. To achieve a uniform SNR, divide the upper spatial sensitivity function of Fig. 8.5a by $\rho_\sigma$. The bottom part of Fig. 8.5a shows the resulting uniformly weighted spatial sensitivity function.

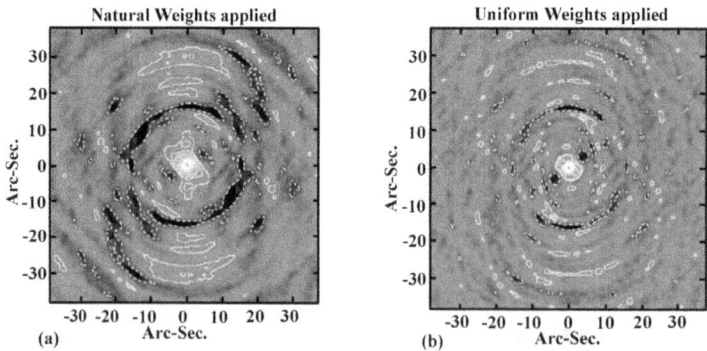

**FIGURE 8.6:** Images formed using (a) natural weights, and (b) uniform weights. Uniform weights give higher resolution at the cost of reduced sensitivity.

In Fig. 8.5c, observe that the main beam resulting from the uniform weights is relatively narrower. The neighboring side lobes are also relatively smaller than those produced by natural weights. The distant side lobes, however, increase with uniform weights. These effects are visible on the images (see Fig. 8.6).

For performing FFT, one requires a rectangular data set. Because the $u$, $v$ data set is usually composed of circles and ellipses, a rectangular data set must be generated by padding zeros outside the area of the actual data, as shown in Figs. 8.4a and 8.5a.

### 8.4.2.3 Imaging Using Tapered Weights

Uniform weights generate strong side lobes close to the main beam, which may reduce the details of images that have a low-intensity profile. We may reduce these near-in side lobes by making some compromises with the increase in the width of the synthesized main beam. A tapered weightage $w_i$ of the form shown in Eq. (8.39), also called *Gaussian tapering,* may be used with the spatial sensitivity function $W(u, v)$, where $t$ represents the tapering parameter, $i$ is the sample number, and $n_d$ represents the total number of discrete visibilities (see Eq. (8.32)). Because $u$ and $v$ axes are in wavelengths, $t$ also uses the same dimension.

$$w_i = \exp\left[-\frac{u_i^2 + v_i^2}{t_i^2}\right], \text{ where, } i = 1, 2, .n_d \tag{8.39}$$

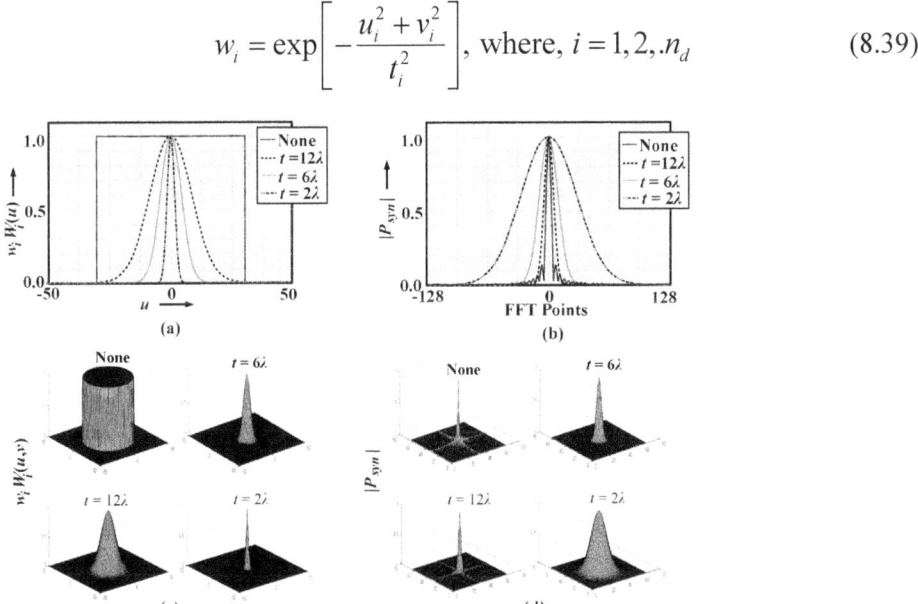

**FIGURE 8.7:** Tapering and its effects. (a) Tapering applied to $W(u, v)$ by modifying $w_i$. (b) Magnitude of the resulting synthesized beam ($|P_{syn}|$). (c) Three-dimensional view of the tapered $W(u, v)$. (d) Three-dimensional plot of $|P_{syn}|$. (Simulated by the author.)

Figs. 8.7a shows a tapering of the spatial sensitivity function $W(u, v)$ by modifying the weighting function $w_i$ using Eq. (8.39) in two dimensions. Fig. 8.7b shows the magnitude of the resulting synthesized beam pattern ($|P_{syn}|$). Note that the side lobes reduce as tapering increases (decreasing $t$), but at the expense of increased main beam width. The same effects are illustrated in Figs. 8.7c and 8.7d in three dimensions.

Images formed by different tapering levels are compared with a natural weighted image in Fig. 8.8. From the tapering magnitudes, we may guess the observed spatial frequencies are in thousands of wavelengths. As tapering increases (decreasing $t$), the images become cleaner, but the extent of the image (near the center) increases due to increasing the synthesized main beam width.

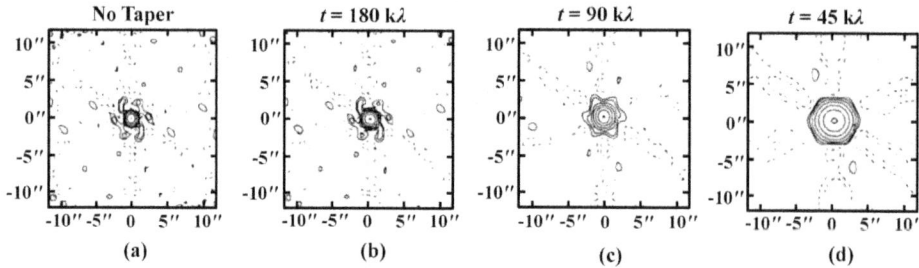

**FIGURE 8.8:** Images from non- or Gaussian-taperings. (a) No tapering. (b) Tapering with $t = 180 \times 10^3 \lambda$. (c) Tapering with $t = 90 \times 10^3 \lambda$. (d) Tapering with $t = 45 \times 10^3 \lambda$. As the tapering increases (decreasing $t$), the images become cleaner, but the extent of the image increases due to the increasing synthesized main beam width.

Gaussian taperings are most commonly used because of its ability to better reduce neighboring side lobes. Tapering functions like the cosine may be used with some compromises between the main beam width and the side lobes.

### 8.4.2.4 Imaging Using Robust Weights

Yet another weighting function called *robust weighting* is gaining popularity in image reconstruction. It is based on a compromise between uniform and natural weights based on the local SNR of the visibilities across the $u$-$v$ plane. When the area density $\rho_\sigma$ is less than a chosen threshold level, natural weights are used, otherwise, uniform weights are applied. The algorithm for applying robust weights is similar to Eq. (8.40), where a filter $S$ is controlled by the robust parameter $R$, and $\sigma$ represents effective (rms) noise as obtained in Eq. (8.30). The constant $k_1$ may be around 5 or so.

$$w_i = \frac{1}{S^2 + \sigma_i^2}, \text{ where } S = k_1 \times 10^{-R}, \text{ and } \sigma = \epsilon_{eff} \qquad (8.40)$$

By a proper selection of the robust parameter $R$, it may be possible to apply natural weights when the area density $\rho_\sigma$ is small, and to otherwise apply uniform weights.

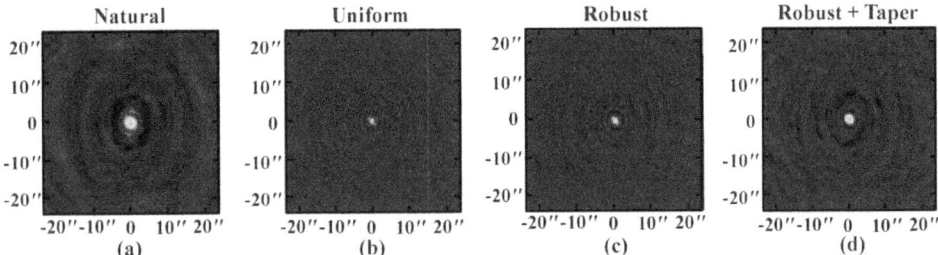

**FIGURE 8.9:** Beam shapes with different weights. (a) Natural weights, $\sigma_i^2 = 1$, HPBW $= 1.4'' \times 1.7''$. (b) Uniform weights, $\sigma_i^2 = 2.50$, HPBW $= 0.7'' \times 0.9''$. (c) Robust weights, $\sigma_i^2 = 1.64$, HPBW $= 0.8'' \times 1.0''$. (d) Combined tapered and robust weights, $\sigma_i^2 = 1.70$, HPBW $= 1.4'' \times 1.7''$.

Examples of beam shapes obtained with different weights and tapers applied to the same data of a spatial sensitivity function are shown in Fig. 8.9. Observe that: (*a*) With natural weights, the HPBW (half-power beam width) is maximum, and the noise variance is minimum. Side lobes are prominent especially near the main beam. (*b*) With uniform weights, the beam-width is minimum, but the noise variance is maximum. (*c*) With robust weights, the beam-width slightly increases, but the noise variance is relatively small. (*d*) Applying taper on robust weights, the beam-width is brought back to natural, but the noise variance has increased. The nearby side lobes, however, have been reduced.

### 8.4.3 Gridding Visibilities for FFT

Spacing between the visibility data samples on the *u-v* plane varies with the changing shape of the projected baselines due to the Earth's rotation. Hence, the spacings are not equal (see Fig. 8.3). However, for performing FFT, the sampled visibility data points should be equally spaced along both the *u-* and *v*-coordinates. Hence, the nonuniformly spaced visibilities $\mathcal{V}(u_i, v_i)$ are convolved with a continuous function and the resulting continuous function is again sampled with equal spacings to fill the *u-v* plane. Generally, the sampling is done by a comb function. Note that this interpolation process is only an approximation, but an unavoidable one.

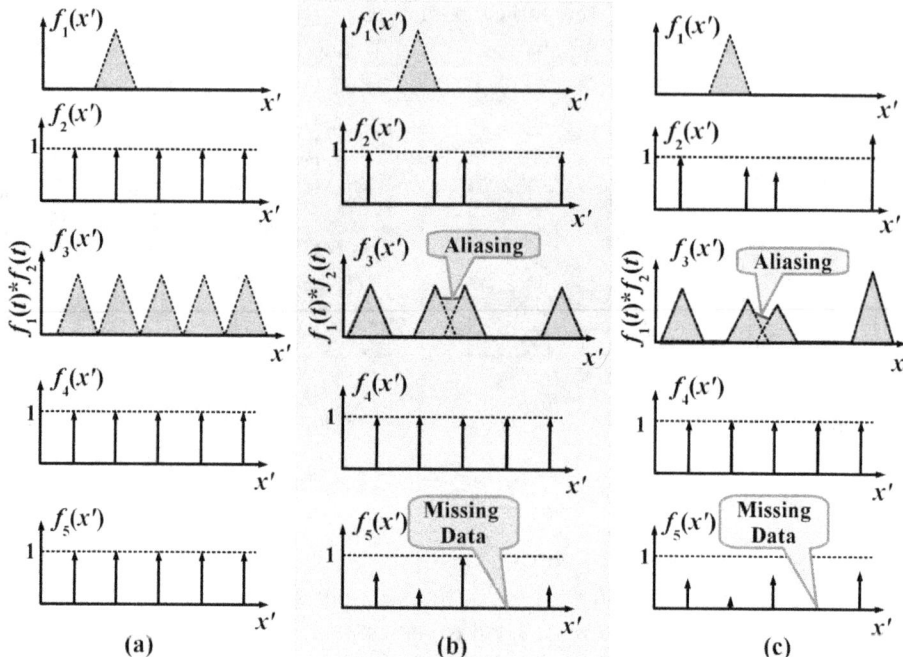

**FIGURE 8.10:** Illustration of the gridding process in one dimension. Here, $f_1(x')$ represents the convolving function, and $f_2(x')$ represents the amplitudes of the data points as a function of some variable $x'$. The function $f_3(x')$ is the convolved product of $f_1$ and $f_2$. The gridded function $f_5(x')$ is the result of sampling $f_3$ at equal intervals of $x'$ by another sampling function $f_4(x')$. (a) $f_2(x')$ has uniform magnitude and is equally spaced. The resulting function $f_5(x')$ resembles $f_2(x')$. (b) $f_2(x')$ has uniform magnitude but is not equally spaced. Due to aliasing, $f_5(x')$ does not resemble $f_2(x')$. Note that the amplitude of $f_5(x')$ at grid points is not uniform, and some data points are missing. (c) $f_2(x')$ has nonuniform magnitude and is also not equally spaced. Here also, $f_5(x')$ does not resemble the function $f_2(x')$. The amplitude of $f_5(x')$ is not uniform, and some data points are missing.

### 8.4.3.1 Gridding Approach

The gridding process in one dimension is illustrated graphically in Fig. 8.10 for three different cases ((a), (b), and (c)). The convolving function is $f_1(x')$. The data points amplitude $f_2(x')$ is shown as a function the variable $x'$. Convolution is performed between $f_1$ and $f_2$, and its result is $f_3(x')$. Gridding is performed next by resampling $f_3$ at equal intervals of $x'$ using the sampling function $f_4(x')$. The result of gridding is represented by $f_5(x')$. The three different cases are now explained.

**Case (a):** If the data points $f_2(x')$ are at equal intervals and have constant amplitude, then the resulting $f_5(x')$ resembles the data $f_2(x)'$, as shown in Fig. 8.10a.

**Case (b):** If the data points $f_2(x')$ are not at equal intervals but have a constant amplitude, then the resulting $f_5(x')$ does not resemble the data $f_2(x')$ due to aliasing, as shown in Fig. 8.10b. Note that some data points are missing and the amplitudes of $f_5(x')$ are not uniform.

**Case (c):** If the data points $f_2(x')$ are not equally spaced and also have varying amplitudes, then the resulting $f_5(x')$ does not resemble the data $f_2(x')$ due to aliasing, as shown in Fig. 8.10c. Here we also find some data points are missing, and the amplitudes of $f_5(x')$ are not uniform.

The same kinds of difficulties exist in two-dimensional gridding. We now describe the gridding process in two dimensions analogous to Fig. 8.10. Let $C(u, v)$ represent the convolving function. For simplicity, the most basic model (without a weight factor) of measured and calibrated visibility $V_{mc}(u, v)$ is used, which is shown in Eq. (8.41) to be a product of the true visibility $V(u, v)$ and the spatial sensitivity function $W(u, v)$ (see Fig. 7.11 of Chapter 7).

$$V_{mc}(u,v) = V(u,v)W(u,v) \tag{8.41}$$

The convolving function is shown as (8.42) and is analogous to Fig. 8.10. Similarly, the measured and calibrated visibilities are analogous to $f_2(u, v)$, as shown in Eq. (8.43). Also the convolution of the two functions would result in $f_3(u, v)$, as shown in Eq. (8.44).

$$f_1(u,v) = C(u,v) \tag{8.42}$$

$$f_2(u,v) = V_{mc}(u,v) = V(u,v)W(u,v) \tag{8.43}$$

$$f_3(u,v) = C(u,v) * [V(u,v)W(u,v)] \tag{8.44}$$

Let $\Delta u$ and $\Delta v$, respectively, represent the minimum (gridding) distances along $u$ and $v$. A comb function similar to Eq. (2.83) of Chapter 2 can be used in analogy to $f_4(u, v)$. This is shown in Eq. (8.45), where $p$ and $q$ are integers ranging from $-\infty$ to $+\infty$. Basically, $\delta_{II}(u - p\Delta u, v - q\Delta v)$ is a two-dimensional delta function shifted by $p\Delta u$ along the $u$-axis and by $q\Delta v$ along the $v$-axis. It looks like as a bed of nails of unit height having separations of $\Delta u$ and $\Delta v$, respectively, along the $u$- and $v$-axes.

$$f_4(u,v) = \mathrm{comb}(\Delta u, \Delta v) = \sum_{p=-\infty}^{\infty} \sum_{q=-\infty}^{\infty} \delta_{II}(u - p\,\Delta u,\ v - q\,\Delta v) \tag{8.45}$$

Sampling using the comb function results in $f_5(u, v)$, as shown in Eq. (8.46). This is the same as the gridded visibilities $V_{grd}(u, v)$.

$$f_5(u,v) = \mathrm{comb}(\Delta u, \Delta v)\Big[ C(u,v) * \{V(u,v)W(u,v)\} \Big] = V_{grd}(u,v) \tag{8.46}$$

### 8.4.3.2 Effects of Convolution Gridding on the Image

Let $I_{grd}(l, m)$ represent the image resulting from the gridded visibilities. It is obtained from the Fourier transform of $V_{grd}(u, v)$, as shown in Eq. (8.47). Here, $V_{grd}$, $V$, and $W$ represent the Fourier transforms of $V_{grd}$, $V$, and $W$, respectively.

$$I_{grd}(l,m) = \overline{V}_{grd}(l,m) = \left[\frac{\text{comb}(\Delta l, \Delta m)}{\Delta u \, \Delta v}\right] * \left[\overline{C}(l,m)\left\{\overline{V}(l,m) * \overline{W}(l,m)\right\}\right] \quad (8.47)$$

The details of the function comb $(\Delta l, \Delta m)$ are shown in Eq. (8.48), where $p'$ and $q'$ represent integers ranging from $-\infty$ to $+\infty$.

$$\text{comb}(\Delta l, \Delta m) = \left\{\begin{array}{c} \displaystyle\sum_{p'=-\infty}^{\infty}\sum_{q'=-\infty}^{\infty} \delta_{\text{II}}\left(u - p'\,\Delta l, \; v - q'\,\Delta m\right) \\[2mm] \text{where, } \Delta l = \dfrac{1}{\Delta u}, \; \Delta m = \dfrac{1}{\Delta v} \end{array}\right\} \quad (8.48)$$

Note that the intensity distribution of the sky is $I(l,m) = \overline{V}(l,m)$. Rearranging Eq. (8.47), we obtain $I_{grd}(l, m)$, as shown in Eq. (8.49).

$$I_{grd}(l,m) = \left[\frac{\text{comb}(\Delta l, \Delta m)}{\Delta u \, \Delta v}\right] * \left[\overline{C}(l,m)\left\{I(l,m) * \overline{W}(l,m)\right\}\right] \quad (8.49)$$

Most of the image information is hidden in the shape of $I(l,m) * \overline{W}(l,m)$, which is a convolution between the spatial intensity distribution of the sky $I(l, m)$ and the beam pattern $\overline{W}(l,m)$. Greater similarity between the shapes of $I(l,m) * \overline{W}(l,m)$ and $\overline{C}(l,m)\{I(l,m) * \overline{W}(l,m)\}$ produces less distortion in the image $I_{grd}(l, m)$. Hence $\overline{C}(l,m)$ should have a flat shape so that it uniformly scales $I(l,m) * \overline{W}(l,m)$ without changing its shape. Again, as the function $\overline{C}(l,m)\{I(l,m) * \overline{W}(l,m)\}$ grows wider along the $l$- and $m$-axes, the aliasing chances increase during convolution with comb $(\Delta l, \Delta m)$. Hence, ideally, $\overline{C}(l,m)$ should have a rectangular shape to restrict the width of $I(l,m) * \overline{W}(l,m)$.

### 8.4.3.3 Gridding Functions

If the function $C(u, v)$ can be generated from two independent functions $C_1(u)$ and $C_2(v)$ as shown in Eq. (8.50), then they become separable and can be applied with more ease.

$$C(u,v) = C_1(u)\, C_2(v) \quad (8.50)$$

Some one-dimensional convolving functions usable with Eq. (8.50) are described next. These functions are (*i*) rectangular, (*ii*) Gaussian, (*iii*) Gaussian sinc, and (*iv*) spheriodal.

(*i*) **Rectangular Function:** A rectangular function is described in Eq. (8.51). Using this, we may define $C_1(u)$ as shown in Eq. (8.52). We may also define $C_2(v)$ in a similar way by replacing $u$ with $v$.

$$C_{\text{Rect}}(x) = \begin{cases} 1, |x| \le \dfrac{1}{2} \\ 0, |x| > \dfrac{1}{2} \end{cases} w_i \qquad (8.51)$$

$$C_1(u) = \frac{1}{\Delta u} C_{\text{Rect}}(u) \qquad (8.52)$$

The response of this function is the sinc function given in Eq. (8.53). It is obtained from the Fourier transformation of (8.52).

$$\overline{C}_1(l) = \frac{\sin(\pi \Delta u \, l)}{\pi \Delta u \, l} = \text{sinc}(\pi \Delta u \, l) \qquad (8.53)$$

Fig. 8.11 shows the magnitude variations of $\overline{C}_1(l)$ using dark lines. The sinc function does not fall off sharply. It extends to infinity, resulting in a significant amount of aliasing. A Gaussian function may be a better choice. It is described next.

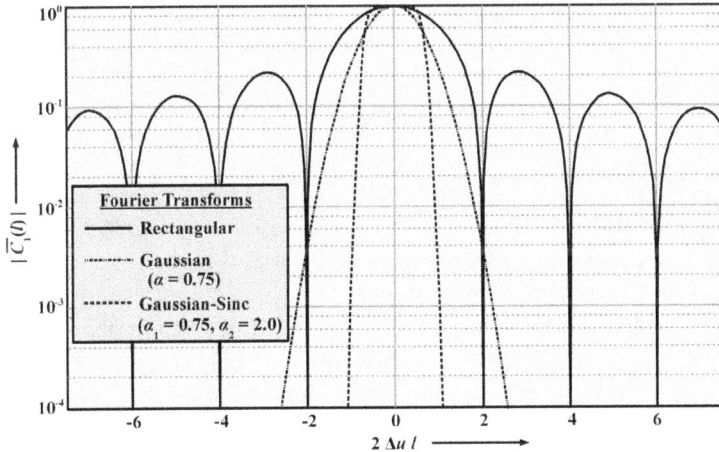

**FIGURE 8.11:** Magnitude of the Fourier transforms of some convolving functions. (Simulated by the author.)

(*ii*) **Gaussian Function:** A Gaussian function is shown in Eq. (8.54). Eq. (8.55) shows its Fourier transform.

$$C_1(u) = C_{\text{Gauss}}(u) = \frac{1}{\alpha \, \Delta u \sqrt{\pi}} \exp\left[ -\left( \frac{u}{\alpha \Delta u} \right)^2 \right] \qquad (8.54)$$

$$\bar{C}_1(l) = \exp\left[-\left(\pi\alpha\Delta u\,l\right)^2\right] \tag{8.55}$$

The desired width of the function $C_1(u)$ can be obtained by varying the value of $\alpha$. Small values of $\alpha$ make $C_1(u)$ more narrow. Hence, contributions come from only those data that are close to the grid points. When $\alpha$ is large, $C_1(l)$ can severely taper the map. Hence the preferred choice of $\alpha$ is an in-between value. Generally, $\alpha$ is given a value close to 0.75. Magnitude variations of $C_1(l)$ for $\alpha = 0.75$ are shown using dashed-dotted lines in Fig. 8.11.

Although a Gaussian function performs better than a rectangular function, it is not ideal for producing rectangular shapes in the spatial $(l, m)$ domain. A further-improved function may be a product of Gaussian and sinc functions. It is described next.

(*iii*) **Gaussian Sinc Function:** Eq. (8.56) shows a Gaussian sinc function. Its Fourier transform is shown in Eq. (8.57), which is a convolution between the Fourier transforms of the sinc and Gaussian functions. It may be seen as a convolution of a rectangular function with a Gaussian function in the spatial $(l, m)$ domain.

$$C_1(u) = C_{\text{GaussSinc}}(u) = \frac{\text{sinc}\left[\dfrac{\pi u}{\alpha_1 \Delta u}\right]\exp\left[-\left(\dfrac{u}{\alpha_2 \Delta u}\right)^2\right]}{\alpha_1 \Delta u} \tag{8.56}$$

$$\bar{C}_1(l) = \left[C_{\text{Rect}}(\alpha_1 \Delta u)\right]*\left[\left(\sqrt{\pi}\alpha_2 \Delta u\right)\exp\left\{-\left(\pi\alpha_2 \Delta u\,l\right)^2\right\}\right] \tag{8.57}$$

With $\alpha_1 = 0.75$ and $\alpha_2 = 2.0$, the magnitude of $\bar{C}_1(l)$ is plotted with a dashed line in Fig. 8.11. Note the central region around the peak. It is flat. The fall is sharp and resembles a distorted rectangular function in the spatial domain.

Convolving functions discussed so far may not be optimal choices for gridding. Based on the desired gridding properties, a completely different approach may be taken. The major requirements are: (*i*) $C_1(l)$ must fall off sharply, and (*ii*) it must be flat around the center. A class of functions called *spheroidal* functions may be a better choice. These are described next.

(*iv*) **Spheriodal Functions:** These are solutions to Laplace's equation. These are found by expressing the equation in spheroidal coordinates and then applying the technique of separation of variables.

The utility of a convolving function is judged from the placement and magnitude of the aliasing side lobes after the image is corrected for roll-off at the edges.

Let $\eta_{\text{sup}}$ represent a measure of the effectiveness of suppressing aliases. It may be described by Eq. (8.58).

$$\eta_{\text{sup}} = \frac{\iint_{\text{map}} \left[ \overline{C}(l,m) \right]^2 dl \, dm}{\int_{\infty}^{\infty} \int_{\infty}^{\infty} \left[ \overline{C}(l,m) \right]^2 dl \, dm}, \text{ where } \eta_{\text{sup}} \leq 1 \qquad (8.58)$$

Observe that the same integrands appear in both the numerator and denominator, but with different limits. The numerator gives the total sum of $[C(l,m)]^2$ within the map. The denominator gives the same over the whole (within and outside the map). Hence, $C(l, m)$ should be such that $\eta_{\text{sup}}$ is nearly unity. Instead of prolate spheroidal functions that are difficult to calculate, a simple approximation using a Kaiser-Bessel window function is shown in Eq. (8.59). Here, $I_0$ represents a zero-order modified Bessel function of the first kind.

$$C_1(u) = \frac{1}{L} I_0 \left[ B \sqrt{1 - \left( \frac{2u}{L} \right)^2} \right], u < \frac{L}{2} \qquad (8.59)$$

The quantities $B$ and $L$ are constants. They are used to determine the window characteristics in the inverse transformation as shown in Eq. (8.60).

$$\overline{C}_1(l) = \frac{\sin\left( \sqrt{\pi^2 L^2 l^2 - B^2} \right)}{\sqrt{\pi^2 L^2 l^2 - B^2}} \qquad (8.60)$$

In two dimensions, the function is obtained as $C(u, v) = C_1(u)C_2(v)$. The values of $B$ are chosen using Eq. (8.61). The first nulls of $\overline{C}_1(l)$ occur at $|l| = \pm\sqrt{l_1^2 + 1/L^2}$ and $|m| = \pm\sqrt{m_1^2 + 1/L^2}$ .

$$\left. \begin{array}{l} B_l = \pi L_l l_1 \\ B_m = \pi L_m m_1 \end{array} \right\} \qquad (8.61)$$

The inverse operation using a discrete Fourier transform will be performed on a set of discrete points in such a way that the points $|l| = l_1$ and $|m| = m_1$ become the folding points in the transform. To do this, we have $l_1 = 1/(2u_0)$ and $m_1 = 1/(2v_0)$. By making $\overline{C}_1(l)$ and $\overline{C}_2(m)$ cover the $L_l$, $L_m$ points in the Fourier domain, we get the values of $B_l$ and $B_m$, as shown in Eq. (8.62). The grid interval is kept unity.

$$\left. \begin{array}{l} B_l = \pi L_l / 2 \\ B_m = \pi L_m / 2 \end{array} \right\} \qquad (8.62)$$

The Kaiser-Bessel window function and its transform for representative values of the window function extent $L$ are shown in Fig. 8.12. It can be seen that as $L$ increases, the performance of the function improves, as measured by rejection beyond the folding frequency.

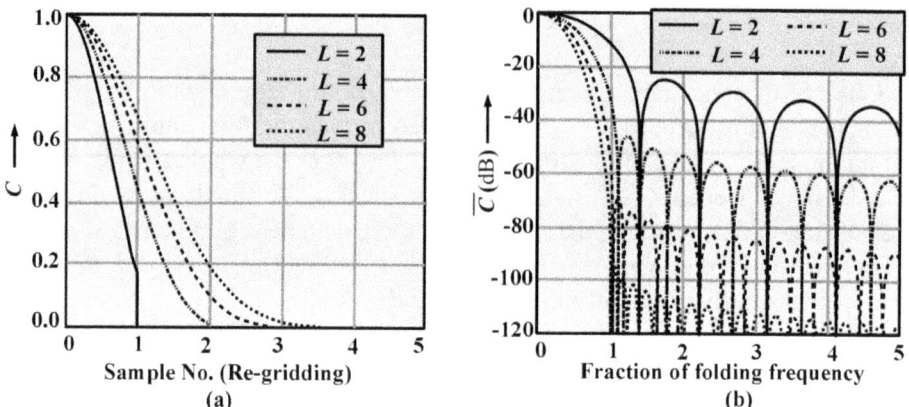

**FIGURE 8.12:** (a) Kaiser-Bessel convolving function with different extents $L$. (b) Fourier transform of (a).

### 8.4.4 Dirty Image from Inverse Fourier Transform

We know that weighting functions are used to obtain (*i*) the uniform effective area density, as well as (*ii*) tapering. We now denote them as $w_{den}$ and $w_{tap}$ for density and tapering, respectively. By substituting them in Eq. (8.22), we get Eq. (8.63).

$$\left.\begin{aligned}\mathcal{V}_{mcw}(u_i,v_i) &= \mathcal{V}_{mc}(u_iv_i)\Big[w_{den}(u,v)\,w_{tap}(u,v)\Big]\\&= \big[\mathcal{V}(u_i,v_i)W(u,v)\big]\Big[w_{den}(u,v)\,w_{tap}(u,v)\Big]\end{aligned}\right\} \tag{8.63}$$

The data contain measured and calibrated visibilities only at fixed locations $(u_i, v_i)$, where $i = 1, 2, 3, ...., N_{vis}$ and $N_{vis}$ represents the number of data pairs. After gridding (after convolving with $C(u, v)$ and sampling with comb $(\Delta u, \Delta v)$ and applying weights (density weighting and tapering), the measured, calibrated, weighted, and gridded visibility $V_{mcwg}(\Delta u, \Delta v)$ can be mathematically shown as Eq. (8.64).

$$\mathcal{V}_{mcwg}(\Delta u,\Delta v) = \Big[C(u,v)*\big\{\mathcal{V}(u_i,v_i)W(u,v)\,w_{den}\,w_{tap}\big\}\Big]comb(\Delta u,\Delta v) \tag{8.64}$$

We have experienced that visibilities appear in pairs that are symmetric about the origin of the $(u, v)$ axes. The actual number of visibilities obtained from the telescope is $N_{vis}$. An equal number of data points are generated that lie exactly on the opposite side on the $u$-$v$ plane. For imaging, this generated data is used together with the measured data. Thus, Eq. (8.64) takes the form shown in Eq. (8.65).

$$\mathcal{V}_{mcwg}(\Delta u, \Delta v) = \text{comb}(\Delta u, \Delta v) C(u,v) * \left[ \{\mathcal{V}(u_i, v_i) + \mathcal{V}(-u_i, -v_i)\} \right. \\ \left. \{W(u,v) w_{den}(u,v) w_{tap}(u,v)\} \right] \tag{8.65}$$

The dirty image $I_D(\Delta l, \Delta m)$ is obtained by taking the inverse FFT of Eq. (8.65) and reorganizing it, as shown in Eq. (8.66). Here, $I(l, m)$ represents the actual intensity distribution of the sky. Its relationship with the true visibility $\mathcal{V}(u, v)$ is shown in Eq. (8.67). The synthesized dirty beam $P_{syn}$ is given by Eq. (8.68).

$$I_D(\Delta l, \Delta m) = \left[ \frac{\text{comb}(\Delta l, \Delta m)}{\Delta u \, \Delta v} \right. \\ \left. * \left[ I(l,m) \{ \bar{C}(l,m) * \bar{W}(l,m) * \bar{w}_{den} * \bar{w}_{tap} \} \right] \right] \\ = \left[ \frac{\text{comb}(\Delta l, \Delta m)}{\Delta u \, \Delta v} \right] * \left[ I(l,m) P_{syn}(l,m) \right] \tag{8.66}$$

$$I(l,m) = \sum_{i=1}^{N_{vis}} \left[ \mathcal{V}(u_i, v_i) e^{j2\pi(lu_i + mv_i)} + \mathcal{V}(-u_i, -v_i) e^{-j2\pi(lu_i + mv_i)} \right] \tag{8.67}$$

$$P_{syn}(l,m) = \bar{C}(l,m) * \bar{W}(l,m) * \bar{w}_{den}(l,m) * \bar{w}_{tap}(l,m) \\ = \int_{-\infty}^{\infty} \int_{-\infty}^{\infty} C(u,v) W(u,v) w_{den}(u,v) w_{tap}(u,v) e^{j2\pi(lu+mv)} \, dl \, dm \tag{8.68}$$

Observe that the dirty beam $P_{syn}(l, m)$ has been expressed as the Fourier transformation of the product of (*i*) the convolution function $C(u, v) = C_1(u) C_2(v)$, (*ii*) the spatial sensitivity function $W(u, v)$, and (*ii*) the density and tapering weights ($w_{den}$ and $w_{tap}$).

Let us look at Eq. (8.65) once more. We define $\mathcal{V}_{grd}(\Delta u, \Delta v)$ as the convolved, sampled, and gridded true visibility on a uniform grid having a hole size $\Delta u \times \Delta v$, as shown in Eq. (8.69). Substituting this into Eq. (8.65), we obtain Eq. (8.70).

$$\mathcal{V}_{grd}(\Delta u, \Delta v) = \text{comb}(\Delta u, \Delta v) \left[ C(u,v) * \{\mathcal{V}(u_i, v_i) + \mathcal{V}(-u_i, -v_i)\} \right] \tag{8.69}$$

$$\mathcal{V}_{mcwg}(\Delta u, \Delta v) = \mathcal{V}_{grd}(\Delta u, \Delta v) W(u,v) w_{den}(u,v) w_{tap}(u,v) \tag{8.70}$$

By performing an inverse FFT operation on Eq. (8.70) the discrete image $I_D(l_i,$ $m_i)$ is obtained as Eq. (8.71). Here $P_{syn}(l_i, m_i)$ represents an equally spaced (gridded), weighted, discritized, synthesized dirty beam.

$$
\left.
\begin{aligned}
I_D(l_i, m_i) &= I_{grd}(l_i, m_i) * P_{syn}(l_i, m_i),\ l_i, m_i = 0, \pm1, \pm2. \\
&= \sum_{l_i'=-\infty}^{\infty} \sum_{m_i'=-\infty}^{\infty} \left[ I_{grd}(l_i, m_i)\, P_{syn}(l_i - l_i', m_i - m_i') \right]
\end{aligned}
\right\}
\qquad (8.71)
$$

### 8.4.5 Image Improvement by Deconvolution

However well the tapers and weights are applied, side lobes will always exist in the synthesized beam, resulting in artifacts on several regions of the dirty image. Hence, they must be removed by additional processes. This involves several operations of deconvolution and convolution on the dirty image. In the most simplified form, shown by Eq. (8.72), the dirty image $I_D$ is a convolution between the true image $I$ and the dirty beam $P_D$.

$$
I_D = P_D * I
\qquad (8.72)
$$

To recover the image $I$, a deconvolution process of the form shown by Eq. (8.73) may be used, where the operators $\mathcal{F}$ and $\mathcal{F}^{-1}$ represent the Fourier and inverse Fourier transforms, respectively.

$$
I = \mathcal{F}^{-1}\left[ \frac{\mathcal{F}[I_D]}{\mathcal{F}[P_D]} \right]
\qquad (8.73)
$$

This process is known as *linear* deconvolution. This method gives a good understanding of the noise properties, and it is computationally cheap. It is, however, not suitable for radio interferometry because the u-v plane has a high population of zeros due to the limited baselines. It is almost impossible to use Eq. (8.73) for image inversion. The alternative solution is to use nonlinear deconvolution algorithms.

In nonlinear deconvolution, the unmeasured parts of the Fourier plane are filled with a plausible distribution based on the properties of the real sky. The logic comes from the appearance of the sky, which does not show any features of the dirty beam, such as rings, spokes, negative regions, or finite extent. Two currently popularly algorithms are (*i*) *CLEAN*, and (*ii*) *MEM*, and they are described below.

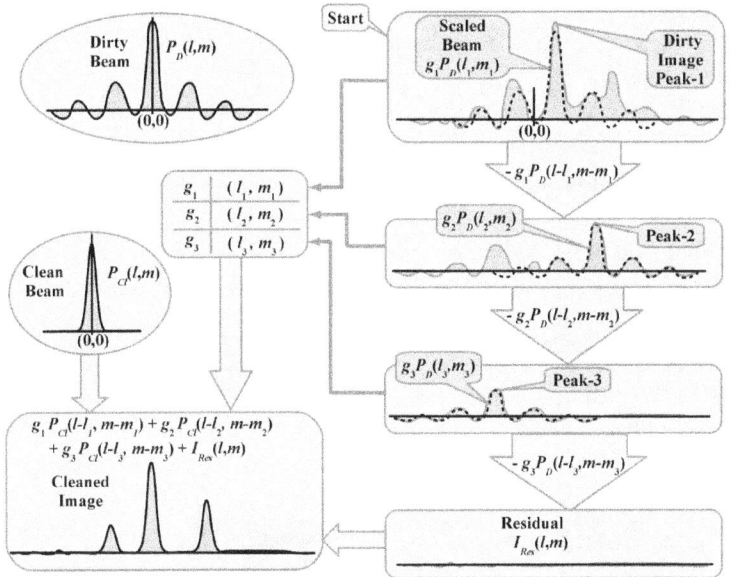

**FIGURE 8.13:** The basic CLEAN algorithm. A dirty beam $P_D$ is scaled by $g_1$ and shifted to $l_1$, $m_1$ on the peak intensity point in the dirty image and subtracted. Values of $g_1$ and $l_1$, $m_1$ are noted. The same operations are repeated until the peak residual from the dirty image is below a specified value. A Gaussian clean beam $P_{cl}$ is convolved with the stored values, and the residuals are added to get the restored CLEAN image.

### 8.4.5.1 The CLEAN Algorithm

In the CLEAN algorithm, a radio source is treated as a composition of several point sources. The sky is treated as mostly empty, but with occasional peaks. The response of a telescope when focussed to a point source is known as the *point spread function* or PSF. The PSF is the synthesized beam in this case. The image is treated as being composed of several point sources and each of these sources can be constructed with an individual PSF. Iterations involve locating these positions, finding their strength, and applying attenuation, as is illustrated in Fig. 8.13 and described in the following steps:

(*i*) In every iteration, the absolute strength and position ($l_0$, $m_0$) of the most brilliant peak in the dirty image $I_D(l, m)$ are determined.

(*ii*) The dirty beam $P_{syn}(l, m)$ is multiplied by a gain factor $g$, which usually has a value lying between 0.1 and 0.25. The result is a weighted dirty beam $gP_{syn}(l, m)$.

(*iii*) The weighted dirty beam $gP_{syn}(l, m)$ is shifted to $l_0$, $m_0$ for alignment with the peak of the dirty map, and then subtracted from it. In the process, the side lobes are also subtracted.

(*iv*) The amplitude and position of the removed component are recorded separately in the CLEAN model. This is done by introducing a delta function (having the same gain) at that location.

(*v*) Iteration steps (*ii*), (*iii*), and (*iv*) are repeated until the remaining peak goes below a specified value.

(*vi*) The CLEAN model obtained from the accumulation of point sources by the iterations is convolved with an ideal beam having no side lobes (usually a Gaussian beam) that produces the CLEAN image.

(*vii*) Residuals from the subtractions in the dirty image are then added to the CLEAN image.

The CLEAN algorithm can be easily understood from Fig. 8.13. Effectively, it removes the image contents produced by the side lobes. A radio image to which CLEAN has been applied is shown in Fig. 8.14. The quality of a CLEAN image depends on the user.

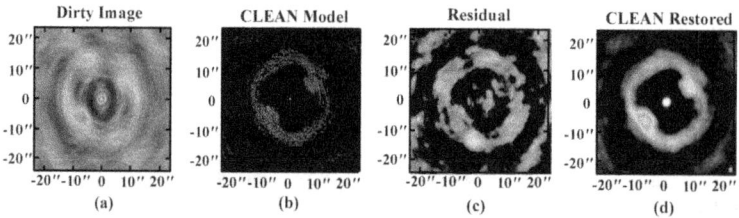

**FIGURE 8.14:** Application results of the CLEAN algorithm. (a) Original dirty image. (b) Model of point source chosen for applying CLEAN. (c) Residuals from the image after applying CLEAN. (d) Restored CLEAN image.

### 8.4.5.2 The MEM (Maximum Entropy Method) Algorithm

Unlike the entropy described by information theory, which is a measure of the uncertainty associated with a random variable, the entropy described for use in imaging is different. Let $I_M(l, m)$ represent a sky intensity distribution obtained using the MEM, and let $F(I_M)$ be its entropy, as shown in Eq. (8.74). Note that $F(I_M)$ has also been defined in a slightly different way by other users.

$$F(I_M) = -\int I_M(l,m)\log\big(I_M(l,m)\big)\,dl\,dm \qquad (8.74)$$

The basic idea is to maximize $F(I_M)$ of the optimized image. An initial model of the source is chosen with the assumption that the sky is uniform and positive. The process starts with this model. Let $k$ represent all the discrete visibility points in the data. Let the measured visibilities be represented as $V_k^{\text{meas}}(u_k, v_k)$, and let $V_k^{\text{model}}(u_k, v_k)$ represent those from the model. Let $\zeta^2$ be a measure of their mean square differences, as shown in Eq. (8.75), where $\sigma_k^2$ represents the noise variance.

$$\xi^2 = \sum_k \frac{\left| V_k^{\text{meas}} - V_k^{\text{model}} \right|^2}{\sigma_k^2} \tag{8.75}$$

Many optimization algorithms have been used. Eq. (8.76) shows one such expression, where $\alpha$ and $\beta$ are Lagrange multipliers, and $S_{\text{model}}$ represents the model's total flux-density. To get an optimized solution, the quantity $J$ expressed in Eq. (8.76) should be maximized.

$$J = F(I_M) - \alpha \xi^2 - \beta S_{\text{model}} \tag{8.76}$$

The values of $\alpha$ and $\beta$ are adjusted during the iteration so that $\xi^2$ and $S_{\text{model}}$ are the same as their expected values. Details of a radio image with MEM applied are shown in Fig. 8.15.

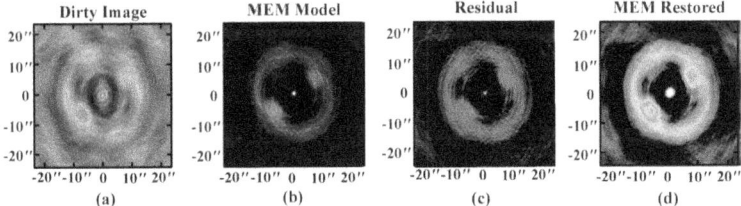

**FIGURE 8.15:** An example of an applicaiton of the MEM. (a) The original dirty image. (b) Model point sources chosen for applying the MEM. (c) Residual of the image after applying the MEM. (d) The MEM restored image.

### 8.4.5.3 CLEAN vs. MEM vs. Other Methods

The CLEAN method works well on compact sources, while the MEM is more useful for extended sources. Unlike the MEM, CLEAN is easy to understand and robust. However, when applied to extended sources, it introduces corrugations in the deconvolved image if there are strong side lobes in the dirty beam. Short baselines give more information about the source intensity if it is large or extended. The absence of short baselines can create a negative bowl (explained later) because the zero level in the image varies with position. This makes CLEAN unsuitable for large structures. The MEM handles these very well.

Some of these problems are solved using the *Multi-Resolution CLEAN* (MRC) technique, in which the image is smoothed to a lower resolution, and a difference is created between the dirty image and the smoothed image. These are separately deconvolved using the smoothed and difference beams, and the results are related so that the fine-scale structures are retained.

In another approach, dirty images of different resolution are made using wavelet transforms. These are then separately deconvolved and restored, as in MRC. However, this has a relatively rigorous framework compared to MRC.

In another approach for deconvolution, a linear system of equations $AX = B$ can be formed, where $A$ represents the samples of the dirty beam, and $B$ represents the samples of the dirty image, while $X$ represents the components of the reconstructed image. Solution methods from linear algebra can be adopted for finding $X$. However, because $A$ is singular, additional information is required for a solution. Methods like *nonnegative least squares* (NNLS) may be used for a solution. Unlike iterated deconvolutions, this approach gives a direct solution.

## 8.5 IMAGING USING WIDE FREQUENCY BANDS

In order to use wide bandwidths in imaging, we have to understand their functionality. We first try to understand the effects of wide bandwidths.

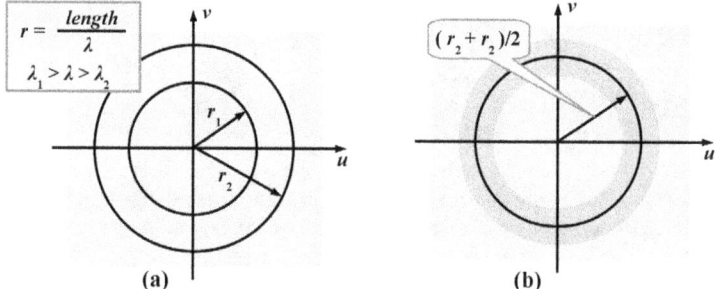

**FIGURE 8.16:** Effect of the different frequency components on the projected baseline length across the *u-v* plane. (a) Two different baseline lengths $r_1$ and $r_2$ are formed as a result of observing a common source at two different wavelengths, $\lambda_1$ and $\lambda_2$, respectively. (b) The effect of integration over the entire band results in merging the data and placing it on the average baseline line $(r_1 + r_2)/2$.

### 8.5.1 Effects of Wide System Bandwidth

Let us observe a source at two wavelengths $\lambda_1$ and $\lambda_2$ that have a common baseline distance $d$. Let the source be at the CNP such that the data appears as circles on the *u-v* plane. The radii of these circles will be $r_1 = d/\lambda_1$ and $r_2 = d/\lambda_2$, as shown in Fig. 8.16a. Note that for the two frequencies, the baselines are different. Let the system have a nonzero bandwidth for which the minimum frequency is $c/\lambda_1$, the highest frequency is $c/\lambda_2$, and $c$ is the speed of light. Hence the visibility will contain the information from all the frequencies lying between $c/\lambda_1$ and $c/\lambda_2$, which should actually appear as a ring of nonzero width on the *u-v* plane, as shown in Fig. 8.16b. However, depending on the hardware configuration, the correlator may produce a single integrated output corresponding to the whole band. Hence, all the individual frequency information is lost because it is combined as a whole. Instead of a ring, we therefore obtain a single circle of radius $(r_1 + r_2)/2$ on the *u-v* plane. In other words, the visibility function is smeared. This effect, known as *bandwidth smearing*, can severely affect the image, especially for large system bandwidths.

### 8.5.2 Multi-Frequency Synthesis

If the primary bandwidth of the system is divided into several narrow bands, and correlations are separately done for each of these, then many different visibilities result corresponding to each subband. Their positions on the *u-v* plane are also different, and the smearing of the visibilities can be reduced or avoided. This, however, requires modification of the hardware.

The above method is known as *multi-frequency synthesis*. It helps to populate the *u-v* plane with more data points, thereby improving the resolution of the image within certain limits.

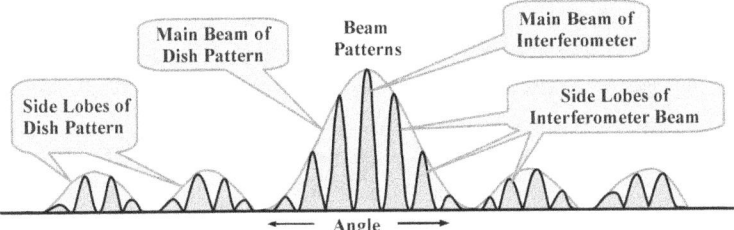

**FIGURE 8.17:** The primary beam of a single dish is larger than the interferometer beam. The sensitivities in both cases decrease with increasing angular distance from the center.

## 8.6   WIDE-FIELD IMAGING

The field of view of the primary beam of a dish antenna is determined by its physical aperture area or diameter. The case is similar for a synthesized aperture using visibilities obtained from an interferometer array where the main beam is determined from the *u-v* coverage. For an interferometer, the angular resolution depends on the projected distance of separation between the two antennas. In all three cases, the sensitivity decreases with an increase in angular distance from the center of the field of view. Fig. 8.17 shows this for an interferometer and a single dish.

### 8.6.1 Absence of Short Baselines

The minimum possible distance $d$ between two antennas of an interferometer is equal to the diameter $D$ of a single dish (assuming both the dishes have the same diameter). Hence the smallest baseline possible in any interferometer array is $D$.

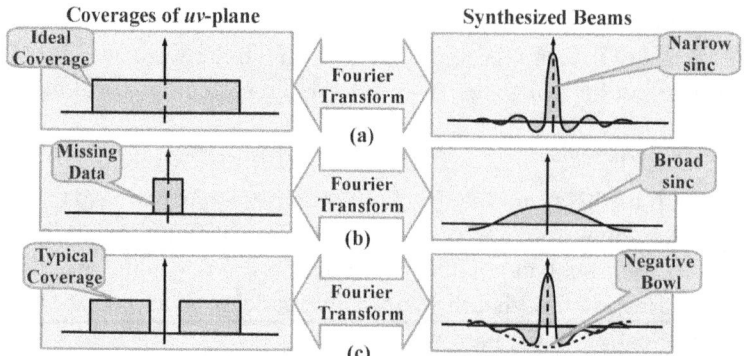

**FIGURE 8.18:** Problems due to the absence of short baselines. (a) Ideal *u-v* coverage produces a synthesized beam like a narrow sinc function. (b) Missing small baseline data and its synthesized beam response. (c) Typical *u-v* coverage. The synthesized beam is a wide sinc subtracted from a narrow sinc.

The expected *u-v* plane coverage from an interferometer array for the full observation time looks like a wide pulse in one dimension, as shown in Fig. 8.18a, where the width of the pulse is equal to the largest baseline in the array. The synthesized beam from this data would be a narrow sinc function.

The spatial frequency components produced from a single dish is from 0 to $\Delta/\lambda$. However, they are not separable and appear as a single value. The smallest interferometer (separated by $D$) produces spatial frequencies ranging from 0 to $\Delta/\lambda$, which should look like a narrow pulse, as shown in Fig. 8.18b. Because they are inseparable, they appear at the edges of the pulse. In other words, the pulse is missing. Hence the broad sinc component produced by this in the synthesized beam is also missing.

The actual interferometer *u-v* coverage looks like Fig. 8.18c. Note that the central part (short spacing) data is missing. The effects are undesirable as the synthesized beam contains a modulated sinc function. The narrow sinc function is modulated by the wide sinc function. Observe that in the central region, the beam has been pulled in the negative direction. This effect is known as a *negative bowl*. This problem is also known as the *short-spacing problem*.

### 8.6.1.1 Solution to the Short-Spacing Problem

As said before, the spatial frequency components available from a dish of diameter $D$ are from zero to $\Delta/\lambda$. If in some way we are able to separate these components so that they actually cover the *u-v* plane with spatial frequencies from zero to $\Delta/\lambda$, we may combine this data with that from an interferometer array. From a practical point of view, a separate dish having a diameter greater the $D$ is required for the reasons described below.

If $d_{\lambda(\min)}$ is the shortest baseline of an interferometer array, then the theoretical minimum diameter of the additional single dish required is $D_{sd} = d_{\lambda(\min)}$. However, we are aware that practical dish illuminations by antenna feeds are nonuniform and reduce at the edges of the dishes (see Fig. 4.4b of Chapter 4). This results in attenuation of the higher spatial frequency components (towards the edges of the dishes). Hence for all practical purposes, a larger diameter dish is chosen, usually with a diameter $D_{sd} = 2d_{\lambda(\min)}$. This gives an overlap region between the single-dish data and the interferometer data. This overlap is useful for cross-calibrations between the two.

We now describe (*i*) how to use a single dish as an interferometer for (*ii*) generating short-baseline data, and (*iii*) combine them with data from an interferometer array.

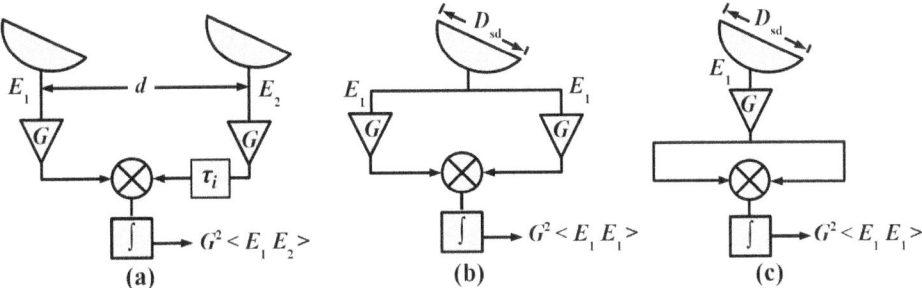

**FIGURE 8.19:** Data analogy between a multiplicative interferometer and a single dish.

(*i*) **Single Dish as Multiplicative Interferometer:** The single-dish data may be visualized as an autocorrelation component when it is combined with interferometer data. This can be understood from Fig. 8.19. A two-element interferometer is shown in Fig. 8.19a , and its output is $G_2 < E_1 E_2 >$, where $G$ is the gain of each amplifier, and $E_1$ and $E_2$ are the antenna signal voltages. The symbol $<>$ represents correlated data with time integration. The instrumental time delay for phase adjustment is represented by $\tau_i$. As a special case, where $E_1 = E_2$, the same configuration takes the form shown in Fig. 8.19b. This may be further simplified as shown in Fig. 8.19c.

(*ii*) **Generating *u,v* data from a single dish:** Single dishes can be used to observe extended objects whose angular size is greater than $1.22\lambda/D_{sd}$, where $D_{sd}$ is the diameter of the dish. Let us consider a square region of the sky containing an extended source, as shown in Fig. 8.20a. Divide this region by the beam width of the single dish so that equally spaced squares are created. The center of each of square is marked with a "+" sign. The dish is now pointed at each of these locations, and

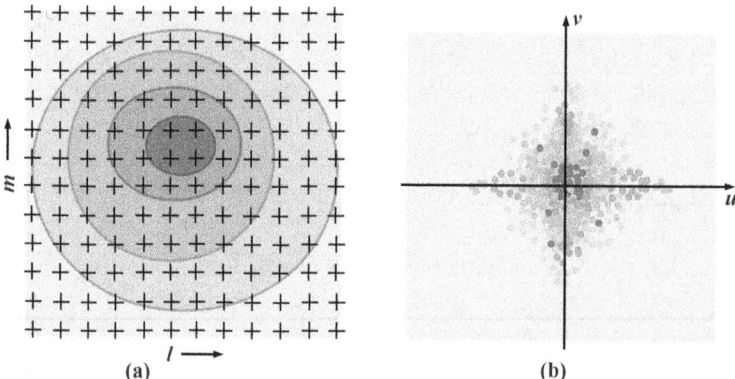

(a)  (b)

**FIGURE 8.20:** (a) Scanning an extended source using a single dish at discrete locations (+). (b) Pseudo $u, v$ data generated from scanned data.

data is taken. Hence we have scanned the extended object at angular intervals of the beam width of the single dish based on the sampling theorem of the observing angle (see Section 2.11). The Fourier transform of this data generates visibility components on the $u$-$v$ plane, as shown in Fig. 8.20b. Note that the data reside in the midsection of the $u$-$v$ plane, which means they are the lower spatial-frequency components.

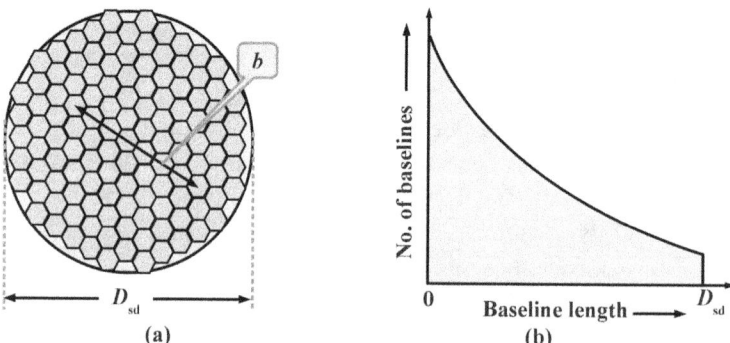

(a)  (b)

**FIGURE 8.21:** (a) The dish aperture may be thought to be composed of a finite number of small apertures. (b) Number distribution of these baselines.

In order to have a better understanding, let us divide the aperture area of the single dish into a finite number of small areas, as shown in Fig. 8.21a. Each of these areas can be treated as the aperture of a small independent antenna. Interferometer-like baselines can be formed between any two such elements. This is shown with a double-headed arrow in the figure. The maximum size of the baselines will be the diameter $D_{sd}$ of the single dish. The number distribution of these baselines is shown in Fig. 8.21b.

If $P_{sd}(l', m')$ represents the beam pattern of the single dish, then the observed sky brightness $I_{sd}(l, m)$ using the dish can be shown as Eq. (8.77), where $I(l, m)$ is the true intensity distribution on the sky. The dummy coordinates $(l', m')$ are used to denote the movement of the antenna while scanning the sky at discrete angular intervals.

$$I_{sd}(l,m) = I(l,m) * P_{sd}(l',m') \qquad (8.77)$$

Applying a Fourier transformation to Eq. (8.77) we obtain the observed single-dish visibilities $V_{osd}(u, v)$, as shown in Eq. (8.78), where $V_{sd}(u, v)$ are the true visibilities, and $W_{sd}(u, v)$ is the spatial sensitivity function of the single dish.

$$V_{osd}(u,v) = V_{sd}(u,v)W_{sd}(u,v) \qquad (8.78)$$

Note that determination of $I(l, m)$ from $I_{sd}(l, m)$ has to be done by deconvolution. However, no interpolations are required on $V_{osd}(u, v)$ because it is continuous across the $u$-$v$ plane.

**(*iii*) Combining Single Dish Data with Interferometer Data:** We present the outline for combining the visibility data obtained from a single dish with that from an interferometer array. As already stated, the $u$-$v$ data obtained from an interferometer always contains a hole in the center, as shown in Fig. 8.22a. The $u$, $v$ data obtained using a single dish covers the central region of the $u$-$v$ plane, as shown in Fig. 8.22b. The size of the single dish is chosen such that there is an overlapping region between the two data sets, as shown in Fig. 8.22c. Relative calibrations between the two are made in this region.

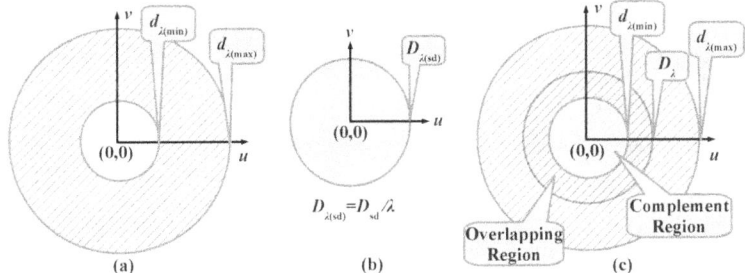

FIGURE 8.22: (a) Interferometer data distribution across the $u$-$v$ plane. The minimum and maximum projected baselines are shown as $d_{\lambda(min)}$ and $d_{\lambda(max)}$, respectively. (b) Single-dish data distribution. (c) Calibration between the two are performed in the overlapping region.

Let $K_{cal}$ be the calibration ratio of the flux densities of an unresolved source in the maps obtained by an interferometer and a single dish, as shown in Eq. (8.79), where $S_{int}$ and $S_{sd}$, respectively, represent the flux densities obtained from an interferometer and a single dish.

$$K_{cal} = \frac{S_{int}}{S_{sd}} \tag{8.79}$$

For a source having a brightness $I$, the intensities measured by both the single dish ($I_{sd}$) and the interferometer ($I_{int}$) should be the same within this region. The calibration ratio $K_{cal}$ can now be shown as Eq. (8.80).

$$K_{cal} = \frac{I_{int}}{I_{sd}} \tag{8.80}$$

The beam widths of an interferometer ($\Omega_{int}$) and a single dish ($\Omega_{sd}$) are not the same. Hence, for extended sources, the quantities $I_{sd}$ and $I_{int}$ are usually expressed in units of flux density per beam width (Jansky/beam) instead of flux density per solid angle (Jansky/steradians). The resolution ratio $\alpha$ between the two is calculated as shown in Eq. (8.81).

$$\alpha = \frac{\Omega_{int}}{\Omega_{sd}} \tag{8.81}$$

To find the value of $K_{cal}$, the following steps are performed:

(*i*) Using $\alpha$, scale the single-dish data, and eliminate the difference in brightness due to different beam widths.

(*ii*) Apply Fourier transformations to the images obtained from both the single dish and the interferometer. One needs to be careful about the edge effects. These may, however, be reduced by applying a smooth gradual tapering of the image intensities towards the edges.

(*iii*) Obtain the true visibilities $V_{sd}$ by deconvolving the single-dish data. This is done by dividing the observed visibilities $V_{osd}$ by the spatial sensitivity function $W_{sd}$ (see Eq. (8.78)). Note that $W_{sd}$ is the Fourier transform of the single-dish beam ($P_{sd}$). Hence, sufficient knowledge about the beam of the single dish is required before performing this step.

(*iv*) Finally, compare the visibilities in the overlapping region of spatial frequencies in the *u-v* plane. This requires sufficient width of the overlapping region. Assuming a Gaussian tapered illumination of the dish by the antenna feed, and considering a cut-off level of 0.2, we may express the minimum diameter $D_{sd(min)}$ as shown in Eq. (8.82), where $d_{min}$ is the shortest projected baseline spacing distance in the interferometer data. The SNR in visibilities derived using a single dish within the overlapping region should be comparable to those obtained from the interferometer, otherwise, the combined map may show a degradation of quality. After the intercalibrations are complete, the data may be combined.

$$D_{sd(min)} > 1.5\,d_{min} \tag{8.82}$$

### 8.6.1.2 Mosaicing

Mosaicing is a technique by which a portion of the sky larger than the primary beam of the interferometer array elements can be mapped. Consider a small square region of the sky having a nonuniform temperature distribution, as shown in Fig. 8.23a. Divide this region by the primary beam width of the interferometer elements so that equally spaced squares are created (the centers are marked with a "+" sign). The array is pointed to each of these locations in turn, and data is collected. This principle is based on the sampling theorem of the observing angle (see Section 2.11 of Chapter 2). For observing an extended source or arbitrary shape, the scanning can be done as shown in Fig. 8.23b, where the circles represent the extent of the primary beams. The entire process of mosaicing can be described in three steps, as follows.

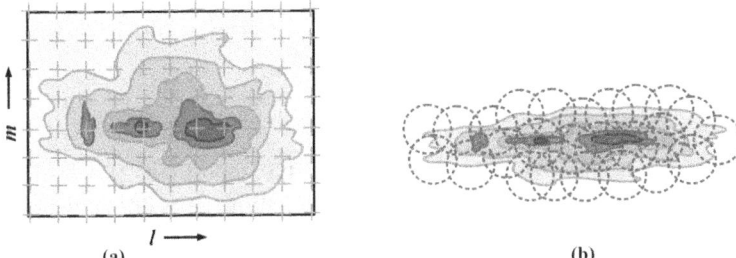

(a)                                                                 (b)

**FIGURE 8.23:** Scanning for mosaicing. (a) Scanning a portion of the sky. (b) Selective scanning of an extended source having an arbitrary shape.

(*i*) Obtain the visibilities for an appropriate series of pointing centers.

(*ii*) Produce a series of maps by independently reducing the data. Each of these maps cover approximately the primary beam solid angle of an antenna element. Nonlinear deconvolution algorithms like CLEAN or MEM may be used for removing the effects of side lobes in the individual images.

(*iii*) Combine these maps together like mosaics to obtain the complete scanned view.

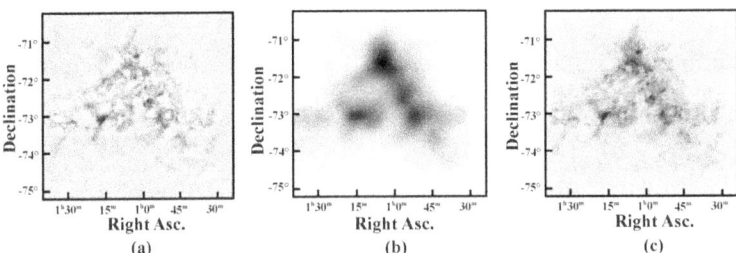

(a)                                  (b)                                  (c)

**FIGURE 8.24:** Some images obtained by mosaicing. The dark regions represents higher intensities. (a) A mosaiced image obtained by using only interferometer data. (b) A single-dish image. (c) Image formed by combining the single-dish and interferometer data. (Stanimirovic et al., 1999).

Fig. 8.24 shows three images of a large field obtained by mosaicing. The first image is constructed from interferometer data alone. The second one is obtained using a single dish. The third image is formed from combined data (interferometer and single dish). The darker regions represent higher intensities. Observe that the second image provides the information lost in the negative bowl. However, finer structures are not resolved. The interferometer image, on the other hand, provides only the finer details. Hence the combined image contains information form both sources, resulting in a more accurate image.

### 8.6.2 Wide-Field Imaging with Noncoplanar Baselines

We have already seen that by mosaicing we can cover a large field of view in the sky. There are also other methods by which one can image a wide region of the sky. Before we go into detail, we need to understand some concepts related to the antenna array, such as planar and nonplanar baselines. We also need to understand the differences, and why the conventional imaging method may not be a good choice for wide-field imaging.

#### *8.6.2.1 Coplanar vs. Noncoplanar Baselines*

By coplanar baselines, we mean baselines that remain in a fixed plane during the entire observation period. Note that we are talking only about baselines and not projected baselines. If the antennas are positioned strictly along east-west lines, only then will we obtain coplanar baselines for the entire observation period irrespective of the source position in the sky. One can easily imagine this by looking at Fig. 7.5 under Section 7.4.2 of Chapter 7, where a single interferometer observing a source at the CNP is discussed. If the source is located in any other position, the $u$-$v$ plane will no longer remain parallel to the equatorial plane, and an ellipse will be traced instead of a circle. However, the actual baselines remain coplanar.

The disadvantages of using east-west baselines is that 2-D (two-dimensional) Fourier mapping becomes difficult if the source is near the celestial equator (see Section 7.4.4 of Chapter 7). Hence, most interferometer arrays are made such that there exists some baseline component along the north-south direction. However, this makes the baseline noncoplanar because, due to the Earth's rotation, it does not remain on a fixed plane as a function of time.

#### *8.6.2.2 Problem With the w Component*

Fig. 8.25a shows an interferometer pointed to a source at the zenith. The projected baseline is co-planar to the baseline because both lie on the $u$-$v$ plane. For simplicity, we have not shown the $v$-axis, which is perpendicular to the page. Let the source be extended in a way that is measurable using $\theta$ or $l$. We may use $l = \sin\theta$,

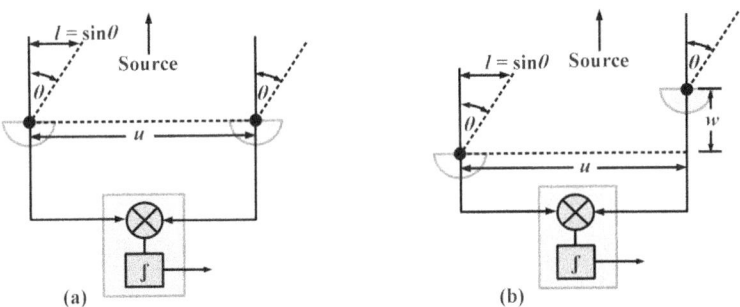

**FIGURE 8.25:** (a) An interferometer pointed to an extended source at the zenith. It has no $w$-component. (b) An interferometer pointed to an extended source at the zenith, but with a $w$-component due to a difference in heights.

as shown in the figure. Hence, $n = \sin\theta$. The $w$-component is absent in this case because the plane of the antennas coincides with the plane of the projected baseline. If we consider the phase of the visibility as zero towards the source, then its phase $\phi$ at an angle $\theta$ is given by Eq. (8.83).

$$\phi = 2\pi u l \tag{8.83}$$

Now consider a slightly different case, where the baseline is not on the $u$-$v$ plane because one antenna is at a higher altitude. It has a $w$-component as shown in Fig. 8.25b. In this case, the phase $\phi$ at an angle $\theta$ from the zenith is given by Eq. (8.84).

$$\phi = 2\pi\left[ul + w(n-1)\right] \tag{8.84}$$

Note that $\phi$ is not the same except when $\theta$ is zero. The additional component $\delta\phi = w(n-1)$ is dependent both on $w$ and $\theta$. The correct phase of the source extent is given by Eq. (8.83), and this means that we should have baselines coplanar with the $u$-$v$ plane. This situation seldom occurs because there exist north-south baseline components in most radio arrays, and the antennas are also usually not at the same altitude.

### 8.6.2.3 Modification of Conventional Imaging Process

We have seen that the entire imaging process discussed so far is based on the van Cittert-Zernike equation shown in Eq. (8.2), which is an approximation of the original equation linking the visibilities with intensity (see Eq. (8.1)). The component $n$ was approximated to zero and eliminated. This is valid for small fields of observation. The approximated van Cittert-Zernike equation does not hold if the source or field of view is larger than the distortionless imaging diameter, and the image obtained will be distorted. Under such circumstances, we have to use Eq. (8.1), which is reproduced below as Eq. (8.85).

$$\mathcal{V}(u,v,w) = \int_{-\infty}^{\infty}\int_{-\infty}^{\infty} \frac{P_n(l,m)\,I(l,m)}{n}\,e^{-j2\pi[ul+vm+w(n-1)]}dl\,dm \qquad (8.85)$$

Note that this general relationship is not a Fourier transformation and thus does not have an immediate inversion to the 2-D brightness distribution. However, it may be considered as a 3-D (three-dimensional) Fourier transformation of $\mathcal{V}(u, v, w)$, giving a 3-D image volume $F(u, v, w)$ that may be somehow related to the intensity $I(l, m)$. The expression of $F(u, v, w)$ is shown in Eq. (8.86).

$$\left.\begin{array}{l} F(u,v,w) = \iiint \mathcal{V}_0(u,v,w)\exp\left[j2\pi\left(ul+vm+wn\right)\right]du\,dv\,dw \\[2mm] \text{where,}\quad \mathcal{V}_0(u,v,w) = \exp\left(-j2\pi w\right)\mathcal{V}(u,v,w) \end{array}\right\} \quad (8.86)$$

The modified visibility $\mathcal{V}_0(u, v, w)$ is actually the observed visibility without phase compensation for the delay distance $w$. It is the visibility with reference to the vertical direction. The relation between the 3-D image volume $F(u, v, w)$ and the intensity distribution $I(l, m)$ is given by Eq. (8.87).

$$F(u,v,w) = \frac{I(l,m)}{\sqrt{1-l^2-m^2}}\,\delta\!\left(l^2+m^2+n^2-1\right) \qquad (8.87)$$

Because $l^2 + m^2 + n^2 = 1$, the delta function $\delta(l^2 + m^2 + n^2 - 1)$ exists on a spherical surface of unit radius. This implies $F(u, v, w)$ is empty everywhere except on this surface of the celestial sphere. Hence the correct intensity distribution is $I(l, m)/n$, which is the value of $F(u, v, w)$ on this unit surface. If $\delta_0$ is the reference declination and $\Delta\alpha$ is the offset from right ascension, we may express the $l$, $m$, $n$ coordinates as shown in Eq. (8.88).

$$\left.\begin{array}{l} l = \cos\delta \sin\Delta\alpha \\[2mm] m = \sin\delta \cos\delta_0 - \cos\delta \sin\delta_0 \cos\Delta\alpha \\[2mm] n = \sin\delta \sin\delta_0 + \cos\delta \cos\delta_0 \cos\Delta\alpha \end{array}\right\} \qquad (8.88)$$

The entire imaging process is now the same as the conventional imaging method. The only difference is we have to deal in 3-D with the effects of finite sampling, dirty beam (called dirty *ball* beam), maximum and minimum baselines, deconvolution, etc. However, these are not straightforward and consume a great deal of computing power.

## 8.7  OBSERVING SPECTRAL LINES

Spectral-line emissions occur under many different circumstances that are useful in astronomy. For example, the hydrogen atom generates a 21 cm line at a frequency of 1420.405 MHz due to a transition in hyperfine levels of its ground state. The line widths are a result of Doppler shifts due to different motions of particles within the gas-like thermal motion. The gas volume itself may be moving away from the observer. Hence, the spectral lines are broadened and their observed frequencies are different from their rest frequency (emission frequency). Additionally, the spectral line also has a natural width of its own imposed by the uncertainty principle, which, however, is more or less dominated by the processes mentioned before. Systematic rotation, gas clouds, and expansions can cause significant deviation of the lines from their expected positions. More details on radio spectral-line sources are presented in Appendix D.

The observation of spectral lines requires higher resolution in frequency to resolve the lines. Hence the number of received channels are generally increased in the range of 100 to 1000. Calibration is performed over the entire bandpass response of the instrument, and is later subtracted from the spectral-line data (containing continuum data from other sources). What is left is the pure spectral-line data. In the following sections, we explain these processes step by step.

### 8.7.1 Rest vs. Observing Frequency

The actual frequency (rest frequency) $v_{rest}$ of the line can be different from that observed $v_{obs}$ due to the relative motion between the observer and the source. These are related with the relative velocities $\vec{V}_{radio}$ for radio and $\vec{V}_{optical}$ for optical (between the source and the observer) as expressed in Eqs. (8.89) and (8.90), where $c$ is the speed of light.

$$\frac{\left|\vec{V}_{radio}\right|}{c} = \frac{v_{rest} - v_{obs}}{v_{rest}} \tag{8.89}$$

$$\frac{\left|\vec{V}_{optical}\right|}{c} = \frac{v_{rest} - v_{obs}}{v_{obs}} \tag{8.90}$$

The above relationships are valid when (*i*) the relative velocity amplitude is much less than $c$, and (*ii*) the angle between the velocity vector and the radiation wave vector is much less than $\pi/2$. These velocities are defined with respect to the chosen rest frame of observation as shown in Table 8.2.

**TABLE 8.2:** Rest frames and maximum relative velocity amplitudes (standardized by IAU).

| Rest Frame | Velocity $|\vec{V}_{radio}|$ |
|---|---|
| **Topocentric** (no corrections) | 0 km/s |
| **Geocentric** (Earth rotation) | < 0.5 km/s |
| **E/M Barycentric** (Earth/Moon barycenter) | < 0.013 km/s |
| **Heliocentric** (Earth around Sun) | < 30 km/s |
| **SS Barycentric** (Sun/planets barycenter) | < 0.012 km/s |
| **Local Standard of Rest** (Sun peculiar motion) | < 20 km/s |
| **Galactocentric** (Galactic rotation) | < 300 km/s |

The position of the observer is *topocentric* and requires no correction. However, in most of the cases, the observer would like to use a different rest frame, such as the *local standard of rest*. Hence, he has to compensate for the velocities of *Geocentric, E/M Barycentric, Heliocentric, SS Barycentric,* and *local standard of rest* from the table. The exact velocity values depend on (*i*) the geocentric latitude of the observer, (*ii*) the source equatorial coordinates, (*iii*) the source ecliptic coordinates, (*iv*) the Sun's longitude, (*v*) the source hour angle, etc.

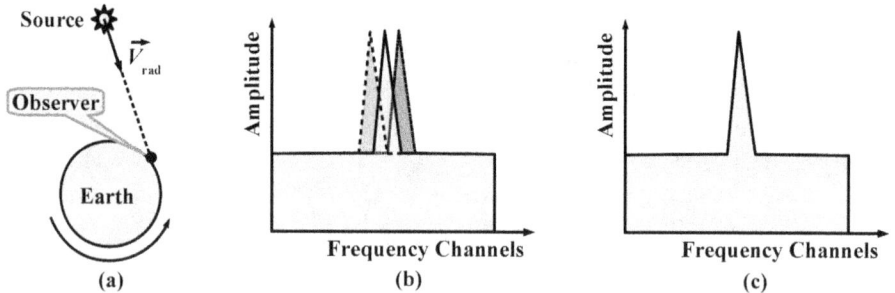

**FIGURE 8.26:** (a) A simple demonstration of change of radial velocity between the source and observer due to the Earth's rotation. (b) Doppler shifts result in positional movement of the spectral line across the frequency channels. (c) Corrections applied in postprocessing.

### 8.7.2 Doppler Tracking

Because of the Earth's rotation, the radial velocity $\vec{V}_{rad}$ between the source and the observer keeps changing, as shown in Fig. 8.26a. Hence, the apparent frequency of the spectral line continuously changes. For a twelve-hour observation, this velocity can vary between ±0.5 km/s (see geocentric velocity in Table 8.2). Thus the spectral line drifts across the frequency band of observation, as shown in Fig. 8.26b. An integration in time will result in broadening of the line and a decrease in its peak. To avoid this problem, the observation frequency is continuous-ly changed such that the line appears in the same channel of the frequency band. This is known as *Doppler tracking*. Wide bandwidths with negligible integration time can be used to avoid Doppler tracking. The shift in frequency can be adjusted during data processing, as shown in Fig. 8.26c.

### 8.7.3 Calibration of Visibilities

The total response of the telescope must be known to determine the true visibili-ties $\mathcal{V}(v, t)$ as a function of frequency $v$ and time $t$ from the observed visibilities $\mathcal{V}_{obs}(v, t)$, as shown in Eq. (8.91), where $G(v, t)$ is the complex gain of the system.

$$\mathcal{V}_{obs}(v,t) = G(v,t)\mathcal{V}(v,t) \tag{8.91}$$

The complex gain $G(v, t)$ can be split into two components: (*i*) an overall com-plex gain $G'(t)$ for a reference RF within the observing band that varies with time, and (*ii*) gain variations $B(v, t)$ across channels (IF band shape), as shown in Eq. (8.92). Hence the telescope gain is a combination of the RF gain calibration and the IF band shape.

$$G(v,t) = G'(t)\,B(v,t) \tag{8.92}$$

#### 8.7.3.1 Gain Calibration
The gain calibration is similar to that for continuum observations. An unresolved source (calibrator) near the direction of the source is observed to estimate the gains of individual antennas in an array. For $N$ antennas we estimate $N$ amplitudes and $N$ phases from which all the visibilities can be calibrated. A certain number of spectral channels are averaged to improve the SNR. The calibration also takes care of atmospheric offsets because the calibrator is chosen in a direction close to the source. For more details, see Section 8.3.

#### 8.7.3.2 Bandpass Calibration
A bright unresolved source containing no spectral lines can be used for bandpass (band shape) calibration. However, the calibrator source need not be in a direction close to the target source. We may simply apply Eq. (8.93) for obtaining $B(v, t)$, where $B_{obs}(v, t)$ is the observed flux density variation across the band shape, and $S_{cal}$ is the flux density of the calibrator source (which is known).

$$B(v,t) = \frac{B_{obs}(v,t)}{S_{cal}} \tag{8.93}$$

This, however, requires a very high SNR, such that the corrected spectrum is not degraded. In other words, one should obtain an intrinsically flat spectrum band shape. In general, one of the two methods are employed in bandpass calibration:

(*i*) Position Switching: This is done while observing the target source. Intermittently, the telescope is pointed to the calibrator once within twenty minutes to an hour, depending on the stability of the telescope properties.

(*ii*) Frequency Switching: The presence of gases, such as galactic neutral hydrogen (HI), may introduce spectral-line features in the path of the calibrator, depending on the frequency of interest. Again, if the target source is brighter than the calibrator, the later has to be observed for longer duration (for the same settings of the telescope) to achieve a good SNR. Under these conditions, it may be useful to note the response at neighboring frequencies outside the spectral line. These adjacent band shapes may be used for calibration of the observed spectrum. Hence, by switching between the adjacent frequencies, one can estimate the band shape that is largely decided by the shape of the baseband filter.

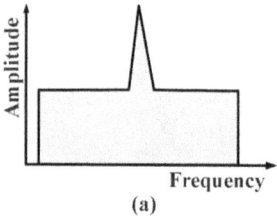 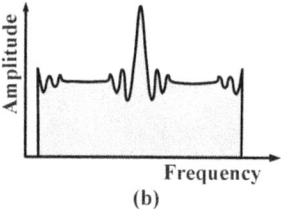

(a)  (b)

**FIGURE 8.27:** (a) Expected spectrum. (b) Measured spectrum.

### 8.7.4 Smoothing the Spectrum

Fig. 8.27a shows the structure of the expected spectrum obtained from spectral-line observations. However, due to the limited availability of the number of correlation products and due to finite time, we cannot measure an infinite number of Fourier components. In other words, a spectrum having a bandwidth $\Delta \nu$ and $N$ channels is a result of cross-correlations between signals sampled at time intervals of $\tau = 1/\Delta \nu$ in the range $-N\tau$ to $(N-1)\tau$ produced by the telescope. This is equivalent to truncation in the time range by a rectangular window. The result is due to Gibbs phenomenon, as shown in Fig. 8.27b. Note that the sharp edges of the spectrum have a ringing effect. This effect may be viewed as a convolution of the true spectrum with a sinc function.

Several different smoothing functions can be used to minimize this unwanted ringing, but at the expense of spectral resolution. The most commonly used is the Hanning window, which helps reduce the first side lobes from 22% to less than 3%.

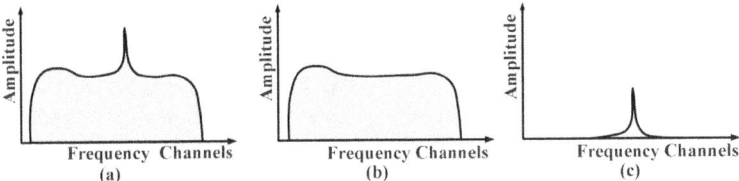

**FIGURE 8.28:** (a) Spectral line with continuum. (b) Bandpass response from continuum source. (c) Spectral line obtained after subtracting (b) from (a).

### 8.7.5 Subtraction of Continuum

Due to emissions from nearby sources in the field of view towards the target source, the data often contains continuum signals, as shown in Fig. 8.28a. This complicates the detection and analysis of the line data. Hence these unwanted signatures must be removed. A model of a continuum is created using frequency channels having no line features, as shown in Fig. 8.28b. This model is subtracted from all channels, and the residue contains only the spectral information, as shown in Fig. 8.28c.

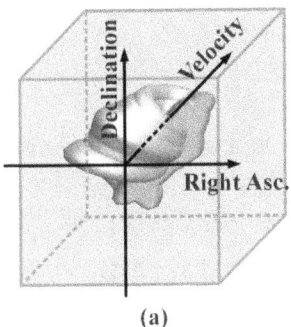
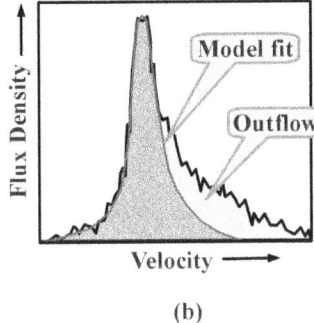

(a)                                          (b)

**FIGURE 8.29:** (a) A pseudo data cube showing variations in position and velocity. (b) Pseudo flux density as a function of velocity for a hot-gas distribution having an outflow.

### 8.7.6 Spectral-Line Features

We give here some basic ideas for understanding the derived features from a spectral-line observation. In reality, the spectral line is an extremely narrow line. But due to the internal motions within the gas due to heat, etc., the contents within a volume of gas are in motion that results in an increase in the line width. The line width is thus a function of the velocity distribution of the particles within the gas volume. Instead of actual velocity, we see it as frequency variations in radio-telescope data. These are converted into a velocity profile before plotting. Thus we have three coordinates, two for position (right ascension and declination) and one for velocity. These kinds of plots are also known as *spectral data cubes*, and an *example* is shown in Fig. 8.29a. One can make slices in the RA-Dec plane at fixed velocities to see the spread for that velocity. On the other hand, if the flux density of a hot-gas distribution is plotted against the velocity, as shown in Fig. 8.29b, one can fit a model curve to find the outflow of gases, if any. The hatched region shows the broadening of the line due to thermal motion of the gas particles. The excess velocity is determined from the gray region on the right.

Before representing the data in the cube format, one has to undergo the imaging process as explained under conventional and or wide-field imaging for each velocity. Other methods for data reduction can be found in the relevant papers.

## REVIEW QUESTIONS

1. Why is the approximated van Cittert-Zernike equation so important in conventional imaging? What are its drawbacks?

2. Why do you need to calibrate the data before imaging?

3. What is meant by flagging bad data?

4. What is the significance of long-term and short-term calibrations?
   [*Hint:* See Sections 8.3.1 and 8.3.2]

5. What is the difference between a gain calibrator and a phase calibrator?
   [*Hint:* See Section 8.3.2.1]

6. Why do you require a point source for phase calibrations?

7. Under what circumstances can a gain calibrator be used as phase calibrator? Explain your reasoning.
   [*Hint:* See Section 8.3.2.1]

8. What is meant by (*i*) phase closure, and (*ii*) gain closure, in self-calibration? [*Hint:* See Section 8.3.2.2(*i*) and (*ii*)]

9. How does a redundant baseline array help in self-calibration?
   [*Hint:* See Fig. 8.2]

10. What is the reason behind applying weights to the visibilities during an imaging procedure?

11. What is the difference between natural weights and uniform weights?
    [*Hint:* See Sections 8.4.2.1 and 8.4.2.2]

12. Why are the tapered weights preferable to natural or uniform weights?
    [*Hint:* See Section 8.4.2.3]

13. Why are robust weights the most preferable?   [*Hint:* See Section 8.4.2.4]

14. Explain the purpose of gridding the visibilities using uniform spacing across the *u-v* plane. [*Hint:* See Section 8.4.3]

15. Explain the gridding procedure in one dimension using a figure.

[*Hint:* Use Fig. 8.10]

16. What should be the ideal shape of the Fourier transform $\overline{C}(l,m)$ of the convolution function $C(u, v)$? [*Hint:* See Section 8.4.3.2]

17. How does expressing the convolution function $C(u, v)$ as a product of two functions $C_1(u)C_2(v)$ help in the gridding process?

[*Hint:* See Section 8.4.3.3]

18. What is meant by a dirty image?

19. Why do you need to perform deconvolution on the dirty image?

20. Using a diagram, explain the basic CLEAN deconvolution algorithm.

[*Hint:* Use Fig. 8.13]

21. What is meant by bandwidth smearing? Using a diagram to explain its effects on the *u-v* plane. [*Hint:* See Section 8.5.1 and use Fig. 8.16]

22. What is meant by multi-frequency synthesis? What are its advantages?

[*Hint:* See Section 8.5.2]

23. What is the basic difference between conventional imaging and wide-field imaging?

24. What is meant by the short-spacing problem, especially in wide-field imaging?

25. Using a diagram, explain the negative bowl effect on the synthesized beam.

[*Hint:* See Section 8.6.1 and use Fig. 8.18]

26. Using a diagram, show the data analogies between a single dish and a multiplicative interferometer. [*Hint:* Use Fig. 8.19]

27. Using a diagram, explain the process for generating *u*, *v* data using a single dish in wide-field imaging. [*Hint:* Use Fig. 8.20]

28. Using a diagram, explain how the negative bowl effect is removed by combining single-dish data with interferometer data. [*Hint:* Use Fig. 8.22]

29. Using a diagram, explain the principle of mosaicing. [*Hint:* Use Fig. 8.23]

30. What is the difference between coplanar and noncoplanar baselines?

[*Hint:* See Section 8.6.2.1]

31. Explain the phase problem arising due to the *w*-component while observing a source at the zenith using an interferometer having two antennas at different heights. [*Hint:* Use Fig. 8.25 and Eqs. (8.83) and (8.84)]

32. What are the differences between a dirty beam and a dirty ball beam? Where are they used? [***Hint:*** See Section 8.6.2.3]

33. What are spectral lines? State some of the possible reasons for spectral broadening.

34. Explain the meaning of the rest and observing frequencies of a spectral line. [***Hint:*** See Section 8.7.1]

35. Give reasons for Doppler tracking in spectral-line observations? [***Hint:*** See Section 8.7.2]

36. What is meant by bandpass calibration? Explain with a suitable diagram. [***Hint:*** See Section 8.7.3.2]

37. Why do you need to subtract the continuum from the data, and how is this done? Explain with a diagram. [***Hint:*** See Section 8.7.5 and use Fig. 8.28]

38. Explain the structure of a data cube using a 3-D diagram. [***Hint:*** Use Fig. 8.29]

# CHAPTER 9

# PROPAGATION EFFECTS IN RADIO ASTRONOMY

## 9.1 INTRODUCTION

The only way to study the Universe is by means of analyzing the radiation reaching us from distant objects. The electromagnetic waves reach us after having travelled extremely large distances in time and space. As the waves propagate, various properties of the interstellar medium modify the original signals, resulting in attenuation, absorption, change of path, change of polarization, etc. Effects like spectral absorption, Faraday rotation, and magnetohydrodynamic waves are seen. The Earth's ionosphere also plays a big role in modifying some of the wave properties by effects like scintillation, Faraday rotation, etc. In this chapter, we shall study the behavior of electromagnetic waves in various media.

## 9.2 THE ELECTROMAGNETIC WAVE EQUATION

Let us begin with the wave equation. It can be derived from the four Maxwell's equations, given in Eq. (9.1), where $\vec{H}, \vec{E}, \vec{D}, \vec{B}$, and $\vec{J}$ represent the magnetic field, the electric field, the electric flux density, the magnetic flux density, and the current density, respectively. For time-harmonic fields, we have

$$\nabla \times \vec{H} = \vec{J} + \frac{\partial \vec{D}}{\partial t} = \vec{J} + j\omega\vec{D} \left.\begin{array}{c} \\ \\ \\ \\ \\ \\ \end{array}\right\} \quad (9.1)$$

$$\nabla \times \vec{E} = -\frac{\partial \vec{B}}{\partial t} = -j\omega\vec{B}$$

$$\nabla \cdot \vec{D} = \rho$$

$$\nabla \cdot \vec{B} = 0$$

The symbols $\rho$ and $\omega$, respectively, represent the electric charge density and the angular frequency. $B = \mu H$, $D = \epsilon E$, and $J = \sigma E$, where $\mu$, $\epsilon$, and $\sigma$ are, respectively, the permeability, the permittivity, and the conductivity of the medium. We may derive the wave equation by taking the curl of the second of Maxwell's equations. The procedure is shown in Eq. (9.2).

$$\nabla \times \nabla \times \vec{E} = -j\omega\mu\left(\nabla \times \vec{H}\right) \left.\begin{array}{c} \\ \\ \\ \\ \end{array}\right\} \quad (9.2)$$

$$= -j\mu\omega\left(\vec{J} + j\omega\vec{D}\right)$$

$$= -j\mu\omega\left(\sigma\vec{E} + j\omega\varepsilon\vec{E}\right)$$

$$\therefore \nabla \times \nabla \times \vec{E} + j\mu\omega\left(\sigma\vec{E} + j\omega\varepsilon\vec{E}\right) = 0$$

Substituting $j\mu\omega(\sigma + j\omega\epsilon) = \gamma^2$ into Eq. (9.2), we obtain Eq. (9.3) as shown below, where $\gamma$ is known as the *propagation constant*.

$$\nabla \times \nabla \times \vec{E} + \gamma^2\vec{E} = 0 \quad (9.3)$$

We may derive an analogous equation for the magnetic field by taking the curl of the first Maxwell equation and substituting the second Maxwell equation listed in Eq. (9.1)

## 9.3 PLANE WAVES IN LOSSLESS MEDIA

Consider a plane wave propagating along the $z$-direction with $\vec{E}$ along the $y$-direction. $E_x$ and $E_z$ are thus zero. Hence, we may write Eq. (9.3) as shown in Eq. (9.4). Its solution is shown in Eq. (9.5).

$$\frac{\partial^2 E_y}{\partial z^2} - \gamma^2 E_y = 0 \quad (9.4)$$

$$E_y = E_0 \exp\left(j\omega t \pm \gamma z\right) \qquad (9.5)$$

Here, $E_0$ is the amplitude of the wave, and $\omega$ is its angular frequency. The propagation constant $\gamma = \alpha \pm j\beta$, where $\alpha$ is called the *attenuation constant* and $\beta$ is known as the *phase constant*. If the medium is loss less, the conductivity $\sigma$ is zero and so is $\alpha$. Hence we may express $\gamma$ as shown in Eq. (9.6).

$$\gamma = \pm j\beta = \pm j\omega\sqrt{\mu\epsilon} \qquad (9.6)$$

Here, $\beta = 2\pi/\lambda$, where $\lambda$ is the wavelength. We may further expand $\beta$ as $\beta = 2\pi v/v$, where $v$ is the velocity of wave propagation (m sec$^{-1}$) inside the medium. It is given as $v = 1/\sqrt{\mu\epsilon}$, where $\mu$ and $\epsilon$ are the permeability and permittivity of the medium. Because angular frequency is defined as $\omega = 2\pi v$ we may express $\beta = \omega/v = \omega\sqrt{\mu\epsilon}$ as in Eq. (9.6) above. Coming back to our problem of plane waves in a lossless medium, we may now express $E_y$ from Eq. (9.5) as shown in Eq. (9.7).

$$E_y = E_0 \exp\left(j\omega t \pm j\beta z\right) \qquad (9.7)$$

Eq. (9.7) represents the space $z$ and time $t$ variation of $\vec{E}$ for a plane wave propagating along the $z$-direction in a loss less media. With a negative sign on $j\beta z$, the solution is suitable for a wave traveling along the positive $z$-axis. Similarly, if $j\beta z$ has a positive sign, the solution is suitable for a wave traveling along the negative $z$-axis. Because the exponent is a pure imaginary quantity, there is no attenuation during wave propagation.

## 9.4 PLANE WAVES IN CONDUCTING MEDIA

If the medium is conducting, and if the conductivity is sufficiently high to fulfill the condition $\sigma \gg \omega\epsilon$, we may write the following equations:

$$\gamma^2 = j\omega\mu\sigma \qquad (9.8)$$

$$\gamma = (1+j)\sqrt{\frac{\mu\omega\sigma}{2}} \qquad (9.9)$$

Thus we find that the propagation constant $\gamma$ (in Neper m$^{-1}$) in a conducting media has both real and imaginary parts, as expressed in Eq. (9.10), where $\alpha$ is the attenuation constant (in m$^{-1}$) and $\beta$ is the phase constant (in rad m$^{-1}$). We may further expand Eq. (9.10) using Eq. (9.9) and obtain Eq. (9.11).

$$\gamma = \alpha + j\beta \qquad (9.10)$$

$$\alpha + j\beta = \sqrt{\frac{\mu\omega\sigma}{2}} + j\sqrt{\frac{\mu\omega\sigma}{2}} \tag{9.11}$$

In a conducting medium, the solution for a plane wave propagating in the $z$-direction with $\vec{E}$ polarized in the $y$-direction is given by Eq. (9.12).

$$\left. \begin{aligned} E_y &= E_0\, e^{j\omega t \pm \gamma z} \\ &= E_0\, e^{-\alpha|z|}\, e^{j(\omega t \pm \beta z)} \end{aligned} \right\} \tag{9.12}$$

Note that the sign of $\alpha|z|$ is negative because it indicates attenuation, and $\alpha$ is a positive quantity. In astronomical studies, the product $\alpha z$ is sometimes termed the *optical depth*. The required distance for the wave to attenuate to $1/e$ of its original value occurs when $z = 1/\alpha$. This is noted as the unity optical depth. It is also called the *depth of penetration*.

## 9.5 PLANE WAVES IN IONIZED MEDIUM HAVING A MAGNETIC FIELD

Consider a plane wave propagating in an ionized medium in the presence of a steady magnetic field. The plasma medium is assumed to have equal numbers of positive and negative charges. If the plasma is cool, so that the collisions are unimportant, only the interactions of the electrons (negative charges) with the wave are taken into account. This is because the positive ions are much more massive than the electrons and also so that they are unable to interact significantly.

Let the ionized medium be a partially conducting medium where the condition $\sigma \gg \omega\epsilon$ is not valid. Let there be a steady magnetic field so that the medium is anisotropic, which means the propagation variable is also a function of direction. Undoubtedly, the situation is complicated in comparison to an isotropic medium where the propagation is the same in all directions.

Fig. 9.1 shows the geometry of wave propagation and the magnetic field in an $x, y, z$ coordinate system. The wave propagates in the $x$-$z$ plane, and its direction subtends an angle $\phi$ from the $z$-axis. The electric field $\vec{E}$ of the wave is shown in a plane perpendicular to the direction of wave propagation. A steady magnetic field $\vec{B}$ is acting along the $z$-direction with amplitude $B_z$, as shown in Eq. (9.13).

$$|\vec{B}| = B_z \tag{9.13}$$

If $e$ is the charge of an electron, then in the presence of the magnetic field, the force equations due to the magnetic and electric fields may be written as shown

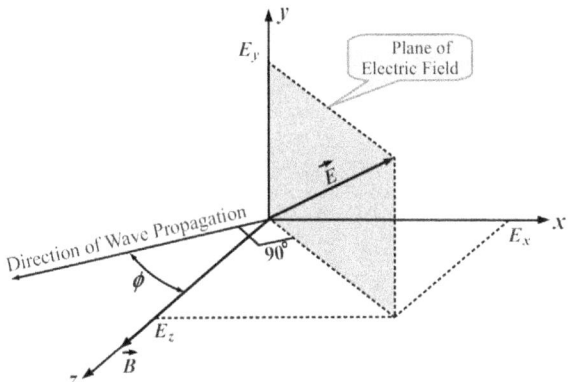

**FIGURE 9.1:** Directions of a magnetic field and wave propagation in an $x$, $y$, $z$ coordinate system. The electric field $\vec{E}$ is in a plane perpendicular to the direction of propagation.

below in Eqs. (9.14) and (9.15), respectively, where $\vec{v}$ is the velocity of the electron.

$$\overline{F_1} = e\left(\vec{v} \times \vec{B}\right) \tag{9.14}$$

$$\overline{F_2} = e\vec{E} \tag{9.15}$$

The total force $\vec{F}$ will the be sum of the above two equations. This can be equated to the product of mass $m$ and acceleration $d\vec{v}/dt$ of the electron using Newton's second law of motion. This is shown below in Eq. (9.16).

$$\vec{F} = e\left(\vec{E} + \vec{v} \times \vec{B}\right) = m\frac{d\vec{v}}{dt} \tag{9.16}$$

Hence the force per unit charge can be expressed as shown below in Eq. (9.17).

$$\frac{\vec{F}}{e} = \vec{E} + \vec{v} \times \vec{B} = \frac{m}{e}\frac{d\vec{v}}{dt} \tag{9.17}$$

We may now express this vector equation in terms of the rectangular component magnitudes as shown in Eqs. (9.18) through (9.20).

$$\frac{F_x}{e} = E_x + v_y B_z = \frac{m}{e}\frac{dv_x}{dt} \tag{9.18}$$

$$\frac{F_y}{e} = E_y - v_x B_z = \frac{m}{e}\frac{dv_y}{dt} \tag{9.19}$$

$$\frac{F_z}{e} = E_z = \frac{m}{e}\frac{dv_z}{dt} \quad \left[\text{since } \left(\vec{v} \times \vec{B}\right)_z = 0\right] \tag{9.20}$$

Let the motion of the charged particles (electrons) be harmonic with the wave. Hence they also oscillate at the same frequency $\omega$ of the wave. If we assume a time-harmonic velocity $\vec{v} = \vec{v}_0 e^{j\omega t}$, then the acceleration is $d\vec{v}/dt = j\omega\vec{v}_0 e^{j\omega t} = j\omega\vec{v}$, where $\omega = 2\pi\nu$ (in rad sec$^{-1}$) represents the angular frequency of the wave in radians, and $\nu$ is its frequency in Hz. Substituting $\vec{v}$ into the above equations and separating the velocity components, we obtain Eqs. (9.21) through (9.23), as shown below.

$$v_x = \frac{j(m\omega/e)E_x + E_y B_z}{B_z^2 - m^2\omega^2/e^2} \tag{9.21}$$

$$v_y = \frac{j(m\omega/e)E_y - E_x B_z}{B_z^2 - m^2\omega^2/e^2} \tag{9.22}$$

$$v_z = -\frac{jeE_z}{m\omega} \tag{9.23}$$

It follows from the above relations that the movement of any charged particle will follow a helical path whose axis is coincidental with the $z$-direction.

### 9.5.1 Gyrofrequency of Radiation

Let us now consider a special case of the magnetic field shown in Eq. (9.24). It occurs when $v_x$ and $v_y$ are very large, such that $B_z^2 - m^2\omega^2/e^2 \approx 0$, as can be obtained from either of Eqs. (9.21) or (9.22).

$$B_z^2 = \frac{m^2\omega^2}{e^2} \tag{9.24}$$

Under this condition, the effect of collisions with other particles may be neglected, and we may call the angular frequency $\omega_g$ of rotation of the electrons around the magnetic field the *angular gyro frequency*. It is measured in rad sec$^{-1}$. The linear gyro frequency $\nu_g$ is given as $\nu_g = \omega_g/2\pi$. The rotation of electrons around the magnetic field emits radiation at the same frequency, called the *radiation frequency*. Rearranging Eq. (9.24), we may write $\omega_g$ as expressed below in Eq. (9.25).

$$\omega_g = \frac{e}{m}B_z \tag{9.25}$$

Substituting Eq. (9.25) into Eqs. (9.21) and (9.22), after simplification we obtain Eqs. (9.26) and (9.27), as shown below.

$$v_x = \frac{e}{m}\frac{\omega_g E_y + j\omega E_x}{\omega_g^2 - \omega^2} \tag{9.26}$$

$$v_y = \frac{e}{m} \frac{-\omega_g E_x + j\omega E_y}{\omega_g^2 - \omega^2} \tag{9.27}$$

## 9.5.2 Plasma Oscillation Frequency

If the electrons are cold, it is possible to show that the charge density oscillates at the plasma frequency. This is known as the *critical frequency*. If $N$ is the number of charged particles per unit volume, then Maxwell's equation $\nabla \times \vec{H} = \vec{J} + j\omega\vec{D}$ may be expressed as shown below in Eq. (9.28).

$$\nabla \times \vec{H} = \vec{J} + j\omega\vec{D} = Ne\vec{v} + j\omega\epsilon_0\vec{E} \tag{9.28}$$

The movement of the charged particles is responsible for the conduction current. We may now express Eq. (9.28) in terms of its three rectangular components, as shown below.

$$\left(\nabla \times \vec{H}\right)_x = j\omega\epsilon_0 E_x + Nev_x \tag{9.29}$$

$$\left(\nabla \times \vec{H}\right)_y = j\omega\epsilon_0 E_y + Nev_y \tag{9.30}$$

$$\left(\nabla \times \vec{H}\right)_z = j\omega\epsilon_0 E_z + Nev_z \tag{9.31}$$

Substituting the values of $v_x$, $v_z$, and $v_z$ from Eqs. (9.26), (9.27), and (9.23) into the above three relations, we get the following equations.

$$\left(\nabla \times \vec{H}\right)_x = j\omega\epsilon_0 E_x \left\{ 1 + \frac{Ne^2}{\epsilon_0 m\left(\omega_g^2 - \omega^2\right)} \right\} + \left\{ \frac{Ne^2 \omega_g E_y}{m\left(\omega_g^2 - \omega^2\right)} \right\} \tag{9.32}$$

$$\left(\nabla \times \vec{H}\right)_y = -\frac{Ne^2 \omega_g E_x}{m\left(\omega_g^2 - \omega^2\right)} + j\omega\epsilon_0 E_y \left\{ 1 + \frac{Ne^2}{\epsilon_0 m\left(\omega_g^2 - \omega^2\right)} \right\} \tag{9.33}$$

$$\left(\nabla \times \vec{H}\right)_z = j\omega\epsilon_0 E_z \left\{ 1 - \frac{Ne^2}{\epsilon_0 m\omega^2} \right\} \tag{9.34}$$

When $\omega \gg \omega_g$, all three relations become identical. The critical or plasma frequency is attained when

$$\frac{Ne^2}{\epsilon_0 m\omega^2} = 1 \tag{9.35}$$

Under this condition, the refractive index and the permittivity of the medium becomes zero. If we symbolize this critical radio frequency (plasma frequency) as $\omega_p$ (in rad Hz$^{-1}$), we may rewrite Eq. (9.35), as shown below in Eq. (9.36).

$$\omega_p^2 = \frac{Ne^2}{\epsilon_0 m} \tag{9.36}$$

If we express $\omega_p = 2\pi v_p$, where $v_p$ is the critical frequency (in Hz), then from Eq. (9.36) we may express $v_p$ as shown in Eq. (9.37), where $m$ is the mass of the particle (in kg), $e$ is the charge of a particle (in Coulombs), and $\epsilon_0$ is the permittivity of free space ($8.854 \times 10^{-12}$ Farad m$^{-1}$).

$$v_p = \frac{e}{2\pi} \sqrt{\frac{N}{\epsilon_0 m}} \tag{9.37}$$

If the charged particles are electrons, then from Eq. (9.37) we get the critical or plasma frequency, as shown below in Eq. (9.38).

$$v_p \approx 9\sqrt{N} \tag{9.38}$$

The wave is totally reflected by the medium at or below the critical frequency. For the Earth's ionosphere, the critical frequency is often about 9 MHz.

### 9.5.3 Propagation Effects Above or Below Plasma Frequency

The Maxwell magnetic curl equation may be written as shown in Eq. (9.39), where $\bar{\epsilon}$ is the tensor for permittivity, whose components satisfy Eqs. (9.32), (9.33), and (9.34). In a generalized form we may express $\bar{\epsilon}$ as shown in Eq. (9.40).

$$\nabla \times \vec{H} = j\omega\bar{\epsilon}\cdot\vec{E} \tag{9.39}$$

$$\bar{\epsilon} = \begin{bmatrix} \epsilon_{11} & -j\epsilon_{12} & \epsilon_{13} \\ j\epsilon_{21} & \epsilon_{22} & \epsilon_{23} \\ \epsilon_{31} & \epsilon_{32} & \epsilon_{33} \end{bmatrix} \tag{9.40}$$

$$= \begin{bmatrix} \left(1 + \dfrac{\omega_p^2}{\omega_g^2 - \omega^2}\right)\epsilon_0 & \dfrac{-j\omega_p^2\omega_g\epsilon_0}{\omega\left(\omega_g^2 - \omega^2\right)} & 0 \\[3ex] \dfrac{j\omega_p^2\omega_g\epsilon_0}{\omega\left(\omega_g^2 - \omega^2\right)} & \left(1 + \dfrac{\omega_p^2}{\omega_g^2 - \omega^2}\right)\epsilon_0 & 0 \\[3ex] 0 & 0 & \left(1 - \dfrac{\omega_p^2}{\omega^2}\right)\epsilon_0 \end{bmatrix}$$

We now study the propagation effects in (*i*) the absence, and (*ii*) the presence, of the external magnetic field $\vec{B}$, as described below.

**Case (*i*):** When the magnetic field is absent ($\vec{B} = 0$), or when the frequency is much higher than the gyrofrequency $\omega \gg \omega_g$, Eq. (9.40) gets simplified as shown below in Eq. (9.41). Using this, Maxwell's magnetic curl can be expressed as Eq. (9.42).

$$\bar{\bar{\epsilon}} = \begin{bmatrix} \left(1-\dfrac{\omega_p^2}{\omega^2}\right)\epsilon_0 & 0 & 0 \\[2em] 0 & \left(1-\dfrac{\omega_p^2}{\omega^2}\right)\epsilon_0 & 0 \\[2em] 0 & 0 & \left(1-\dfrac{\omega_p^2}{\omega^2}\right)\epsilon_0 \end{bmatrix} \tag{9.41}$$

$$\nabla \times \vec{H} = j\omega\left(1-\frac{\omega_p^2}{\omega^2}\right)\epsilon_0\vec{E} \tag{9.42}$$

In the above equation, the term within brackets is the relative permittivity $\varepsilon_r$ of the medium. The refractive index $\eta$ can be obtained as shown below in Eq. (9.43).

$$\left.\begin{aligned} \eta = \sqrt{\epsilon_r} &= \sqrt{1-\left(\omega_p / \omega\right)^2} \\[1em] &= \sqrt{1-\left(v_p / v\right)^2} \end{aligned}\right\} \tag{9.43}$$

From Eq. (9.43), it may be seen that for frequencies much above the plasma frequency ($v \gg v_p$), the refractive index becomes real, and hence the wave can propagate. However, when $\eta$ is less than unity, the refraction is opposite of that which occurs when waves enter into a denser medium. On the other hand, for frequencies much below plasma frequency ($v \ll v_p$), the refraction index becomes imaginary, and thus the waves cannot propagate and get reflected.

Perturbed charged particles in the form of a cloud may tend to oscillate at the plasma frequency. During such plasma oscillations, it is believed that certain types of radiation are produced from the solar atmosphere or the corona. The electron density within the solar corona reduces with height. Eq. (9.38) suggests that the plasma frequency also decreases with decreasing electron density. Hence, at a given frequency $\nu$, radiation takes place from the corona at heights where $\nu = \nu_p$. In reality, most of the radiation comes from a thin layer just above the height corresponding to $\nu = \nu_p$. These conclusions are applicable only for a quiet Sun.

For a disturbed Sun, such as during solar flares, shock waves or high-velocity jets rise into the corona. The plasma oscillations of electron clouds excited by various mechanisms result in relatively narrow-band radiation whose frequency drifts downwards because the disturbances rise through the corona into regions having lower electron density.

**Case (ii):** When the magnetic field $\vec{B}$ is present, we have from Eqns. (9.2) and (9.39)

$$\left.\begin{aligned} \nabla \times \vec{E} &= -j\omega\mu_0\vec{H} \\ \text{or, } \nabla \times \nabla \times \vec{E} &= -j\mu_0\omega\left(j\omega\bar{\epsilon}\cdot\vec{E}\right) \\ &= \omega^2\mu_0\bar{\epsilon}\cdot\vec{E} \end{aligned}\right\} . \tag{9.44}$$

The above gives a wave equation in terms of $\vec{E}$. Ignoring the effect of collisions, let us choose its solution in the form given in Eq. (9.45), where $r$ is the distance measured in the direction of wave propagation and $\beta$ is the phase constant.

$$\vec{E} = E_0 e^{j(\omega t - \beta r)} \tag{9.45}$$

From a rigorous analysis, it may be shown that when the wave propagates parallel to the magnetic field (longitudinal propagation), the phase constant $\beta$ may be expressed as shown below in Eq. (9.46).

$$\beta = \omega\sqrt{\mu_0\left(\epsilon_{11} \pm \epsilon_{12}\right)} \tag{9.46}$$

The above equation indicates that the wave consists of two opposite circularly polarized components.

If the wave propagation is perpendicular to the magnetic field (transverse propagation), the phase constant $\beta$ can be expressed as shown below in Eqs. (9.47) and (9.48).

$$\beta = \omega\sqrt{\mu_0\left(\epsilon_{11} - \frac{\epsilon_{12}^2}{\epsilon_{11}}\right)} \qquad \text{when} \quad \vec{E} \perp \vec{B} \tag{9.47}$$

$$\beta = \omega\sqrt{\mu_0\,\epsilon_{33}} \qquad \text{when} \quad \vec{E} \parallel \vec{B} \tag{9.48}$$

Note that when $\vec{E}$ is parallel to $\vec{B}$, the propagation is the same as the case when no magnetic field is present. The wave is referred to as an *ordinary ray*. When $\vec{E}$ is perpendicular to $\vec{B}$, the wave is referred to as an *extraordinary ray*.

Now we establish more general relationships for quasi-longitudinal and quasi-transverse propagation conditions. In the former case, $\phi$ is sufficiently small so that the terms $\sin^2\phi$ and $\sin^4\phi$ may be ignored. In the latter case, $\phi$ is close to 90°, and so we may ignore $\cos\phi$. Thus for a quasi-longitudinal propagation where $\phi'$ is small, we may express $\beta$ as shown below in Eq. (9.49). Similarly, for the quasi-transverse propagation where $\phi \approx 90°$, we may express $\beta$ as shown below in Eq. (9.50).

$$\beta = \omega\sqrt{\mu_0\left(\epsilon_{11} \pm \epsilon_{12}\cos\phi\right)} \qquad (9.49)$$

$$\beta = \omega\sqrt{\mu_0\left\{\epsilon_{11} - (1\pm1)\frac{\epsilon_{12}^2}{2\epsilon_{11}}\sin\phi\right\}} \qquad (9.50)$$

## 9.6  FARADAY ROTATION

In 1845, Michael Faraday observed that the plane of polarization of light rotates as it passes through a crystal. A linearly polarized wave may result from two circularly polarized waves having identical amplitudes but opposite rotations. If the phase constants of the two circularly polarized waves are different, the plane of polarization of the resulting linearly polarized wave rotates with its propagation.

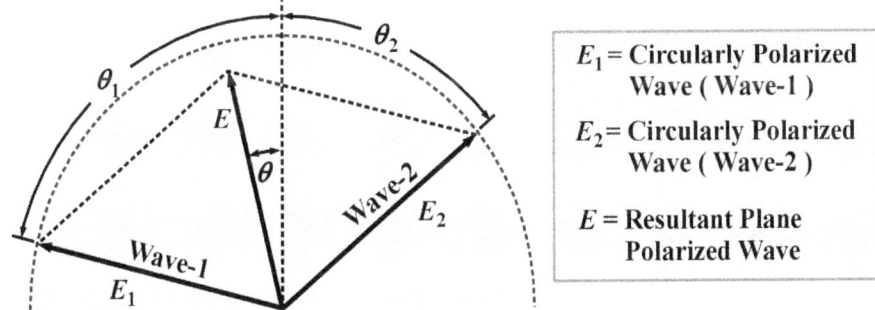

**FIGURE 9.2:** The linearly polarized wave *E* is shown as a resultant combination of two oppositely rotating circularly polarized waves $E_1$ and $E_2$ having the same amplitude.

Let us consider two circularly polarized waves as shown in Fig. 9.2, where the waves 1 and 2 rotate in opposite directions. We now consider two cases of wave propagation: (*i*) quasi-longitudinal, and (*ii*) quasi-transverse.

**Case (i):** Let the waves be traveling in a quasi-longitudinal fashion such that $\phi$ is small. The elemental angular rotation over a propagation distance $dr$ of one wave may be expressed as shown below in Eq. (9.51), while that of other may be expressed as in Eq. (9.52). The values of $\beta^-$ and $\beta^+$ can obtained from Eq. (9.49) with the sign in the parentheses as minus and plus, respectively. The net rotation angle $d\theta$ is shown in Eq. (9.53).

$$d\theta_1 = \beta^- dr \tag{9.51}$$

$$d\theta_2 = \beta^+ dr \tag{9.52}$$

$$d\theta = \frac{\beta^- - \beta^+}{2} dr \tag{9.53}$$

From Eq. (9.41), we substitute the values of $\epsilon_{11}$ and $\epsilon_{12}$ into Eq. (9.49). Taking approximations for $\omega \gg \omega_g$, $\omega \gg \omega_0$, and small $\phi$, we obtain Eq. (9.54), as shown below.

$$d\theta = \frac{Ne^3 \beta \lambda^2 \cos\phi}{8\pi^2 c^3 \epsilon_0 m^2} dr \tag{9.54}$$

The total Faraday rotation for the quasi-longitudinal case can now be obtained by integrating Eq. (9.54) from 0 to $r$ as shown in Eq. (9.55), where $r$ is the propagation distance.

$$\theta = \frac{e^3 \lambda^2}{8\pi^2 c^3 \epsilon_0 m^2} \int_0^r N B \cos\phi\, dr \tag{9.55}$$

If we assume the values $\vec{B}$ and $\phi$ to be unchanging, then Eq. (9.55) can be re-expressed as shown in Eq. (9.56).

$$\theta = \frac{e^3 \lambda^2 B}{8\pi^2 c^3 \epsilon_0 m^2} \int_0^r N dr \tag{9.56}$$

With $\vec{B}$ and $\phi$ known, a measurement of $\theta$ at a particular wavelength $\lambda$ allows us to determine the total number of charged particles within a column having a cross section 1 m$^2$ between the source and the observer. This may be obtained from Eq. (9.57).

$$N_t = \int_0^r N\, dr \tag{9.57}$$

**Case (ii):** For the quasi-transverse case where $\phi \approx 90°$, we have

$$d\theta = \frac{Ne^4 \lambda^3 B^2 \sin^2\phi}{32\pi^3 c^4 m^3 \epsilon_0} dr \ . \tag{9.58}$$

The total Faraday rotation can then be obtained as shown in Eq. (9.59).

$$\theta = \frac{e^4 \lambda^3}{32 \pi^3 c^4 m^3 \epsilon_0} \int_0^r N B^2 \sin^2 \phi \, dr \qquad (9.59)$$

The ratio of longitudinal and transverse rotations may be determined from Eqs. (9.54) and (9.58) as shown in Eq. (9.60).

$$\frac{d\theta \left( \text{longitudinal} \right)}{d\theta \left( \text{transverse} \right)} = \frac{4 \pi c m}{e B \lambda} = \frac{2 m \omega}{B e} \qquad (9.60)$$

If we consider the electrons within the Earth ionosphere at a frequency of 100 MHz, and if we take the Earth's magnetic field as $5 \times 10^{-5}$ Weber m$^{-2}$, the ratio of longitudinal to transverse rotation is on the order of 100.

## 9.6.1 Astronomical Radio Observations

The magnitude of $\vec{B}$ is on the order of $10^{-5}$ Weber m$^{-2}$ in the interstellar medium. As a result, when propagation is parallel to $\vec{B}$, the rotation becomes dominant. This is particularly true if higher frequencies are chosen. Thus we may treat the rotational effects entirely the same as in the quasi-longitudinal case.

Because the position angle $\theta$ of the electric field vector is directly proportional to the square of the wavelength $\lambda$, the graph of $\theta$ vs. $\lambda^2$ exhibits a straight line over a range of wavelengths. If this straight line is extrapolated to $\lambda = 0$, the polarization angle at the source, or the *intrinsic polarization measure,* can be determined by taking the slope of the line, which is $\theta/\lambda^2$. The rotation is positive if the direction of the magnetic field is towards the observer. To avoid ambiguity, the measurements should be within a restricted range where changes in $\theta$ do not exceed $\pi/2$.

During the early 1960s, polarization measurements were made for Taurus A. The results summarized by Kraus are presented in Fig. 9.3. The upper curve shows the variation of position angle with wavelength. The lower curve shows the percentage of linear polarization as a function of wavelength. As shown, when the position angle is extrapolated to zero wavelength, the curve provides the intrinsic polarization angle (about 150° ) for Taurus A (M1), which is also known as the *Crab Nebula.* The rotation measure is approximately −25 rad m$^{-2}$.

If the magnetic field $B$, angle $\phi$ (between the wave direction and $\vec{B}$ ), and the electron density are considered fixed, then Eq. (9.55) reduces to Eq. (9.61), as shown below, where $\Delta r$ is the path-length in meters, $\lambda$ is in meters, $N$ is the number of electrons per m$^3$, $B$ is in Weber m$^{-1}$, and $\phi$ is in radians.

$$\Delta \theta = 2.6 \times 10^{-13} \, N B \lambda^2 \cos \phi \Delta r \quad \text{rad} \qquad (9.61)$$

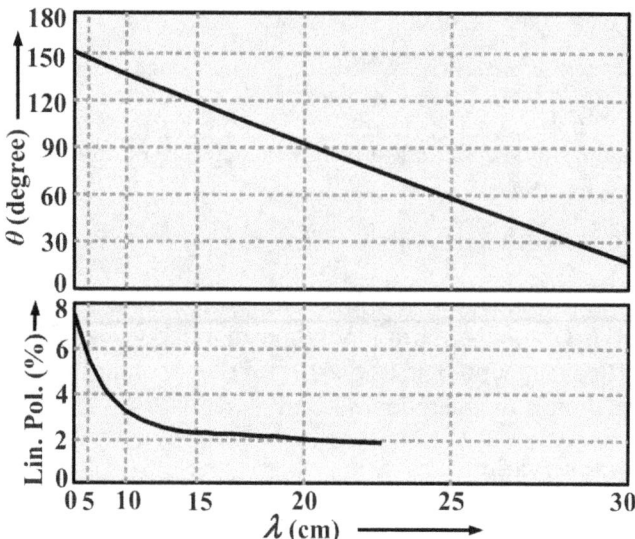

**FIGURE 9.3:** Position angle and degree of linear polarization as functions of wavelength for Taurus A (M 1), the Crab Nebula.

For astronomical calculations, one may prefer the number of electrons $N$ per cm$^3$, magnetic flux density $B$ in Gauss, and path length $\Delta r$ in pc (parsecs). The wavelength $\lambda$ remains in meters. With these units, Eq. (9.61) can be expressed as shown in Eq. (9.62).

$$\Delta\theta = 8.1 \times 10^5\, N\, B\, \lambda^2 \cos\phi\, \Delta r \quad \text{rad} \tag{9.62}$$

For example, if we have $B = 10^{-5}$ gauss, $\phi = 0$, and $\Delta r = 1{,}100$ pc, then for Taurus A ($|\theta/\lambda^2| = 25$), using Eq. (9.62) we find the average electron density of the interstellar medium to be nearly $3 \times 10^{-3}$ electrons/cm$^3$.

The integral $\int N\, B\, \cos\phi\, dr$ taken over different parts of a radio source shows clear differences. These result in Faraday rotation differences (increasing with the wavelength squared), thereby tending to depolarize the radiation that effectively decreases the degree of linear polarization. This is applicable to most polarized radio sources. The depolarization in some sources like Cygnus A is extremely rapid. The degree of linear polarization for Cygnus A diminishes between 8 to 1.5 percent for wavelengths between 3 and 5 cm.

## 9.7    MAGNETOHYDRODYNAMIC WAVES

The words *magneto, hydro,* and *dynamics,* respectively, stand for magnetic field, liquid, and movement. *Magnetohydrodynamics* therefore refers to the dynamics of electrically conducting fluids. Examples of such fluids include plasmas, liquid metals, and salt water. Hannes Alfvén pioneered the study of magnetohydrodynamic waves during the 1960s. We know that magnetic fields induce currents in moving conductors. If instead of a conductor we have a moving conductive fluid in a magnetic field, currents will be induced in the liquid. These currents give rise to mechanical forces that change the state of motion of the liquid. In other words, magnetohydrodynamic waves may be generated within the liquid. These waves are also referred to as *Alfvén waves* and are usually described by a combination of Maxwell's equations of electromagnetism and the Navier-Stokes equations of fluid dynamics. These equations are solved simultaneously by either analytic or numerical methods.

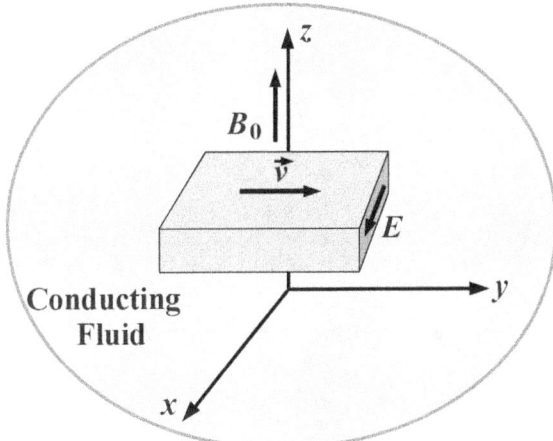

**FIGURE 9.4:** A slab within a conducting fluid is moving with a velocity $\vec{v}$.

As shown in Fig. 9.4, let a small slab-shaped region of a conducting fluid be moved with a velocity $\vec{v}$ in the $y$-direction in the presence of a steady magnetic field $\vec{B}$ ($|\vec{B}| = B_0$) applied in the $z$-direction. Let the extent of the slab be infinite along the $\pm\,y$-direction. Thus an electric field $\vec{E}$ will be induced in the $x$-direction, and it generates a current in the same direction. The current has return paths through the medium (above and below the slab), which in turn produce forces. These forces tend to prevent the motion of the slab and impart an acceleration to the fluid above and below the slab in the $y$-direction. Thus the original motion along the $y$-direction initiates a motion of the fluid along the $z$-direction that propagates as a wave.

Let $\rho$ be the mass density in kg m$^{-3}$, $p$ be the pressure in kg m$^{-1}$ sec$^{-2}$, and $\vec{G}$ be the acceleration involving non-electromagnetic forces in m sec$^{-2}$. The electrodynamic relations are listed in Eqs. (9.63) through (9.66). The first two are Maxwell's curl equations. The third shows the current density $\vec{J}$ in the fluid, and this will be the sum of the induced current density $\sigma(\vec{v} \times \vec{B})$ due to the motion of the slab at velocity $\vec{v}$ and the conduction current density $\sigma \vec{E}$. The fourth equation characterizes the medium.

$$\nabla \times \vec{H} = \vec{J} + \frac{\partial \vec{D}}{\partial t} \tag{9.63}$$

$$\nabla \times \vec{E} = -\frac{\partial \vec{B}}{\partial t} \tag{9.64}$$

$$\vec{J} = \sigma \left( \vec{E} + \vec{v} \times \vec{B} \right) \tag{9.65}$$

$$\vec{B} = \mu \vec{H} \tag{9.66}$$

We now write the hydrodynamic relation as shown in Eq. (9.67), which has a dimension of acceleration.

$$\frac{d\vec{v}}{dt} = \vec{G} + \frac{1}{\rho} \left( \vec{J} \times \vec{B} - \nabla p \right) \tag{9.67}$$

We have to connect the electrodynamic relations given in Eqs. (9.62) through (9.66) with the hydrodynamic relation given in Eq. (9.67). Let us consider the case of plane waves in a fluid of constant density (incompressible fluid). We rearrange Eq. (9.65) as shown in Eq. (9.68).

$$\vec{E} = \frac{\vec{J}}{\sigma} - \vec{v} \times \vec{B} \tag{9.68}$$

From the geometry shown in Fig. 9.4, we have $J_y = J_z = 0$. Thus, we may express the electric-field components as shown below in Eqs. (9.69) and (9.70). From Eqs. (9.64) and (9.66), we obtain Eq. (9.71).

$$E_x = \frac{J_x}{\sigma} - v_y B_0 \tag{9.69}$$

$$E_y = E_z = 0 \tag{9.70}$$

$$\frac{\partial E_x}{\partial z} = -\mu \frac{\partial H_y}{\partial t} \tag{9.71}$$

Substituting the value of $E_x$ from Eq. (9.69) into Eq. (9.71) we obtain Eq. (9.72). Because $G \approx 0$, from Eq. (9.67) we obtain Eq. (9.73).

$$\mu \frac{\partial H_y}{\partial t} = B_0 \frac{\partial v_y}{\partial z} - \frac{1}{\sigma} \frac{\partial J_x}{\partial z} \tag{9.72}$$

$$\frac{\partial v_y}{\partial t} = -\frac{1}{\rho} J_x B_0 \tag{9.73}$$

It may be assumed that $\nabla p$ has no component perpendicular to $z$. From Eqs. (9.63) and (9.73), we obtain Eq. (9.74).

$$\frac{\partial v_y}{\partial t} = \frac{B_0}{\rho} \frac{\partial H_y}{\partial z} \tag{9.74}$$

Eliminating $v_y$ between Eqs. (9.72) and (9.74), we arrive at the equation for a wave propagating in the $z$-direction. This is shown below in Eq. (9.75). We can ignore the last term of Eq. (9.75) if the conductivity is very high. The velocity of the wave can then be expressed as in Eq. (9.76).

$$\frac{\partial^2 H_y}{\partial t^2} = \frac{B_0^2}{\mu \rho} \frac{\partial^2 H_y}{\partial z^2} + \frac{1}{\mu \sigma} \frac{\partial^2 H_y}{\partial t \, \partial z} \tag{9.75}$$

$$v = \frac{B_0}{\sqrt{\mu \rho}} \tag{9.76}$$

The speed $v$ obtained from the above expression is known as the *Alfvén velocity* when expressed as a vector. Here, $v$ is expressed in units of m sec$^{-1}$, $B_0$ is in Weber m$^{-2}$, $\mu$ is in Henry m$^{-1}$, and $\rho$ is in kg m$^{-3}$.

With the Alfvén assumptions, we now consider that the medium has infinite conductivity, such that the magnetic field lines are *held in place*. Thus, both the field and the medium move together, and the field lines may be treated as stretched elastic strings. We may write d'Alembert's wave equation for wave motion on a stretched string towards the $z$-direction as expressed below in Eq. (9.77), where $S$ is the tension of the force in kg m sec$^{-2}$, $y$ is the transverse displacement in meters, and $m$ is the mass per unit length in kg m$^{-1}$.

$$\frac{\partial^2 y}{\partial t^2} = \frac{S}{m} \frac{\partial^2 y}{\partial z^2} \tag{9.77}$$

Analyzing the dimensions of Eq. (9.77), we find that the factor $S/m$ is equal to the square of the wave speed that is expressed below in Eq. (9.78).

$$v = \sqrt{\frac{S}{m}} \tag{9.78}$$

Let $S$ be the force per Weber of flux, in the case of a magnetic field, and let $m$ be the mass per unit length per Weber of flux. We thus obtain the following two equations, Eqs. (9.79) and (9.80), which are the same as we found in Eq. (9.76).

$$m = \frac{\text{kg}}{\text{meter Weber}} \left.\rule{0pt}{8em}\right\} \tag{9.79}$$

$$= \frac{\text{kg}}{\text{meter}^3 \dfrac{\text{Weber}}{\text{meter}^2}}$$

$$\text{or}, m = \frac{\rho}{B}$$

$$v = \sqrt{\frac{H\,B}{\rho}} = \sqrt{\frac{B^2}{\mu\rho}} \ \left[\text{m sec}^{-1}\right] \tag{9.80}$$

To evaluate Eq. (9.76) for a magnetohydrodynamic wave in the Sun's photosphere, we may choose (i) $B_0 = 0.1$ Weber m$^{-2}$, and (ii) $\mu = \mu_0 = 4\pi \times 10^{-7}$ Henry m$^{-1}$. The velocity thus obtained is 6.3 km sec$^{-1}$. Because the density of the medium is much less in the solar corona at about 1 solar radius above the photosphere, the velocity is very close to the velocity of light.

## 9.8  EFFECTS OF IONOSPHERE PLASMA ON RADIO OBSERVATIONS

We now apply some of the concepts developed earlier in this chapter to the observation of sources using radio telescopes. The Earth ionosphere plays an unwanted role in corrupting the observed data, especially below 1 GHz. In Earth's ionosphere and in outer space, gas may be in an ionized state over a long period of time. Radio waves are subject to changes when they pass through these plasmas. The refractive index $\eta$ of a cold neutral plasma is shown below in Eq. (9.81), where $v_p$ is the plasma frequency and $v$ is the frequency of observation.

$$\eta(v) = \sqrt{1 - \frac{v_p^2}{v^2}} \tag{9.81}$$

The plasma frequency $v_p$ is expressed below in Eq. (9.82), where $e$ represents the charge of an electron, $m_e$ is the mass of an electron, $\epsilon_0$ is the permittivity of a vacuum, and $N$ is the electron density (also see Eq. (9.37)).

$$v_p = \frac{1}{2\pi}\sqrt{\frac{Ne^2}{\epsilon_0 m_e}} \approx 9\sqrt{N} \tag{9.82}$$

The refractive index $\eta$ of the medium becomes imaginary at frequencies below $v_0$. Under such conditions, most of the energy within the wave is reflected back. Waves that are not reflected are attenuated exponentially with a certain distance and may not be detectable. Because the electron density of the Earth's ionosphere varies in the range of $10^4 - 10^5$ cm$^{-3}$, the Earth's ionosphere plasma frequency varies within the range of $1 - 10$ MHz. At such low frequencies, radio waves from outer space are unable to reach the Earth's surface and can only be received by space-based telescopes.

The material that fills the solar system and through which all the larger solar-system bodies (such as planets, asteroids, and comets) move is known as the *inter-planetary medium* (IPM). The plasma within the IPM and at the Earth's location has an electron density typically 0.03 cm$^{-3}$, for which the cut-off frequency is roughly 1 kHz. Such low-frequency waves arriving from stars, galaxies, and other objects are more or less impossible to observe even with the aid of space craft. This is because of severe attenuation below the kHz range in the IPM and the interstellar medium (ISM).

The dispersion relationship in a cold plasma may be written as $c^2 k^2 = \omega^2 - \omega_p^2$, where $c$ is the speed of light. We may approximate the magnitude of the phase velocity $\vec{v}_{ph}$ as shown in Eq.(9.83). Assuming $v \gg v_p$, the magnitude of the group velocity $\vec{v}_{gr}$ can be expressed as shown in Eq. (9.84).

$$v_{ph} = \frac{\omega}{k} = \frac{c}{\eta} \approx c\left(1 + \frac{1}{2}\frac{v_p^2}{v^2}\right) \tag{9.83}$$

$$v_{gr} = \frac{d\omega}{dk} = c\eta \approx c\left(1 - \frac{1}{2}\frac{v_p^2}{v^2}\right) \tag{9.84}$$

### 9.8.1 Observing Through Homogeneous Plasma

Many types of propagation effects exist above the cut-off frequency that affects the radio waves passing through the plasma. Let a radio wave be observed through a homogeneous plasma slab having length $L$. In absence of the plasma, the wave is delayed by an amount $\Delta T$, as shown in Eq. (9.85), where $c$ is the speed of light. The magnitude of the propagation delay may also be expressed as in Eq. (9.86).

$$\Delta T = \frac{L}{v_{gr}} - \frac{L}{c} = \frac{L}{c}\left(\frac{1}{\eta} - 1\right) = \frac{Lv_p^2}{2cv^2} \tag{9.85}$$

$$|\Delta T| = \frac{L}{c} \times \frac{4 \times 10^6}{v^2} N \tag{9.86}$$

The equivalent excess path length $\Delta L$ created by the delay is equal to $c\Delta T$. Because $(v_{gr}/c - 1)$ and $(v_{ph}/c - 1)$ differ in sign alone, the magnitude of the excess phase $[2\pi v(L/v_{ph} - L/c)]$ can be obtained as $\Delta\phi = 2\pi v\Delta T$. Because propagation delay is a function of frequency $v$, different frequency components get delayed accordingly. Thus, near the slab's far end, the incident pulse gets smeared out as it propagates through the slab, which is known as *dispersion*. If a magnetic field runs through the plasma, then it becomes birefringent, where the refractive index is different for right and left circularly polarized waves. Thus, the two circularly polarized components are phase shifted by different amounts due to Faraday rotation. As a result, there is a rotation of the plane of polarization. The amount of angular rotation $\Delta\theta$ is given by Eq. (9.87), where $RM$ is the rotation measure, wavelength $\lambda$ is in meters, $N$ is in cm$^{-3}$, $B_0$ is in micro-Gauss, and the length $\Delta r$ of the region has units of parsecs (see also Eq. (9.62)).

$$\Delta\theta = RM\,\lambda^2 = 0.81\lambda^2\Delta r\,N\,B_0 \tag{9.87}$$

### 9.8.2 Observing Through Parallel and Curved Ionospheres

While using an interferometer, one must have a knowledge about the (*i*) time delay $\delta t = \Delta t_1 - \Delta t_2$ between the signals reaching the two arms, where $\Delta t_1$ and $\Delta t_2$ are the propagation delays in the two interferometer arms, and (*ii*) the phase difference $\delta\phi = (2\pi/\lambda)(\Delta L_1 - \Delta L_2)$ between the two signals reaching the two arms, where $\Delta L_1$ and $\Delta L_2$ are the excess path lengths in the two arms at a wavelength $\lambda$. In general, $\delta t$ is small in comparison with the coherence bandwidth of the signal.

Consider a parallel ionosphere with refractive index $\eta$, as shown in Fig. 9.5. From Snell's law, we have $\eta\sin\theta_0 = \sin\theta_1$. The observed geometric delay $\tau_g$ is given by $\tau_g = \eta d\sin\theta_0/c$, where $d$ is the separating distance, $c$ is the speed of light, and $c/\eta$ is the magnitude of the group velocity. Applying Snell's law, we get $\tau_g = d\sin\theta_1/c$. Thus, a homogeneous plane parallel ionosphere has no net effect over the visibilities. Even if the apparent position of the source is changed so that the interferometer is located outside the slab, no change is observed in the apparent position or phase.

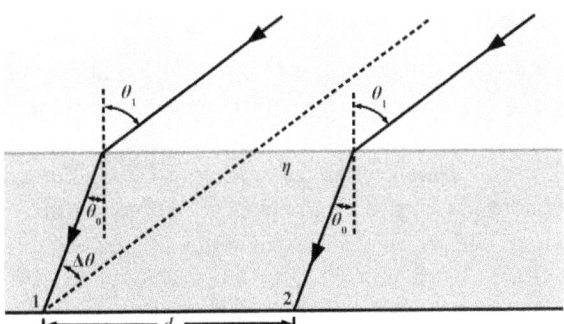

**FIGURE 9.5:** Refraction within a parallel ionosphere.

In reality, the ionosphere is curved and has a radial variation of electron density. Thus the change in apparent position and $\delta\phi$ are nonzero even outside the ionosphere. The situation is shown in Fig. 9.6. The arrival of rays from a distant source appear in a direction different from the actual direction of arrival.

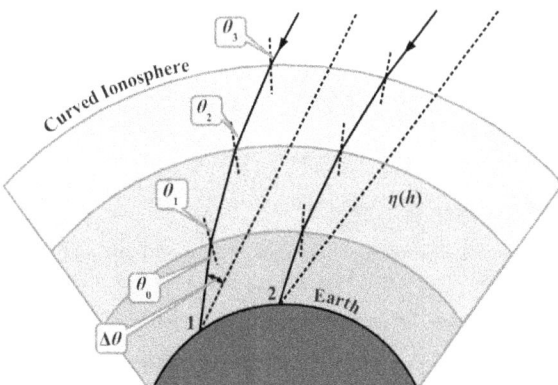

**FIGURE 9.6:** Refraction within a curved ionosphere.

The difference between the true and the apparent directions $\Delta\theta$ is given by Eq. (9.88), where $K$ is a constant, $\theta_0$ is the observed zenith angle, $h$ is the height above the Earth's surface, $r_0$ is the radius of the Earth, and $\eta(h)$ is the refractive index at height $h$. If the baseline uses $u$, $v$ coordinates, the phase difference for the apparent position change of the source may be expressed as in Eq. (9.89), where $\Delta\theta_{EW}$ and $\Delta\theta_{NS}$, respectively, represent the components of $\Delta\theta$ along the east-west and north-south directions.

$$\Delta\theta = \frac{K\sin(\theta_0)}{r_0}\int_0^\infty \left[\frac{\alpha^2\eta(h)}{1-\alpha^2\sin^2(\theta_0)}\right]dh \tag{9.88}$$

$$\Delta\phi = 2\pi\left(u\,\Delta\theta_{EW} + v\,\Delta\theta_{NS}\right) \tag{9.89}$$

### 9.8.3 Observing Through Nonhomogeneous Ionosphere

The Earth's ionosphere shows density fluctuations on large time scales and lengths. Above the Earth's surface at a height $h$, a density fluctuation of length $l$ corresponds to a fluctuation in an angular scale of $l/h$. Typically, for $l = 10$ km at a height $h = 200$ km, the angular scale (phase change) is roughly 3°. Hence an astronomer requires an unresolved source (for use as a phase calibrator) within 3° of the radio source under observation. The problem worsens with a decrease in the observation frequency because the ionospheric phase is proportional to the inverse of the square of the frequency. This phase may be combined with the electronic phase of the receiver system and can be solved by self-calibration, provided the excess ionospheric phase remains constant over the field of view within that period.

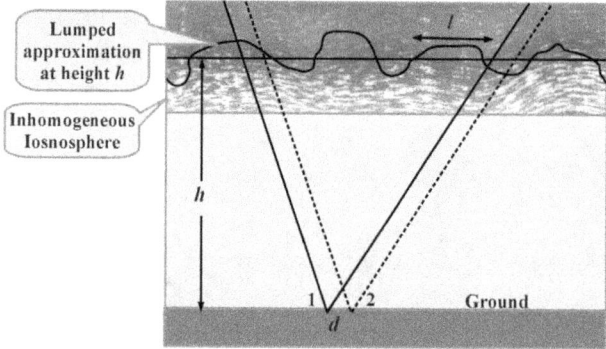

**FIGURE 9.7:** Propagation effects in a nonhomogeneous ionosphere with a short baseline $d$. Both arms of the interferometer receive essentially the same excess phase because they look through almost the same ionosphere.

The field of view of an antenna increases inversely with frequency. The ionospheric phase variation over the field of view is known as *non-isoplanaticity*. The excess ionospheric phase increases rapidly with decreasing frequency. Hence, a self-calibrating algorithm cannot be applied over long baselines. If the interferometer baseline length is smaller than the typical length scale of ionospheric density fluctuations, the excess phase is almost identical at both ends of the baseline. This is shown in Fig. 9.7. Because interferometers are sensitive only to the phase differences (between the two antenna signals), the isoplanatic assumption is still applicable. When both the baselines and the field of view are sufficiently large, the non-isoplanaticity situation arises.

### 9.8.4 Broadening of Sources Due to Scattering From Ionosphere

The Earth's ionosphere consists of plasma irregularities that affect the waves passing through them. As the waves progress, they are scattered in various directions depending on the plasma irregularities. Hence the observer sees an angular broadening of the source, as shown in Fig. 9.8.

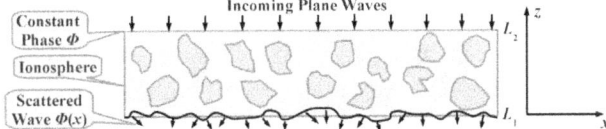

**FIGURE 9.8:** Signals are scattered in various directions from an ionosphere layer containing plasma irregularities. The observer sees an angular broadening of the source.

Small-scale fluctuations of ionospheric electron density contribute excess phase to radio waves propagating through it. If the change in the refractive index $\Delta\eta$ is due to the electron density fluctuation, then the excess phase can be expressed as in Eq. (9.90), where $\Delta N(x,z)$ is the fluctuation of electron density at the point $(x,z)$, $\lambda$ is the wavelength, and $C$ is a constant. As shown in Fig. 9.8, the integration is performed over the entire path traversed by the ray.

$$\phi(x) = \frac{2\pi}{\lambda} \int_{L_1}^{L_2} \Delta\eta \, dz = C\lambda \int_{L_1}^{L_2} \Delta N(x,z) \, dz \qquad (9.90)$$

Let $\phi(x)$ be a Gaussian process having zero mean, as shown in Eq. (9.91), where $\sigma$ is the standard deviation. The autocorrelation function $R_{\phi\phi}$ will be as shown in Eq. (9.92), where $\phi_0 = 1/\sqrt{2\pi\sigma^2}$ and $\rho(r) = \exp\left(-r/(2\sigma^2)\right)$.

$$\phi(x) = \frac{1}{\sqrt{2\pi\sigma^2}} \exp\left(-\frac{x}{2\sigma^2}\right) \qquad (9.91)$$

$$R_{\phi\phi} = \left\langle \phi(x)\phi(x+r) \right\rangle = \frac{1}{2\pi\sigma^2} \exp\left(-\frac{r}{2\sigma^2}\right) = \phi_0^2 \, \rho(r) \qquad (9.92)$$

From Eq. (9.90), we find that $\phi_0$ is proportional to $\lambda^2 (\Delta N)^2 L$, where $L$ is the total path length of the ionosphere. Consider a plane wave front produced from an extremely distant point radio source. Let the wave-front be incident at the top of the ionosphere. The wave reaching the surface of the Earth would be a plane wave if the ionosphere was absent. For this case, the visibility or correlation function of the electric field is obtained as $\left\langle E_i(x)E_i^*(x+r) \right\rangle = E_i^2$, which is a constant and independent of $r$. In reality, the ionosphere exists. Thus, different parts of the wave front acquire different phases. Hence the emerging wave front is not planar. If $E(x)$ is the electric field at some point within the emergent wave, we may write $E(x) = E_i e^{-j\phi x}$. Because $E_i$ is a constant, the correlation of the emergent field can be written as in Eq. (9.93). From the statistical nature of $\phi(x)$, we can modify Eq. (9.93) to obtain Eq. (9.94).

$$\left\langle E_i(x)\, E_i^*(x+r)\right\rangle = E_i^2 \left\langle e^{-j(\phi(x)-\phi(x+r))}\right\rangle \tag{9.93}$$

$$\left\langle E_i(x)\, E_i^*(x+r)\right\rangle = E_i^2 e^{-2\phi_0^2(1-\rho(r))} \tag{9.94}$$

Because $(1-\rho(r))$ increases with increasing $r$, when $\phi_0^2$ is extremely large, the exponent tends to zero. Hence it may be adequate to evaluate Eq. (9.94) with small values of $r$. This may be done by expanding $\rho(r)$ using a suitably truncated Taylor series, such as $\rho(r)=1-r^2/2a_\phi^2$. Eq. (9.94) can then be re-expressed as in Eq. (9.95).

$$\left\langle E_i(x)\, E_i^*(x+r)\right\rangle = E_i^2 e^{-\phi_0^2\frac{r^2}{\sigma^2}} \tag{9.95}$$

Thus we conclude that the emergent electric field has a finite coherence length. In view of the van Cittert-Zernike equation, it may be again concluded that the original unresolved point source has blurred out to a source of finite size. This type of blurring of point sources is known as *scatter broadening* or *angular broadening*. If we define $a=\sigma/\phi_0$, the visibilities follow a Gaussian distribution and can be expressed as $e^{-ir^2}/a^2$. In other words, the characteristic angular size $\theta_{\text{scat}}$ of the scatter broadened source may be roughly shown as $\theta_{\text{scat}} \sim \lambda/a \propto \lambda^2\sqrt{(\Delta N)^2 L}$ where $\theta_{\text{scat}}$ is called the *scattering angle*. For small $\phi_0^2$, the exponent in Eq. (9.94) is expanded using a Taylor series in Eq. (9.96). This represents the visibilities of an unresolved core having a flux density $E_i^2(1-2\phi_0^2)$, surrounded by a faint halo.

$$\left\langle E_i(x)\, E_i^*(x+r)\right\rangle = E_i^2\left[1-2\phi_0^2(1-\rho(r))\right] \tag{9.96}$$
$$= E_i^2\left[(1-2\phi_0^2)+2\phi_0^2 e^{-\frac{r^2}{2\sigma^2}}\right]$$

### 9.8.5 Scintillations Produced by Ionosphere

The ionosphere considered so far had random density fluctuations with position, but not with time. In reality, the density varies both with position and time. Temporal variations are a result of both (*i*) intrinsic variations, and (*ii*) traveling disturbances. Temporal fluctuations in density variations cause the coherence function to vary with time. This is known as scintillation in radio astronomy. The scintillation could be weak or strong, depending on both the height of the scattering layer from the Earth and the scattering angle.

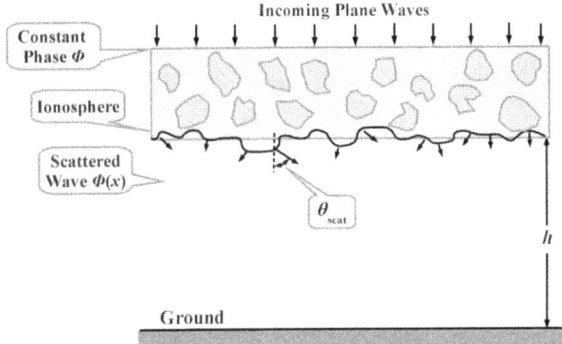

**FIGURE 9.9:** Scintillations caused by the ionosphere.

The rays get scattered by an angle $\theta_{\text{scat}}$ when they pass through an irregular ionosphere. Let the scattering take place at a height $h$ above the antennas. As shown in Fig. 9.9, the scattered rays traverse a distance $h$ before being detected. Then the distance traversed by a scattered ray is roughly $h\theta_{\text{scat}}$. For very small lengths compared to the coherence length $a$, the scattered rays (due to different irregularities) do not intersect before hitting the ground. This condition is $h\theta_{\text{scat}} < a$. Mathematically, $h\theta_{\text{scat}} < \lambda/\theta_{\text{scat}}$, or $h\theta_{\text{scat}}^2 < \lambda$. Under these conditions, at any instant of time, the observer finds an undistorted image of the source, the position of which is shifted because of refraction. The density fluctuations vary with time, causing the image to appear as wandering across the sky. The image created from long exposure is essentially the average of many such wanderings. Hence the source appears to have a scatter-broadened size $\theta_{\text{scat}}$. This effect may be corrected only by self-calibration on a time scale smaller than the time scale over which the image wanders. When $h\theta_{\text{scat}}^2 > \lambda$, the intersecting rays from different density fluctuations interfere with one another. More than one image is seen by the observer, and the amplitude of the received signal fluctuates with time due to interference. This is *amplitude scintillation*. At frequencies below 500 MHz, scintillation can be very strong. On very short time scales, the source flux can vary by a factor of two or more. One cannot reliably eliminate this effect from data. Hence, observations are avoided during powerful amplitude scintillations.

| BOX 9.1 | **Plane Waves in Conducting and Dielectric Media** |
|---|---|

In Section 9.3, we considered plane waves in a lossless media where $\sigma = 0$. Plane waves in conducting media with $\sigma \gg \omega\epsilon$ were treated in Section 9.4. Plane waves in ionized media with magnetic fields were treated in Section 9.5. We seek the general plane wave solution to the wave equations

$$\left. \begin{array}{l} \nabla^2 \vec{E} - \gamma^2 \vec{E} = 0 \\ \nabla^2 \vec{B} - \gamma^2 \vec{B} = 0 \end{array} \right\} \qquad (9.97)$$

in media with nonzero $\sigma$ and $\epsilon$, but with zero background magnetic field. Here we solve for the real and imaginary parts of the complex propagation constant

$$\gamma = \sqrt{j\omega\mu\sigma - \omega^2\mu\varepsilon} \,. \qquad (9.98)$$

We write $\gamma = \alpha + j\beta$ and square it to solve for $\alpha$ and $\beta$, giving

$$\gamma^2 = \alpha^2 - \beta^2 + 2j\alpha\beta = j\omega\mu\sigma - \omega^2\mu\varepsilon \,. \qquad (9.99)$$

Equating the real and imaginary parts of Eq. (9.99) gives

$$\alpha^2 - \beta^2 = -\omega^2\mu\varepsilon \qquad (9.100)$$

$$2\alpha\beta = \omega\mu\sigma \qquad (9.101)$$

Squaring (9.101) and substituting $\beta^2$ from (9.100) and solving for $\alpha$ gives

$$\alpha = \omega \left\{ \frac{\mu\varepsilon}{2} \left[ \sqrt{1 + \left(\frac{\sigma}{\omega\varepsilon}\right)^2} - 1 \right] \right\}^{1/2} \,. \qquad (9.102)$$

Substituting $\beta^2$ from (9.100) and solving (9.101) for $\beta$ gives

$$\beta = \omega \left\{ \frac{\mu\varepsilon}{2} \left[ \sqrt{1 + \left(\frac{\sigma}{\omega\varepsilon}\right)^2} + 1 \right] \right\}^{1/2} \,. \qquad (9.103)$$

If the waves are polarized with $\vec{E} = (0, E_y, 0)$ and $\vec{B} = (B_x, 0, 0)$ propagating in the direction of $\vec{E} \times \vec{B}$ along the z-axis, we have

$$E_y = E_0 \exp(-\alpha|z|)\exp(j\omega t \pm j\beta z) \qquad (9.104)$$

$$B_x = B_0 \exp(-\alpha|z|)\exp(j\omega t \pm j\beta z) \,. \qquad (9.105)$$

These waves are attenuated with distance as in Eq. (9.12), except here the attenuation constant is given by Eq. (9.102).

## REVIEW QUESTIONS

1. A plane electromagnetic wave has its electric field aligned along the x-axis and propagates along the z-axis through a lossless medium. If the angular frequency is $\omega$ and the phase constant is $\beta$, write an equation representing the wave in time and space. [***Hint:*** See Eq. (9.9)]

2. Modify the above equation for a conducting medium having an attenuation constant $\alpha$.

3. How are the attenuation phase constants related to the propagation constant? What is meant by depth of penetration?     [**Hint:** See Section 9.4]

4. Calculate the force produced by an electron travelling with a velocity $10^6$ m/s perpendicular to a magnetic field of 5 mT. Assume the charge is $-1.602 \times 10^{-19}$ C.                    [**Hint:** Use Eq. (9.14)]

5. A proton travels with a velocity $10^6$ m/s perpendicular to a magnetic field of 5 mT. Find the radius of curvature of the proton's path. Assume the mass of the proton is $m = 1.673 \times 10^{-27}$ kg and its charge is $1.602 \times 10^{-19}$ C.

   [**Hint:** Centrifugal force $F = \dfrac{mv^2}{R}$, where $R$ is the radius of curvature.

   Balance this with Eq. (9.14), and solve for $R$. Ans: 20.8 m]

6. An electron rotates in a magnetic field of 5 mT. Calculate the gyrofrequency, given its mass is $9.10938 \times 10^{-31}$ kg and its charge is $-1.602 \times 10^{-19}$ C. What is the radiation frequency?

                    [**Hint:** Use Eq. (9.25) and then convert to $v_g$.]

7. Calculate the plasma frequency of the ionosphere, given the number of electrons as $10^{12}$ per m$^3$.                    [**Hint:** Use Eq. (9.38)]

8. What is the significance of critical frequency, and how is it related to the plasma frequency of the ionosphere?

9. Explain the concept of Faraday rotation using a diagram and equations.
                    [**Hint:** Use Fig. 9.2 and Eq. (9.53)]

10. Calculate the total Faraday rotation at a frequency of 150 MHz, taking the density of electrons in the ionosphere as $10^{12}$ per m$^3$, the magnetic field as $10^5$ gauss, the angle between the direction of propagation and the magnetic field as 30°, and the length of the ionosphere as 1000 km.

                    [**Hint:** calculate the wavelength and use Eq. (9.61)]

11. Explain the principle of magnetohydrodynamic wave generation in a conducting liquid in the presence of a magnetic field using a diagram. [Hint Use Fig. 9.4]

12. Given a plasma frequency of 9 MHz, calculate the refractive index of the ionosphere at the following frequencies: (*i*) 150 MHz, (*ii*) 233 MHz, (*iii*) 327 MHz, (*iv*) 610 MHz, and (*v*) 1420 MHz.     [**Hint:** Use Eq. (9.81)]

304 • Radio Astronomy

13. Calculate the group and phase velocities for problem no. 10 at all the given frequencies.                    [*Hint:* Use Eqs. (9.83) and (9.84)]

14. If the ionosphere is homogeneous and parallel, are there any net effects over the visibilities? If not, why not?          [*Hint:* See Section 9.10]

15. What are the effects of a curved homogeneous ionosphere over the visibilities if the interferometer baselines are long?          [*Hint:* See Fig. 9.6]

16. Why are phase calibrators used within 3° of target source? What happens when phase calibrators are at larger angles from the target source? Explain using a diagram.          [*Hint:* See Fig. 9.7]

17. What are the reasons for broadening of the source? Explain with a diagram.          [*Hint:* See Section 9.8.4 and Fig. 9.8]

18. How are scintillations produced? Is there any remedy for scintillations?

# 10

# *THE GMRT RADIO ARRAY*

## 10.1   INTRODUCTION

In this chapter we shall describe the GMRT (Giant Meterwave Radio Telescope), which is a multiplicative interferometer array consisting of thirty antennas spread within a radius of 15 km.

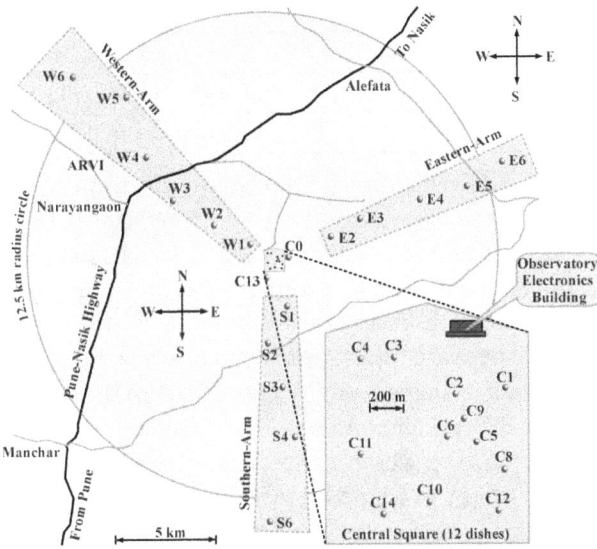

**FIGURE 10.1:** A map of the GMRT antenna array. The observatory building is located near the center. Fourteen antennas form a short baseline array around the observatory building.

**TABLE 10.1:** Position coordinates of the GMRT antennas. Altitude = 650 m.

| Antenna | Lat. (deg) | Lon. (deg) | Antenna | Lat. (deg) | Lon. (deg) |
|---------|-----------|-----------|---------|-----------|-----------|
| C0 | 19.0929 | −74.0570 | E2 | 19.1023 | −74.0772 |
| C1 | 19.0927 | −74.0536 | E3 | 19.1117 | −74.0940 |
| C2 | 19.0931 | −74.0505 | E4 | 19.1209 | −74.1244 |
| C3 | 19.0944 | −74.0469 | E5 | 19.1251 | −74.1474 |
| C4 | 19.0931 | −74.0505 | E6 | 19.1365 | −74.1652 |
| C5 | 19.0908 | −74.0511 | W1 | 19.0988 | −74.0353 |
| C6 | 19.0910 | −74.0502 | W2 | 19.1066 | −74.0210 |
| C8 | 19.0893 | −74.0531 | W3 | 19.1208 | −74.0011 |
| C9 | 19.0917 | −74.0509 | W4 | 19.1415 | −73.9836 |
| C10 | 19.0875 | −74.0489 | W5 | 19.1678 | −73.9734 |
| C11 | 19.0901 | −74.0447 | W6 | 19.1784 | −73.9436 |
| C12 | 19.0871 | −74.0521 | S1 | 19.0664 | −74.0565 |
| C13 | 19.0825 | −74.0444 | S2 | 19.0523 | −74.0470 |
| C14 | 19.0871 | −74.0460 | S3 | 19.0320 | −74.0536 |
| — | — | — | S4 | 19.0074 | −74.0595 |
| — | — | — | S6 | 18.9653 | −74.0470 |

Fig. 10.1 shows the map of the GMRT radio array. There are fourteen antennas, C0 to C14 (with C7 absent), close to the observatory building. Except for C0 and C14, the rest of these antennas are within the boundary area of the observatory. These antennas are called *central square* antennas and form a short baseline array. The electronics building, observatory control room, labs, library, and offices etc. are inside the observatory building. There are three arms for longer baselines that extend approximately 15 km to form a Y-shape, namely (*i*) an eastern arm, (*ii*) a western arm, and (*iii*) a southern arm. The eastern arm consists of five antennas (E2 to E6). Note that there is no E1 antenna. The western arm consists of six antennas (W1 to W6). The southern arm consists of five antennas (S1 to S4 and S6). Note that there is no S5 antenna. The position coordinates of all the antennas are shown in Table 10.1.

Instead of a solid dish, wire mesh is used to reduce the weight and cost, but at the expense of dish leakage at higher frequencies. The mesh also reduces resistance against air flow. The wind meter monitors the speed of the wind. If it is greater than 40 km/h, the antennas are automatically brought to a position facing

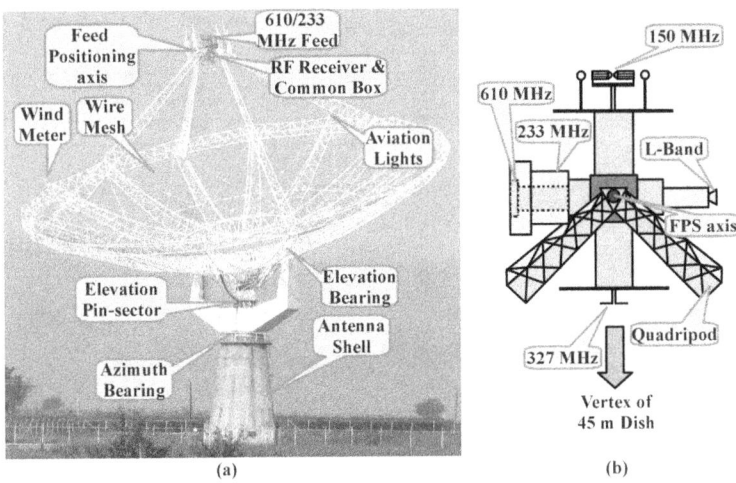

**FIGURE 10.2:** (a) A picture of a GMRT antenna (array element). (b) Mounting geometry of the antenna feeds.

the zenith and all observations are stopped. It is kept locked in this parked position until observing conditions are favorable. The aviation lights are used at night as a precaution for airplanes.

## 10.2 ARRAY ELEMENTS

Each array element is a dish antenna having a diameter of 45 m. The details are shown in Fig. 10.2a. There are four antenna feeds: (*i*) 150 MHz, (*ii*) 327 MHz, (*iii*) 610/233 MHz (dual frequency), and (*iv*) 1000-1450 MHz (L-band), as shown in Fig. 10.2b. The first three antenna feeds have a minimum usable bandwidth of 32 MHz. The third feed has two frequencies, 610 MHz and 233 MHz, that can be operated together. However, due to channel restrictions, only one polarization can be used in dual-mode observations. The feeds are all mounted on a rotating assembly, but at any given time, only one of the antenna feeds can be brought to the focus of the dish. This is done with the help of the FPS (feed-positioning system), as shown in Fig. 10.2b. Each antenna feed has two linear cross polarizations, namely V and H. The frontend receiving system (RF receiver) is fitted near the antenna feeds. Table 10.2 lists the primary HPBW (half-power beam widths) of the dish at different frequencies. These ae calculated using the Rayleigh criterion $1.22\lambda/D$, where $\lambda$ is the wavelength and $D$ is the dish diameter (45m).

**TABLE 10.2:** The half-power beam widths at different frequencies, calculated using the Rayleigh criterion $(1.22\lambda/D)$.

| Freq. | 150 MHz | 233 MHz | 327 MHz | 610 MHz | 1000 MHz | 1200 MHz | 1400 MHz |
|-------|---------|---------|---------|---------|----------|----------|----------|
| **HPBW** | 3.10° | 2.00° | 1.43° | 0.76° | 0.47° | 0.39° | 0.33° |

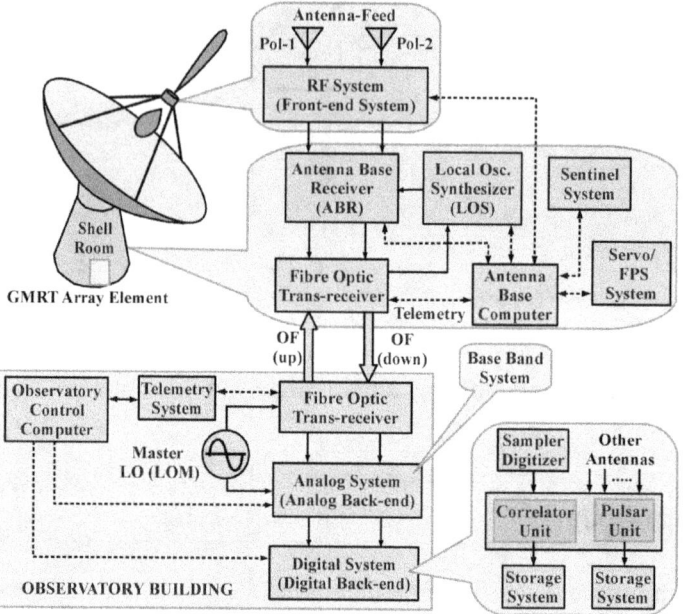

**FIGURE 10.3:** Functional blocks of the GMRT. The telemetry, control, and monitoring signals are shown by dotted lines. The RF, IF, and data signals are shown by dark lines.

Each antenna feed uses a separate frontend RF receiver. The outputs of these are fed into a common receiver (common box) that selects the feed to be used. Signal outputs from these are carried over to the antenna shell, where it is converted to IF and sent over through an OF (optical fiber) to the observatory building. The antenna shell also consists of servo and feed-positioning amplifiers. All of these are controlled from the ABC (antenna base computer) inside the shell. The control commands to the ABC from the observatory are sent through the OF.

## 10.3   FUNCTIONAL BLOCKS OF GMRT

The basic functional blocks of the GMRT are shown in Fig. 10.3. The telemetry, control, and monitoring signals are shown by dotted lines. The RF, IF, and data signals are shown by dark lines. The frontend RF system, consisting of an RF receiver and a common box, is mounted very close to the antenna feeds. Vertical and horizontal polarizations (V and H) are processed separately and simultaneously and are brought into the antenna shell to the ABR (antenna base receiver). The two signals undergo two stages of mixing and are finally downconverted to 130 MHz and 175 MHz IF signals. These are passed to the optical transreceiver, where they are converted to optical signals and sent to the observatory electronics building through the OF downlink (return link). The LO signals for downconversion are generated in a synthesizer that is locked to the LOM (master local oscillator) kept in the observatory building. The LOM signal is passed to all the antennas through the OF uplink (forward link).

Inside the observatory building, the signals from the antennas are converted back to IF from optical and are then fed to the analog backend receiver system. The same LOM is used as a reference oscillator to recover the signals and convert them to BBs (basebands). These are then passed to the digital backend system, which consists of two correlator systems in parallel, namely the GHB (GMRT hardware backend) and the GSB (GMRT software backend). Both the GHB and the GSB function by converting the baseband signals into digital signals, correlating them or processing them for pulsar beam formation, and saving the final results. The difference between the GHB and the GSB lies only in the architecture; the end results are the same. However, the GSB has a greater frequency resolution. The GSB is a new system that is expected to completely replace the GHB.

The servo system is used to control the azimuth and elevation of the antenna dishes. It consists of servo amplifiers for controlling the motor drives. The FPS is also a motor-driven system. It is used to focus the various antenna feeds on the dish. The sentinel system protects the antenna-shell electronics from fire, smoke, and heat. It can automatically turn off systems like the servo when conditions are not favorable. It also sets the alarm under such conditions. The temperature inside the antenna shell is constantly monitored by this system. Each of the systems in an antenna is controlled by telemetry signals reaching the ABC. It also brings monitored data into the control room from systems like the sentinel.

### 10.3.1 The RF Frontend System

Fig. 10.4 shows the blocks of the frontend low-noise receiving units. These are used next to the antenna feeds (before the common box). The system blocks used with 150 MHz, 233/610 MHz, and 327 MHz antenna feeds are similar and are shown in part (a) of the figure. The two linearly cross-polarized outputs (V and H) from the antenna feed are fed to a polarizer for conversion to circular polarization.

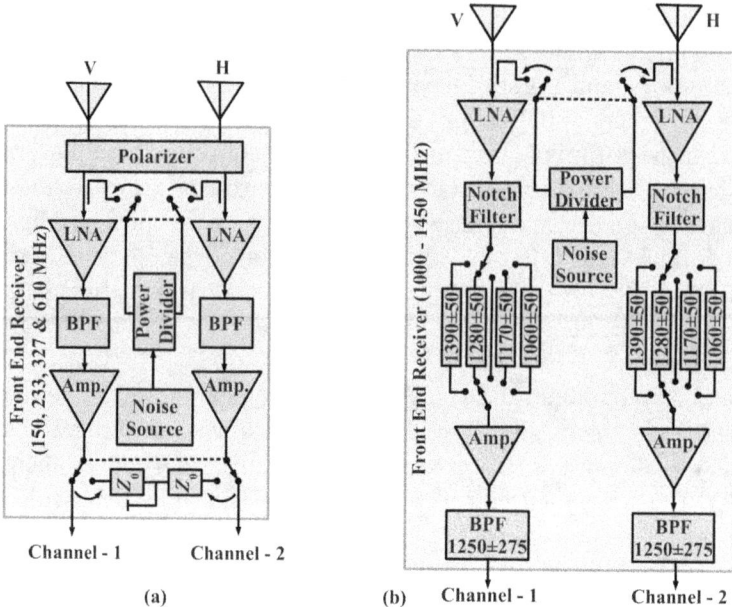

**FIGURE 10.4:** Low-noise receiving units (before the common box) of the RF frontend system. (a) System blocks used with 150 MHz, 233/610 MHz, and 327 MHz antenna feeds. (b) System blocks used with the L-band (1000–1450 MHz) antenna feed.

The two circularly polarized outputs are then sent to two LNAs and then passed through the BPF (bandpass filters) for removing unwanted signals. These are again amplified and output as two frequency channels (1 and 2), which are passed to the common box. These can be terminated to matched loads $Z_0$ using an RF SP2T switch for testing.

Part (b) of the figure shows the same for the L-band antenna feed. Notice that the polarizer is not used here, because ionospheric effects like Faraday rotation, source broadening, scintillation, etc., are less severe at frequencies above 1 GHz. Notch filters are used to prevent the GSM signals from entering into the receiving chain. Any of the four BPFs centered at 1060, 1170, 1280, and 1390 MHz, with a bandwidth of 100 MHz, may be chosen for observations. A direct path is also provided for bypassing these filters. The two polarized outputs appear as two frequency channels (1 and 2), which are passed to the common box.

In both modules, a noise source is used for calibration by coupling equal amounts of coherent noise in both the channels by using directional couplers, as shown. The noise source is turned on, and a SPDT switch is used to connect the outputs to the directional couplers.

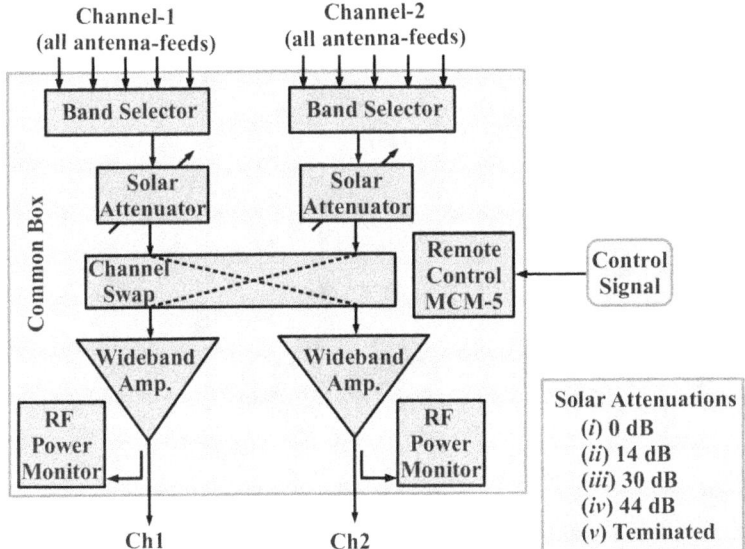

**FIGURE 10.5:** Common box (frequency multiplexer and variable gain/attenuator) following the low-noise receiving units.

The common box chooses the frequency of operation by selecting the antenna feed and setting the various BPFs and the attenuation in the front-end RF unit. The operational commands to the common box are decoded by an MCM (machine control and monitoring) unit, as shown in Fig. 10.5. The common box can swap the polarized channel outputs (Ch1 and Ch2) when this is required. It can provide specified amounts of attenuation (0 dB, 14 dB, 30 dB, and 44 dB) by using the solar attenuators. Sometimes it is necessary to check the authenticity of an interfering signal (to determine if it is generated within the system or coming through the antenna from outside). This is done by terminating the front-end input in the attenuator setup.

### 10.3.2 IF Receiver at Antenna

Fig. 10.6a shows the blocks of the IF receiver, also known as the ABR. The LO signals are obtained from the LOR (LO reference generator), as shown in Fig. 10.6b. The outputs from the common box (Ch1 and Ch2) are taken as input and downconverted to the first IF of 70 MHz. This bandwidth can be set by external control to 16 MHz, 6 MHz, or an unlimited amount. Attenuation ranging from 0 to 30 dB in steps of 0.5 dB can be applied to the first IF, as shown. This is known as *preattenuation*. The 70 MHz first IF are converted to a second IF of 130 MHz and 175 MHz, respectively, by mixing with the LO signals of 200 MHz and 105 MHz.

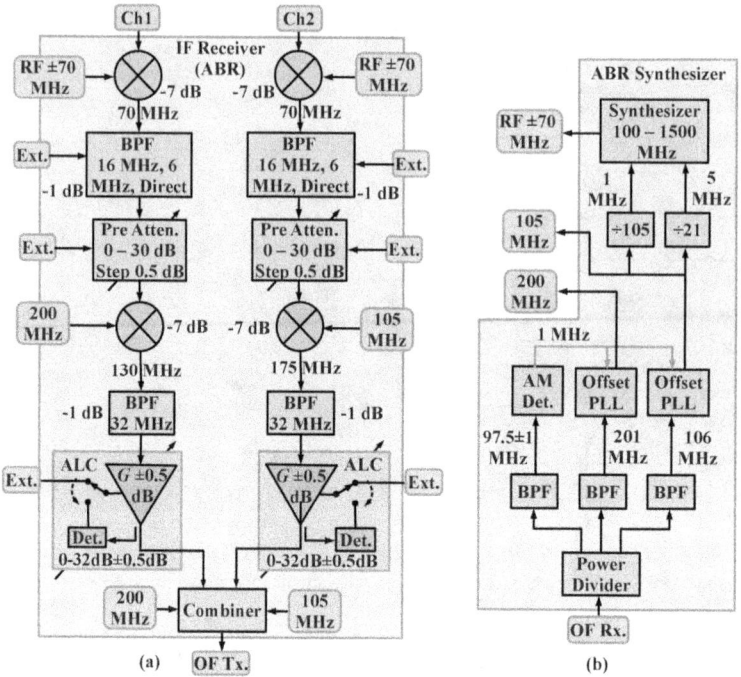

**FIGURE 10.6:** IF receiver at the antenna shell. The externals controls are represented by Ext. (a) Antenna base receiver or ABR. (b) LO reference generation at antenna shell or LOR.

Thus, the final IF for Ch1 and Ch2 are, respectively, 130 MHz and 175 MHz, and their bandwidths are kept at 32 MHz by using the BPF, as shown.

An automatic level control (ALC) system may be used to provide a gain correction within a range of 0 to 32 dB in steps of 0.5 dB. It has two modes: automatic and manual. The automatic mode detects the power level, on which it then bases an increases or decreases in gain. In the manual mode, fixed or changing gains can be set using an external control.

Finally, the two frequency signals and the second LO signals are combined in a power combiner and sent to the optical transmitter, which then sends these signals to the observatory building through the downlink fiber.

The LO signal required for the first IF conversion is simply RF ± IF. However, GMRT frequencies range roughly from 120 MHz to 1450 MHz. Hence, a wide frequency range synthesizer is required. As mentioned above, all the LO frequencies must be generated with reference to the master LO at the observatory building. Three LO signals (generated with reference to the LOM) reach the antenna shell through the uplink fiber. These are set at 201 MHz, 106 MHz, and 97.5 ±

1 MHz The last one is an amplitude modulation of a 97.5 MHz carrier with a 1 MHz signal. These are recovered from the optical receiver and then separated using a power divider and some BPF. The 1 MHz signal is recovered and used with the 206 MHz and 106 MHz signals to generate 205 MHz and 105 MHz LO signals, respectively. Using these two signals, a broad-range synthesizer generates the first LO frequency, which ranges from 100 MHz to 1500 MHz.

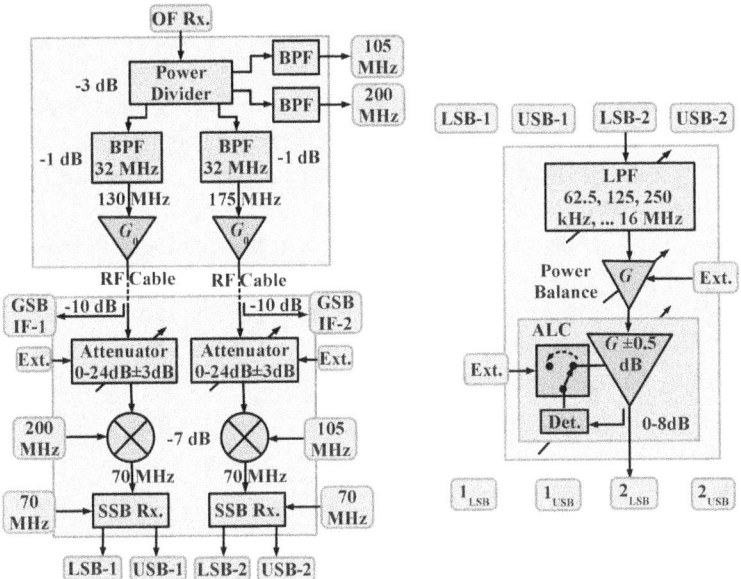

**FIGURE 10.7:** Baseband system at the observatory building. The LO signals comes from the LOM reference generator.

### 10.3.3 Baseband System at the Observatory

Fig. 10.7 shows the functional blocks of the IF-to-baseband conversion. The signals from the antenna (130 MHz and 175 MHz) corresponding to the two polarization channels[1] and their LO signals (200 MHz and 105 MHz) are converted back from optical to radio and then separated using the BPF. The maximum IF bandwidths are 32 MHz, and they are recovered with two BPF centered at 130 MHz and 175 MHz, with 32 MHz bandwidths, as shown. These IF signals are again boosted by fixed gains of $G_0$ and carried over to the next stage using two RF cables. Two signals (IF-1 and IF-2) are tapped out using directional couplers (−10 dB), for use with the GSB. We shall call this *baseband system II*.

---

[1] Channels 1 and 2 with reference to 150, 233, 327, and 610 MHz are right and left circular, whereas in the L-band, the signals are linearly polarized (horizontal and vertical).

Attenuation is applied between 0 to 24 dB in steps of 3 dB to bring the IF signals to a standard power level, using external control. These are then mixed with 200 MHz and 105 MHz LO frequencies to obtain 70 MHz IF (third IF) that are given to SSB (single sideband) receivers (mixers). The LO for the SSB mixers is 70 MHz. Two sidebands (USB and LSB) of 16 MHz width are separated out from each of the IF bands of 32 MHz. A variable bandwidth LPF (62.5 kHz, 125 kHz, 250 kHz, ... 16 MHz) is used to select the required spectrum for output. These bandwidth selections may be required especially during spectral-line observations.

The overall power of the continuum signals is a function of bandwidth. Hence, power-level changes due to bandwidth selection is automatically or manually compensated by power-balancing amplifiers. The output is passed through an ALC circuit that has a gain of between 0 and 8 dB and corrects it in steps of 0.5 dB. The ALC can be switched off by external control. The final outputs are labeled $1_{LSB}$, $1_{USB}$, $2_{LSB}$, and $2_{USB}$, where 1 and 2 represent the polarization of the channels. These are taken to the digital system for conversion to digital and further processing.

It must be remembered that from a 32 MHz observation bandwidth, four baseband outputs of 16 MHz (or less) bandwidths are produced. Further processing is done separately for the USB and LSB in the digital system.

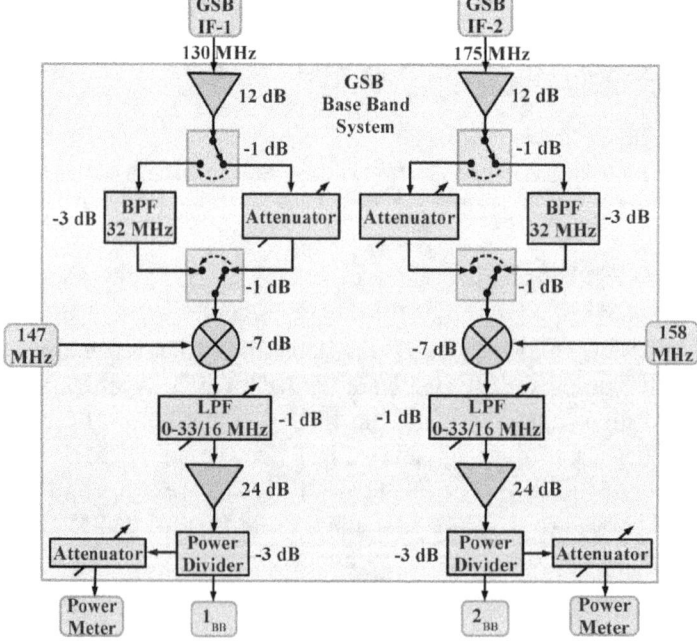

**FIGURE 10.8:** Baseband system II for the GSB correlator and pulsar receiver at the observatory building.

### 10.3.4 Baseband System II at Observatory

Functional blocks of baseband system II for use with the software-based correlator and pulsar receiver (GSB) is shown in Fig. 10.8. The tapped IF (GSB IF-1 and GSB IF-2, respectively, centered at 130 MHz and 175 MHz), with a 32 MHz bandwidth, are passed to baseband system II. After amplification, the signals are passed through a BPF (or an attenuator) using two SPDT switches. The outputs are mixed with 147 MHz and 158 MHz LO to recover the full baseband of 33 MHz. The LPF following this decides the size of the basebands. These filters can be set to either 33 MHz or 16 MHz depending on the user's choice and the type of observation. The outputs are amplified and divided into two for (*i*) power measurement, and (*ii*) use by the software backend. Note that unlike the previous baseband system that generates four outputs per antenna, the present system generates only two baseband outputs ($1_{BB}$ and $2_{BB}$), each with a double bandwidth.

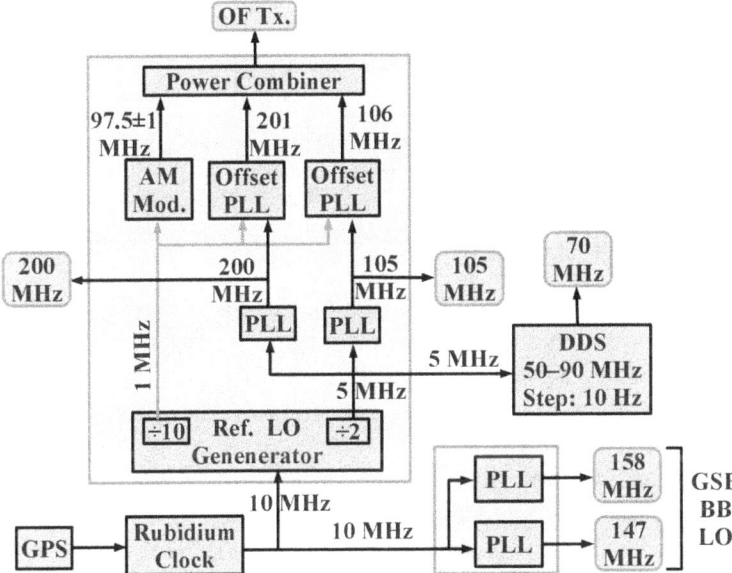

**FIGURE 10.9:** The LOM and reference LO generator at the observatory building.

### 10.3.5 LOM Reference Generator at Observatory

Fig. 10.9 shows the master local oscillator and reference clock (LO) generations. The rubidium clock is used as the standard clock generator and is synchronized with the GPS. The 10 MHz clock output is sent to a reference generator that generates two frequencies (1 MHz and 5 MHz) by division. The 5 MHz signal generates 200 MHz and 105 MHz signals using PLL (phase locked loops). From these frequencies and using offset PLL, 201 MHz and 106 MHz are generated with the 1 MHz signal offset. The 1 MHz signal amplitude modulates a carrier of 97.5 MHz. The modulated output is combined with the 201 MHz and 106 MHz signals and sent to the optical transmitter for transmission to the antennas using the uplink fiber.

A separate DDS (direct digital synthesizer) unit generates a 70 MHz signal from the 5 MHz input signal. The 200 MHz, 105 MHz, and 70 MHz signals are used as LO frequencies by the baseband receiving system in the observatory building. These are also used in every antenna because they are derived from the OF output.

Two additional LO frequencies (158 MHz and 147 MHz) are generated from the PLL for use in baseband system II. The same rubidium clock is used as a reference.

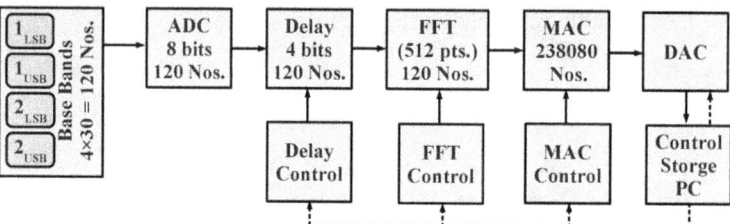

**FIGURE 10.10:** Functional blocks of the hardware correlator and data storage systems at the observatory building.

### 10.3.6 Hardware Correlator and Pulsar Receiver

The correlator and pulsar receivers form the digital backend of the GMRT. Presently, there are two systems based on hardware and software, respectively, known as the GHB (GMRT hardware backend) and the GSB (GMRT software backend). Fig. 10.10 shows the functional blocks of the hardware correlator. It consists of ADC (analog-to-digital convertor) cards, delay-compensating cards, FFT cards, MAC (multiplier-and-accumulator) cards, and data-accumulator cards, which are PC add-on cards. The sampling frequency of the ADC is 32 MHz (double the baseband width). DSP processors are used in control circuits (except for the ADC), which are controlled by the control storage PC, in which the data are saved.

*10.3.6.1 Common Subsystem for Correlator and Pulsar Receiver*
Some portion of the hardware is common to both the correlator and the pulsar-receiving systems. These consist of digitization, delay compensation, and Fourier transformation, as shown in Fig. 10.11a.

All four baseband outputs ($1_{LSB}$, $1_{USB}$, $2_{LSB}$, and $2_{USB}$) from each antenna are sampled and processed. Correlations are performed separately for the USB and

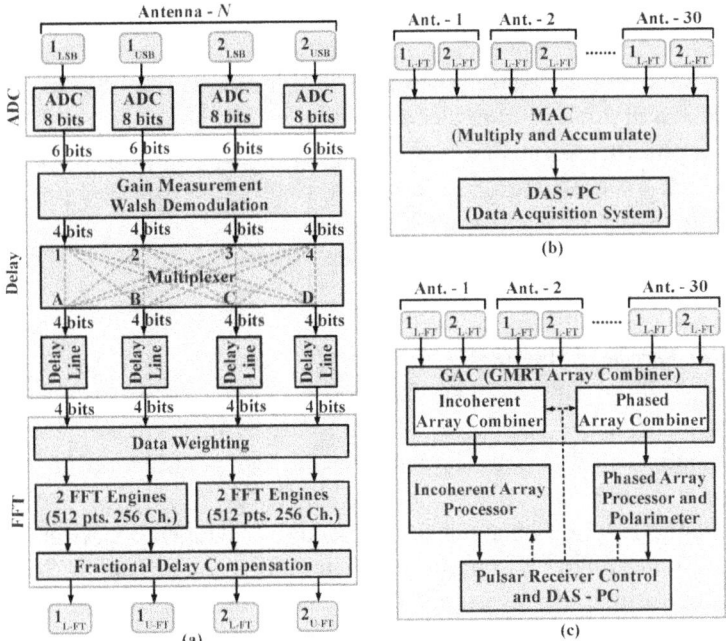

**FIGURE 10.11:** (a) Signals are sampled, delay-compensated, and Fourier-transformed. (b) FX correlation and data acquisition. (c) Pulsar receiver.

LSB between all antennas. The instrumental delays between any two antennas are compensated by the delay cards. Each of the baseband outputs are sampled and digitized to eight bits. The delay compensation card takes the six MSB (most significant bits) of the digitized data and rounds them to four bits for further processing. Provision for (*i*) gain measurement, and (*ii*) Walsh demodulations, exist along with an inbuilt multiplexer. If the baseband outputs have bandwidths less than 16 MHz (due to user selection), the delay card effectively resamples the digitized signal so that the signal appears Nyquist sampled downstream of the delay unit.

The four rounded bits are sent to a multiplexer that has various modes of selection. For example, any of the four inputs (1, 2, 3, 4) can be mapped to all four of its outputs (A, B, C, D). Other mappings for polarization observations with the correlator are also included, such as $A = 1_{LSB}$, $B = 2_{LSB}$, $C = 2_{LSB}$, $D = 1_{LSB}$, and $A = 1_{USB}$, $B = 2_{USB}$, $C = 2_{USB}$, $D = 1_{USB}$. The outputs of the multiplexer are given to the adjustable delay lines that use shift registers to compensate for geometric delay.

The four-bit delay-compensated outputs are sent to the FFT cards. Each FFT subsystem takes two data streams (either A, B or C, D) from the delay card. Each FFT-engine can perform 512 point FFT, giving 256 channels (due to symmetry) across zero frequency. The length can be reduced to 256, 128, or 16 points. The

FFT can also be by-passed if required. A software-selectable weighting function can be applied before performing the FFT. The weighting function can represent one of the windows, such as the Hanning window, to bring down the side lobes of the bandpass response that occurs due to the Gibbs phenomenon, especially in spectral-line observations. After the FFT operation, fine-delay corrections (less than the sampling clock interval) are done to obtain a complete fringe-stopping condition. The outputs of the FFT cards are labeled as $1_{\text{L-FT}}$, $1_{\text{U-FT}}$, $2_{\text{L-FT}}$, and $2_{\text{U-FT}}$, where the subscripts L and U, respectively, represent the LSB and USB of the baseband and FT represents Fourier transform.

### 10.3.6.2 FX Correlator Receiver Subsystem
The outputs from the common subsystem are further processed for correlation and data storage as shown in Fig. 10.11b. These are sent to the MAC card, which can operate with a maximum of 128 channels per sideband per polarization in the regular mode. Hence, two adjacent FFT channels are usually averaged to form one inside the MAC. Thus the 256 FFT channels are reduced to 128 channels. A single MAC card takes the output of two FFT engines. For each side band (USB or LSB), several products can be formed. For example, for the LSB, the self-products from two antennas $m$ and $n$ will be between (*i*) $1^m_{\text{L-FT}}$ and $1^m_{\text{L-FT}}$, (*ii*) $1^m_{\text{L-FT}}$ and $2^m_{\text{L-FT}}$, (*iii*) $2^m_{\text{L-FT}}$ and $2^m_{\text{L-FT}}$, (*iv*) $1^n_{\text{L-FT}}$ and $1^n_{\text{L-FT}}$, (*v*) $1^n_{\text{L-FT}}$ and $2^n_{\text{L-FT}}$, and (*vi*) $2^n_{\text{L-FT}}$ and $2^n_{\text{L-FT}}$. The first three are from the $m$ antenna and the remaining are from the $n$ antenna. The cross products between the $m$ and $n$ antennas will be (*i*) $1^m_{\text{L-FT}}$ and $1^n_{\text{L-FT}}$, (*ii*) $2^m_{\text{L-FT}}$ and $2^n_{\text{L-FT}}$, (*iii*) $1^m_{\text{L-FT}}$ and $2^n_{\text{L-FT}}$, (*iv*) $2^m_{\text{L-FT}}$ and $1^n_{\text{L-FT}}$, which are usually used for imaging. The MAC can be configured into several different modes depending on the observer's requirements. In general, the components 11, 22, 21, and 12, between the antennas are required. The various products formed are integrated continuously over a period of time and sent to the DAS card, which forms a part of the control and storage computer. The integration time is user selectable.

### 10.3.6.3 Pulsar Receiver Subsystem
The functional blocks of the pulsar receiver subsystem are shown in Fig. 10.11c. The system can perform in both: (*i*) an incoherent phased array mode, used for pulsar-search operations, and (*ii*) a coherent phased array mode for specific known pulsars. Theoretically, the incoherent phased array detects the signal power of each antenna using a detector and then adds them. The coherent phased array, however, adds the actual signals from the antennas after a delay correction. An array of $N$ antennas gives a sensitivity $\sqrt{N}$ and $N$ times greater than that of a single antenna for incoherent and coherent phased modes, respectively. Also the beam width of the former remains the same as that of a single antenna (approximately $\lambda/D$, where

$\lambda$ is the wavelength and $D$ is the diameter of a single dish). The beam width of the latter is much narrower.

As shown in Fig. 10.11c, all the polarizations and frequency bands ($1_{L\text{-}FT}$, $1_{U\text{-}FT}$, $2_{L\text{-}FT}$, and $2_{U\text{-}FT}$) of all the antennas are taken into the GAC (GMRT array combiner), which combines the signals both incoherently and coherently (phased) and produces two outputs. The incoherent array combiner takes the FFT spectrum of all the antennas and detects their power (using a square law) for every frequency channel and adds them separately for every polarization and every frequency band. These signals are fed to the incoherent array processor where both the polarization powers are added giving a total intensity from the telescope as a function of frequency. These are finally acquired by the DAS card and saved in the computer.

The GAC also produces a coherent or phased addition that is fed to the phased array processor where the voltages from the two polarizations are fed to a polarimeter that gives the four Stokes parameters as a function of frequency. These are finally acquired by the DAS card and saved in the computer.

Operationally, only one of the modes is selected at a time. The control signals (shown in dotted lines) are provided by the computer.

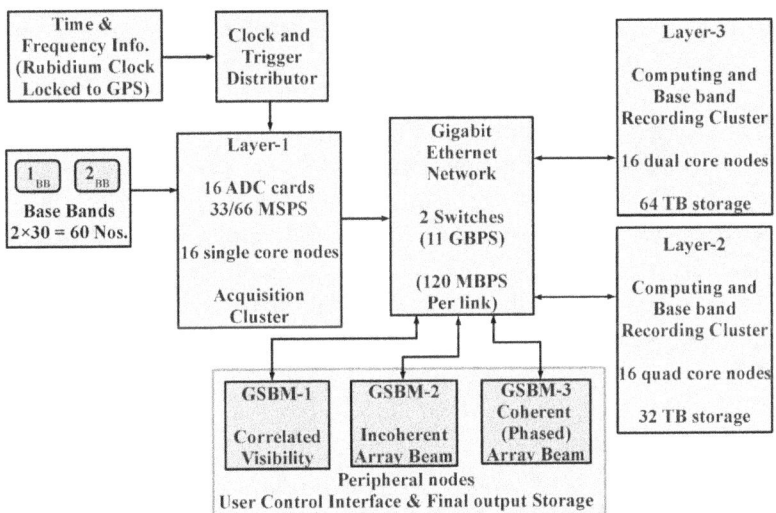

**FIGURE 10.12:** Functional blocks of the software-based correlator and the pulsar receiver at the observatory building.

## 10.3.7 Software Correlator and Pulsar Receiver

The combination of the software correlator and the pulsar receiver is known as the GSB (GMRT software backend). It is functionally the same as the hardware correlator, except that after conversion of the baseband signals from analog to digital, the entire processing is done using sophisticated parallel-pipelined software. Note that the baseband signal for this system is usually a single 33 MHz band per antenna per polarization. Hence, there are only sixty baseband inputs for all thirty antennas marked $1_{BB}$ and $2_{BB}$, respectively, for the two polarization channels from every antenna as shown in Fig. 10.12. This is in contrast to the GHB where 120 baseband signals of 16 MHz bandwidths are taken as input. Here also (as in the GSB) the time and frequency information is derived from the same rubidium clock of the LOM.

The hardware part consists of sixteen ADC cards, fourty-eight Linux PCs, and two independent forty-eight-port giga-bit switches. Three distinct groups of sixteen PCs each are formed, as shown in the figure as Layer 1, Layer 2, and Layer 3. All these PCs are interconnected using the two giga-bit switches. These PCs are also called *nodes*. There are also a few peripheral nodes attached to the system.

The ADC cards are connected to the single-core processor-based first-layer PCs through a PCI bus. Each ADC card can acquire data from four analog inputs, operating in one of the two modes: (*i*) 33 MHz analog bandwidth with four-bit digital conversion, or (*ii*) 16 MHz analog bandwidth with eight-bit digital conversion. Thus, sixteen cards can take sixty-four analog inputs, which is more than sufficient for the present GMRT. These ADC cards convert the baseband signals ($1_{BB}$ and $2_{BB}$) into digital form and transfer this data via a PCI bus to Layer 1 at a rate of 132 MBPS, which results in an effective data rate of 2 GBPS.

Layer 2 contains sixteen quad-core dual-processor PCs. The major part of the computational pipelines is taken up by these nodes.

Layer 3 contains sixteen dual-core dual-processor PCs, each possessing 4 TB of storage capacity. These are used to record the data in raw-dump mode. However, they also assist in computing when computational requirements are higher than the capacity of Layer 2. An example of this is when the baseband signal bandwidth is 33 MHz, and computations are required for all polarizations.

There are three additional nodes (peripheral nodes) attached to the system, namely (*i*) GSBM-1, (*ii*) GSBM-2, and (*iii*) GSBM-3 as shown. GSBM-1 forms the primary interface to the outside world. It provides control and configuration information, including antenna connectivities to acquisition nodes, current frequency and source settings of all antennas, updates of antenna specific gain and phase, etc. It takes commands to start data-recording new scans, and does other related operations. It obtains the final results of computation (visibility) and sends them to another computerfor long-term accumulation and recording.

GSBM-2 receives the pulsar data of incoherent beam formation. Similarly, GSBM-3 receives the pulsar data of coherent beam formation. Both of these nodes record the data on local hard drives after any necessary preprocessing.

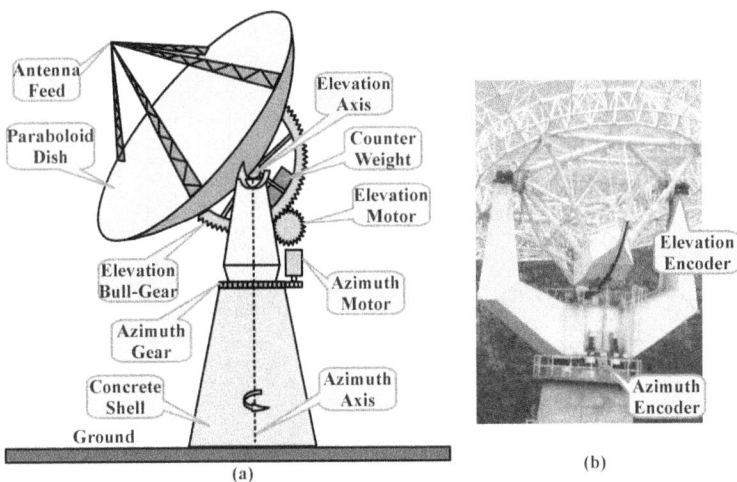

**FIGURE 10.13:** (a) Motor-driven arrangements for azimuth and elevation movements. (b) Locations of azimuth and elevation encoders.

## 10.3.8 Servo System for Antenna Positioning

The center of gravity of the GMRT dishes lies on the elevation axis of rotation. A counterweight is used for this purpose. The details of the azimuth and elevation rotation mechanisms are shown in Fig. 10.13a. Both of these are adjusted using motors connected to the servo system. In order to know the azimuth and elevation positions of the dish, two encoders are fitted as shown in Fig. 10.13b. These are used by the servo system to position the dish correctly to point towards the target (astronomical source). Also there are switches for the azimuth and elevation rotations that inform the servo system to stop the antennas when rotation limits are exceeded either in the azimuth, the elevation, or both. While not in use, by default the brakes are applied to the motors, which are then released temporarily for antenna movements.

Fig. 10.14 shows the functional blocks of the GMRT azimuth and elevation servo control system. Both the elevation and azimuth rotations are performed with motors controlled by servo amplifiers using current loops. The control over the current loops comes from the speed-error feedbacks. There are four motors (two for azimuth and two for elevation movements). Each of these motors are fitted with tachos that generate signals that carry speed information. These are used to control the dish rotation speeds. The two azimuth and elevation encoders send the respective angular positions of the antenna with respect to the antenna shell and horizontal directions, respectively. The interlock system has the final authority to determine whether the antenna should be moved or not.

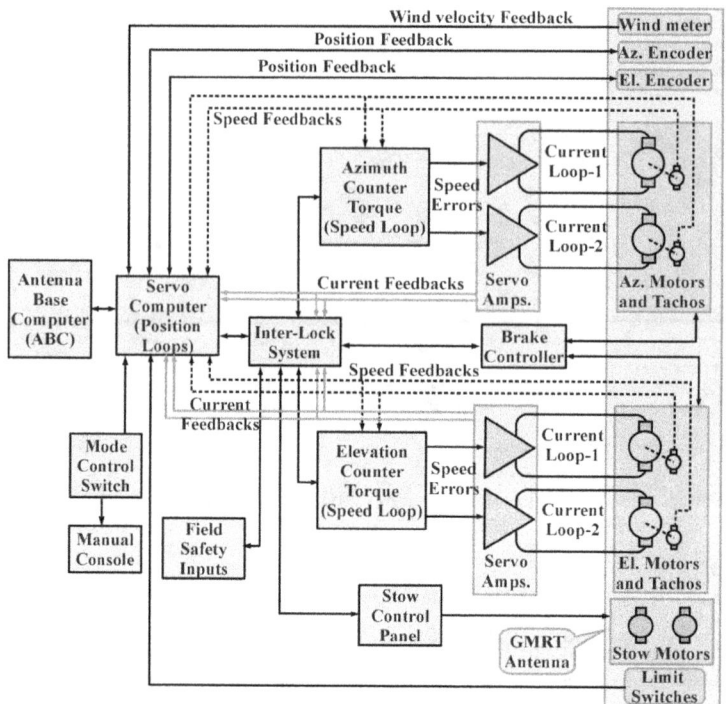

**FIGURE 10.14:** Functional blocks of the GMRT servo system for the control of dish movements.

The ABC (antenna base computer) receives commands from the observatory through the OF and passes them to the servo computer. Based on the commands and various other information (such as speed and position), it communicates with the interlock system that checks that it is safe to rotate the antenna. If all is well, it releases the brakes and sets up the current loops using the azimuth and elevation countertorque systems.

The antenna can also be controlled manually from the shell after by-passing the observatory commands reaching the servo computer. This is usually done during maintenance, using the manual console that sets changes to manual mode using the mode-control switch.

The limit switches restrict the azimuth rotation within 270° and the elevation rotation within 110° relative to the azimuth and elevation reference positions on the antennas. Crossing any of these limits activates signals from the respective limit switch that are sent to the servo computer, which prevents any further rotation in that direction.

During high wind speeds (above 40 km/h), the antennas are parked in a position facing the zenith called the *stow* position, and mechanical locks are applied

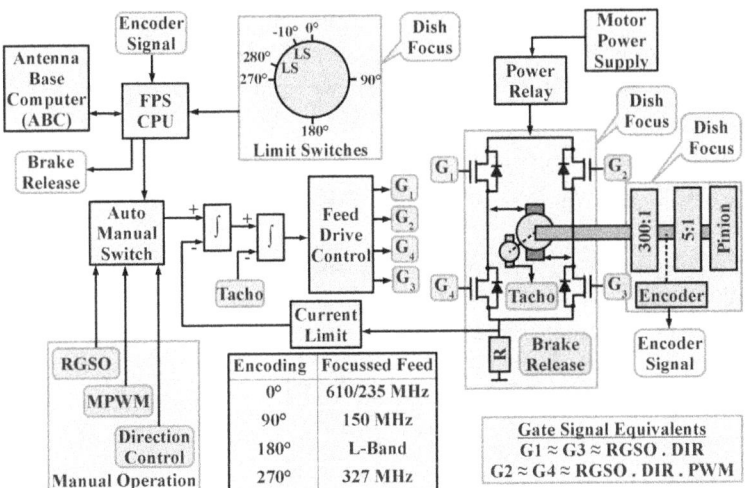

**FIGURE 10.15:** Functional blocks of the feed-positioning system in GMRT antennas.

automatically. This decision is taken by the servo computer and interlock system based on the wind velocity feedback.

### 10.3.9 Feed-positioning System (FPS)

The functional blocks of the GMRT feed-positioning system are shown in Fig. 10.15. Any of the GMRT antenna feeds can be brought to the focus (facing the vertex) of the dish using the FPS. The angular encodings for various antenna feeds are tabulated in the figure. A motor with bidirectional rotation is used, and it is fitted near the dish focus. The direction of rotation and the speed of the motor are controlled by a four-MOS-FET bridge operated by the feed drive control. The diodes in the MOS-FET bridge are for free-wheeling the currents during switching. A resistance at the bottom of the bridge is used to send a feedback voltage proportional to the current through the motor. The motor speed is monitored by the tachometer output in the feedback path. A gear box with a 500:1 gear ratio is used to reduce the rotation speed in the first level that is coupled to the encoder. Another gear box with a gear ratio of 5:1 further reduces the speed (second level). This is applied to the axis of the four-feed fixture at the dish focus. Limit switches prevent the feed from excess rotation in any direction.

Feed-positioning commands from the observatory reach the ABC via an OF that are passed to the FPS CPU (processing unit) through an RS-485 serial line. The CPU takes input from the limit switches and then determines the current position of the antenna feeds from the encoder signal. It then releases the brakes of the motor and signals the feed drive control through two difference integrators. The first integrator generates a difference-integrated output between the driving signal and the current through the motor. The output is again compared with the speed voltage from the tachometer, and their difference is integrated. The result is sent to the feed drive control that controls the MOS-FET bridge using PWM (pulse-width modulation). The gate signals ($G_1$, $G_2$, $G_3$, and $G_4$) are logically dependent on the PWM, the RGSO (remote gate shutoff) and the DIR (direction of motor rotation), as tabulated on the figure.

For manual operation at the antenna shell, the observatory control is bypassed using the automanual switch. Manual checking and maintenance operations like direction control, MPWM (manual PWM), and RGSO can then be safely performed.

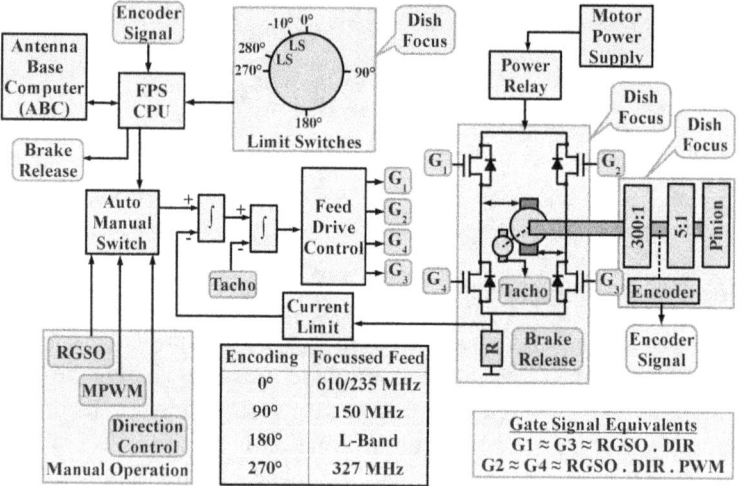

**FIGURE 10.16:** Functional blocks of the control and monitoring (telemetry) system of the GMRT.

### 10.3.10 Control and Monitoring System

The control and monitoring system shown in Fig. 10.16 is used to control the various electronic units, such as the RF system and the FPS, and to monitor various quantities, such as temperature, speed at the antennas. All these are done remotely from the observatory control computer. As usual, two links are set up, one for sending information to the antennas and the other for receiving information from the antennas. There are several MCM (machine control cards) that operate the various units at the antennas and also collect information from them. Their usage is tabulated in the figure. The control and monitoring system sets up a link between the MCM cards and the observatory control computer. The system also supports a voice telephonic link between the control computer and all antennas.

The observatory control computer sends its commands to two communication-handler systems (COMH-1 and COMH-2) through an RS-422 link. One of these handles twenty antennas and the other handles ten antennas. The observatory commands are broadcast by these to all the antennas after conversion to 18 MHz FSK (16 and 19 MHz for 0 and 1, respectively) through the OF system, along with a 201 MHz LO signal, as shown. At the antenna shells, the optical signals are converted back to electrical signals (FSK 18 MHz) and passed to the demodulator module inside the ABC. This recovers the HDLC (high-level data-link control) packets and sends them to then ABC. The ABC checks whether they are meant for the same antenna. If so, it then identifies the signal (whether it is for the servo or the MCM) and channelizes it.

The monitored data from various points in the antenna collected by the MCM cards (including FPS) is then sent back to the ABC through the RS-485 link. They are then converted back to HDLC and modulated to produce a 5 MHz FSK that is upconverted using the 201 MHz LO (collected from the OF uplink) and sent to the OF transmitter. These are received at the observatory and are handled separately using thirty communication receivers (CEB coms). The voice signal from the telephone and the data signals are separated by CEB coms and passed on to the EPABX and multiplexers, respectively. One of the two multiplexers handles twenty antennas and the other handles the remaining ten antennas. The multiplexed signals come back to the communication handlers that send the data to the control computer by using the RS-422 link.

### 10.3.11 Optical Fiber System

The functional blocks of the optical-fiber uplinks (forward) and downlinks (return) between the observatory and the antennas are shown in Fig. 10.17. The telemetry signals (18 MHz FSK), LO signals ($97.5 \pm 1$ MHz, 106 MHz and 201 MHz), are combined in the signal combiner (adder) and sent to the optical-modulator transmitter for transmission to the antennas through the OF uplink (forward) at 1310 nm wavelength. At the antenna shell, these are converted back to RF and separated (telemetry and LO). The downlinks from the antennas to the observatory building follow a similar arrangement as shown. Here, the two LO frequencies, the two IF bands corresponding to two polarizations and the telemetry signals are combined and sent. The power gains and losses occurring at various block levels are marked.

### 10.3.12 Sentinel System

The sentinel system protects the antennas from hazards caused by increases in temperature, fire, and smoke. The functional blocks are shown in Fig. 10.18.

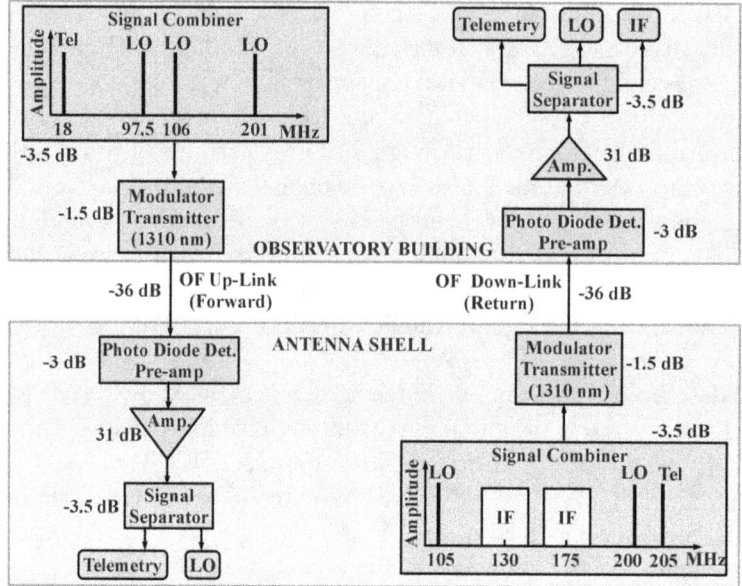

**FIGURE 10.17:** Functional blocks of the optical fiber uplinks (forward) and downlinks (return) between the observatory and the antennas.

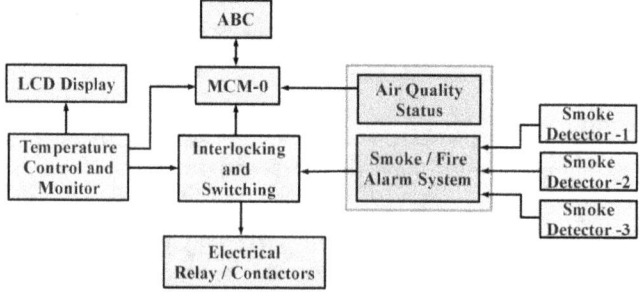

**FIGURE 10.18:** Functional blocks of the GMRT sentinel system at the antenna shell.

Inside each antenna shell there are three smoke detectors at different locations and a temperature monitor. The temperature of the antenna shell electronics room is displayed there on a LCD display. Continuous monitoring of temperature and air quality (smoke free) is monitored and sent to the observatory using the MCM-0 card connected to the ABC. Also, if any of these figures exceeds the allowable safety limits, the interlocking and switching system is activated. This in turn shuts down all the electrical and electronic systems within the antenna shell through relays and electrical contactors. The alarm systems are also simultaneously switched on.

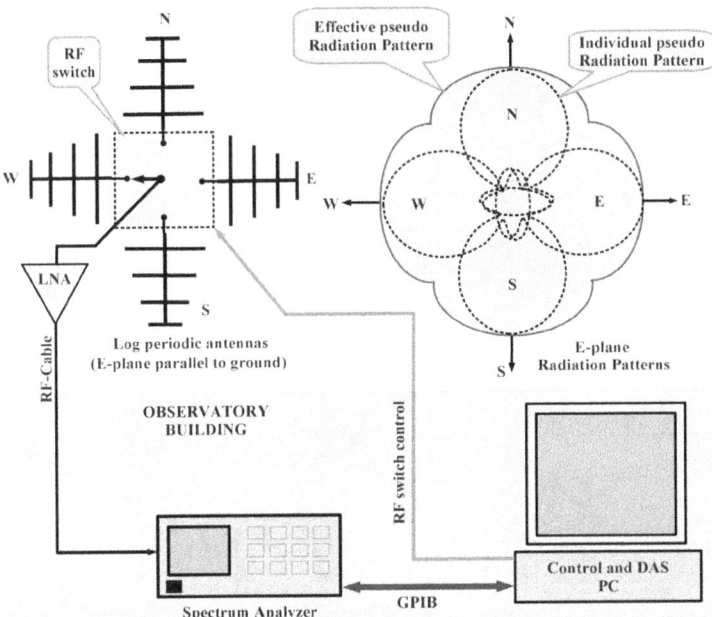

FIGURE 10.19: Omnidirectional Spectrum-monitoring system of the GMRT near the observatory building.

## 10.3.13 Spectrum-monitoring System

Fig. 10.19 shows the functional blocks of the spectrum-monitoring system. It consists of four identically designed log-periodic dipole arrays (100 MHz to 1500 MHz) pointed to the east, west, north, and south, and fitted on top of a tower of 20 m height. The polarizations are parallel to the ground. A 4PST RF switch multiplexes the antenna output and, after amplification (LNA), sends it to a spectrum analyzer inside the observatory building. The control and DAS computer controls the sequence of the switching and also controls the spectrum analyzer (settings and data acquisition) using a GPIB interface. Effectively, if the power amplitude patterns of the antennas are added, a more or less omnidirectional pattern is formed as shown. The data is stored in the control and the DAS PC for further processing, such as direction-finding, omnidirectional spectrum, broad-band and narrow-band spectrum separation, probability of occurrence (spectral occupancy), etc.

## 10.3.14 Online Control from Observatory

All the systems described so far are controlled from the control room of the observatory building. A main Unix online PC forms the heart of the system that is connected to several other supporting Linux PCs, baseband PCs (Microsoft® Win-

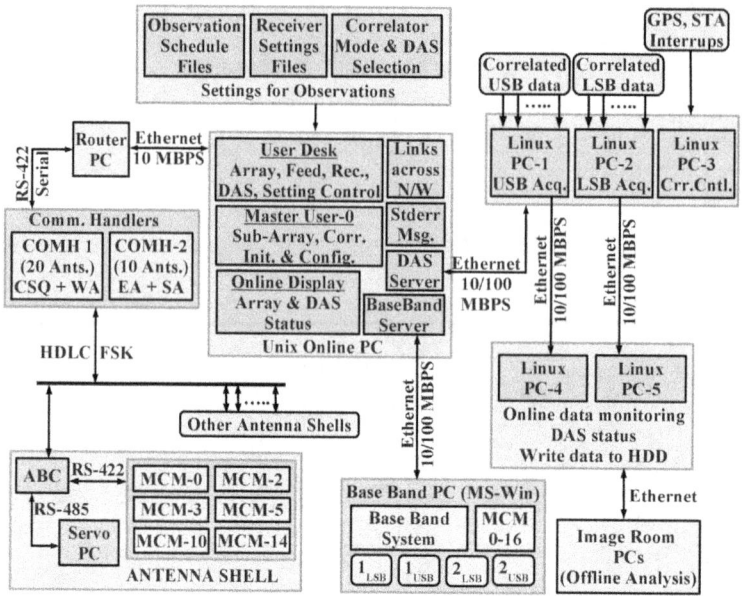

**FIGURE 10.20:** Functional blocks of the online control of the observatory from the control-room.

dows®) and communication handlers through a router PC as shown in Fig. 10.20. The Linux PCs (1 and 2), respectively, are used for receiving correlated USB and LSB data. The third Linux PC controls the correlator. The GPS and STA (short-term acquisition) interruptions occur to all three PCs. Data from PCs 1 and 2 are sent to PCs 3 and 4, respectively, for writing onto the hard drives. These two PCs also monitor online data and report the status of the DAS (data-acquisition system). These are also linked over the ethernet with the terminal-room Linux PCs, which are used for postreduction, processing, and imaging of the data. The baseband PC controls the baseband system using the MCM (0 to 16). All the PCs are interconnected via the ethernet, as shown.

For controlling the antennas, the Unix PC communicates with them via a router PC and the COMH (communication handlers) using a HDLC-FSK network (described earlier in the discussion of control and monitoring). The COMH-1 handles the central square and WA (west-arm) antennas, whereas COMH-2 handles the EA (east-arm) and SA (south-arm) antennas. In each antenna shell, there are several MCMs (0, 2, 3, 5, 10, and 14) that control the various electronics, such as the FPS, the common box, temperature monitors, etc. The servo PC operates on an RS-485 serial link, whereas the MCM operates on an RS-422 link.

The terminals of the five Linux PCs and the baseband PC are opened on the Unix PC and commands are set. The observation settings, receiver settings, and data-acquisition settings are usually supplied as files to the Unix PC. These are used to focus the proper antenna feed, select the proper RF frontend, and to set the different LO, bandwidths, attenuation, ALC (on or off), etc. The observations are continuously monitored for any errors in the antennas, correlator, baseband output, RFI, etc.

## REVIEW QUESTIONS

1. List the number of antennas in the east, west, and south arms of the GMRT, and name them.

2. How many antennas are there in the central square region?

3. How many observing frequency bands and antenna feeds are there in the GMRT?

4. Explain the purpose of using notch filters in the L-band RF frontend.

    [*Hint:* RFI]

5. Explain the purpose of using a polarizer in the 150, 233, 327, and 610 MHz bands. [*Hint:* Ionosphere at lower frequencies.]

6. What is the purpose of using the common box?      [*Hint:* Multiplexing]

7. List the different polarizations for the 150, 233, 327, 610 MHz, and L-band outputs of the common box.

8. List all the IF conversions at the antenna base.

9. How many baseband outputs are there for use with the GHB? Also list their bandwidths.

10. List the baseband outputs used in the GSB and mention their possible bandwidths.

11. How many nodes and layers are there in the GSB? Name them.

12. How many nodes are used for the ADC in the GSB?

13. List the different feedback loops in the servo system (feedback from the antenna).

14. What precautions are necessary before operating the servo system at the antenna base?

15. What are the roles of the limit switches in the servo system and the FPS?

16. List the various MCMs and describe their usage at the antenna shell.

17. How many optical links are there between the observatory and the thirty antennas?

18. What are the RF contents in the optical return links?

19. Under what conditions does the sentinel system shut down the power system at the antenna shell?

20. What are the objectives of spectrum monitoring?

21. How many MCMs are used by the baseband system?

22. What are the objectives of online control?

# COORDINATE SYSTEMS USED IN RADIO ASTRONOMY

## A.1 STRUCTURE OF ASTRONOMICAL COORDINATES

Consider a sphere of infinite radius with the Earth at its center. All celestial objects like stars and galaxies appear to lie on the inner surface of this sphere as observed from Earth. Hence, it is known as the *celestial sphere*. Two angular coordinates are required to locate any point, such as a star, from the Earth. The intersection of the celestial sphere with any plane touching the center forms a circle known as a *great circle*. Planes that do not pass through the center intersect the celestial sphere and form circles known as *small circles*. The diameter of any great circle is equal to the celestial sphere's diameter, whereas the small circle's diameters are always less.

## A.2 THE SPHERICAL COORDINATE SYSTEM

A reference axis may be chosen that passes through the center and intersects the celestial sphere at two points known as the poles of the celestial sphere. A great circle intersecting this axis orthogonally is called the *fundamental great circle* and is used as a reference of measurement. As shown in Fig. A.1a, let $P_1$ and $P_2$ be the poles of the fundamental great circle with C as the center. Great circles joining $P_1$ and $P_2$ are called *secondary circles*. A secondary circle used as a reference of measurement is called a *fundamental secondary circle*.

Let us consider a point $X$ on the celestial sphere. To locate this point, let us draw a secondary circle passing through it that intersects the great circle at point $Y$. The angle $\theta = \angle OCY$ is measured along the fundamental great circle. Similarly, the angle $\phi = \angle YCX$ is measured along the secondary circle of $X$. These are the two coordinates of the point $X$. The loci of constant $\phi$ are small circles parallel to the fundamental great circle. These are also known as the $\phi$-circles. The loci of constant $\theta$ that touch the poles are the secondary circles.

The latitude and longitude of any place can be described in spherical coordinates, where the Earth's equator and the Greenwich meridian can be taken as the two fundamental circles. Here, the longitude represents the angular difference between the Greenwich meridian and the local meridian measured along the equator. Similarly, the latitude represents the angular distance of the place from the equator measured along the local meridian.

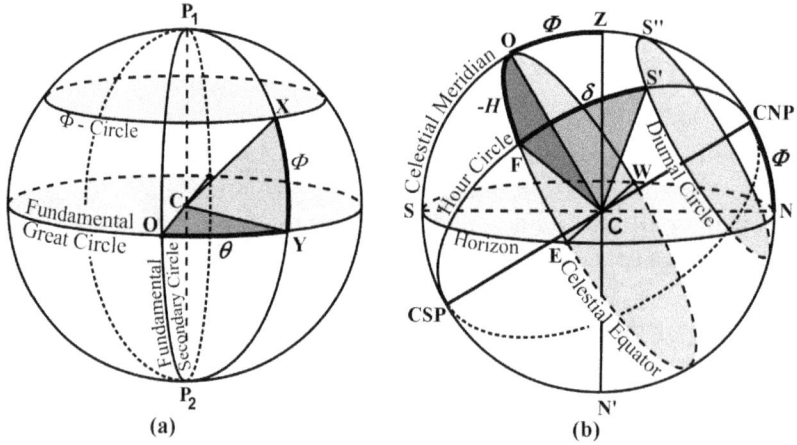

**FIGURE A.1:** Different spherical coordinate systems. (a) A general spherical coordinate system. $(\theta,\phi)$. (b) The local equatorial system $(H,\delta)$.

## A. 3 THE LOCAL EQUATORIAL SYSTEM

The Earth's rotational axis is extended to the celestial sphere touching the celestial north pole (CNP) and the celestial south pole (CSP), as shown in Fig. A.1b. Thus the plane of the Earth's equator cuts the celestial sphere to form a great circle called the *celestial equator*. The small circles parallel to the celestial equator are known as *diurnal circles* or *declination circles*. As shown in the figure, the secondary circles joining the CNP and CSP are known as the *hour circles*. The half of the hour circle that passes through the local zenith $Z$ and the celestial poles is the observer's local meridian.

Consider a star to be observed at $S'$. Let a meridian cross $S'$ and cut the celestial equator at F. The $\angle OCF$ (shown in dark shading) is the hour angle $H$ of the star, which is also a measurement of time. $H$ is measured positively towards the west and negatively towards the east. The $\angle FCS'$ between the star and the celestial equator is called the *declination* $\Delta$. It is measured positively towards the north and negatively towards the south. For example, the declinations of the celestial north and south poles are, respectively, $+90°$ and $-90°$. For a star, the declination remains fixed, whereas its hour angle changes from $-12$ hrs (east) to $+12$ hrs (west) over the course of a day. Note that the hour-angle reference changes with the longitude of the location. For example, if $H_1$ and $H_2$ are the hour angles of a star at a given time measured from two places on Earth having geographical longitudes $l_1$ and $l_2$, respectively, we may relate the differences in the longitudes of the respective places as shown in Eq. (A.1).

$$H_2 - H_1 = l_2 - l_1 \qquad (A.1)$$

Instead of degrees, astronomers prefer to measure the hour angles in hours. This is done using the conversion rates given in Eq. (A.2).

$$\left.\begin{array}{c} 1^h = 15^\circ \\ 1^m = 15' \\ 1^s = 15'' \end{array}\right\} \qquad (A.2)$$

Because the hour angle is measured with respect to the local meridian of the observer, these equatorial coordinates are known as *local* equatorial coordinates.

## A.4  THE UNIVERSAL EQUATORIAL SYSTEM

This system of coordinates is used for providing fixed coordinates to the stars. As shown in Fig. A.2a, the fundamental great circle is the celestial equator. The hour circles are treated in a similar manner. A location is fixed on the great celestial equator, and it is then taken as the reference point. This is one of the two intersecting points of the celestial equator with the ecliptic[1], which is inclined to the celestial equator by $23.5°$. This inclination is known as the obliquity of the ecliptic and is represented by $\epsilon$. The point at which the ecliptic crosses the equator from south to north is the *vernal equinox* and is symbolically represented by $\Upsilon$. The opposite

---

[1] The ecliptic is the apparent path of the Sun in the sky as the Earth revolves around it.

point A, where the ecliptic crosses the equator from north to south is the *autumnal equinox*. Halfway between the equinoxes, the Sun's declinations $\delta$ is either a positive maxima ($\delta = \delta_{max} = +23.5°$), which is the summer solstice, or a negative minima ($\delta = \delta_{min} = -23.5°$), which is the winter solstice. The vernal equinox is accepted as the origin of the universal equatorial system. Hence the zero-hour circle is the hour circle passing through it.

The angle measured from the vernal equinox along the celestial equator from west to east is the *right ascension $\alpha$*. It ranges from 0 to $24^h$, or $0°$ to $360°$. The right ascension and declination $(\alpha,\delta)$ are used for preparing permanent star charts.

The vernal equinox $\Upsilon$ also changes with time due to the precession of the Earth's axis. It is roughly $50.26''$ per year towards the west. Hence, astronomers also specify a date at which $\Upsilon$ is considered for a given right ascension and declination of a star. This is known as the epoch. For example, the epoch J2000.0 refers to the object's position at Greenwich noon (UT) in $1^{st}$ January 2000.

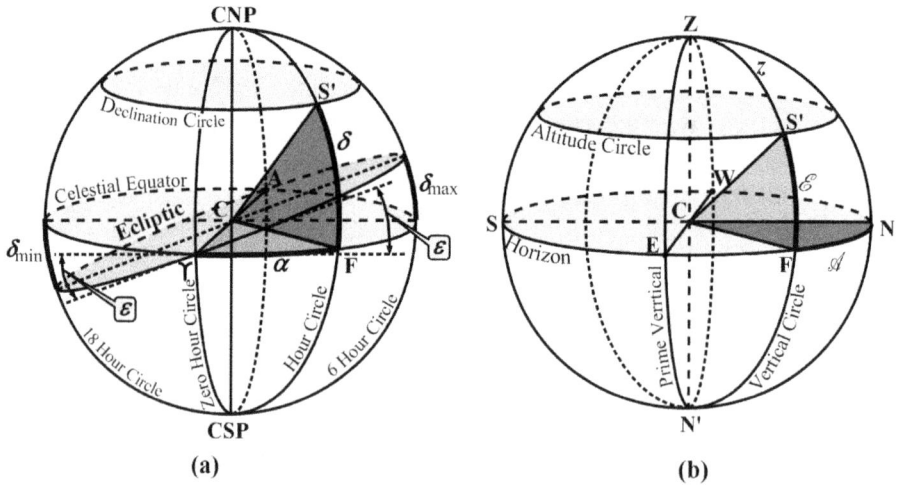

(a)                (b)

**FIGURE A.2:** Different spherical coordinate systems. (a) The universal equatorial system $(\alpha,\delta)$. (b) The horizon or altazimuth system $(\mathcal{A}, \mathcal{E})$.

## A.5 THE ALTAZIMUTH SYSTEM

The altazimuth system is used for continuously pointing a radio telescope antenna and tracking a radio source in the sky. This is shown in Fig. A.2b. The fundamental great circle represents the local horizon. It is formed from the intersection of the horizontal plane of a place on Earth with the celestial sphere. The zenith Z and the nadir N' are the poles of this great circle and are, respectively, the directions directly overhead and underfoot at that place. They may be easily located by using a local plumb line, which points towards the center of the Earth. Any great circle touching the zenith and the nadir is known as a *vertical circle*. A vertical circle joining the east and west (points E and W) is known as the *prime vertical*. Small circles parallel to the horizon are known as the *altitude circles*. The origin of the coordinate system can be either N (north) or S (south). However, N is a popular choice for radio-astronomy antennas.

For locating a star, a vertical circle ZS'FN' is passed through it that intersects the horizon at F. The azimuth angle $\mathcal{A}$ is measured from north towards east along the horizon. The altitude $\angle$FCS' of the star represented as $\mathcal{E}$ is measured from the horizon towards the zenith along the vertical circle ZS'FN'. The compliment angle $\angle$ZCS' represented by $z$ may also be used, and it is known as the *angular zenith distance*.

Although this system is convenient for antenna positioning, it is not useful for preparing charts. The reason is obvious. Because of the Earth's rotation from west to east (resulting in east-to-west apparent motion of the heavenly bodies), both the altitude and the azimuth of a star vary with time. Also, the altazimuth system changes with different positions and so the coordinates of a celestial object differ from place to place at any given time. The conventional coordinates of the heavenly bodies are transformed to this system only for positioning antennas.

## A.6  THE GALACTIC COORDINATE SYSTEM

The galactic coordinate system is useful for studying problems related to the structure and motion of the Milky Way. A fundamental great circle is formed by extending the plane of the Milky Way, which is called *galactic equator*, as shown in Fig. A.3a. The poles are above and below the galactic equator. The galactic pole in the direction closer to the celestial north pole is called the *galactic north pole* (GNP). The opposite one is called the *galactic south pole* (GSP). The secondary half-circles between the poles are known as *galactic longitudes*. The small circles parallel to the galactic equator are called *galactic latitudes*.

The galactic latitude and longitude are zero in the direction from the center C towards the center of the Milky Way, represented by CO. It is called the *galactic center* and is in the constellation of Sagittarius. The longitude $l$ increases positively towards the east from 0° to 360°, represented by $\angle$OCF. The galactic latitude $b$ represented by $\angle$FCS becomes positive towards the north and negative towards the south of the galactic equator. The range of $b$ is within ±90°.

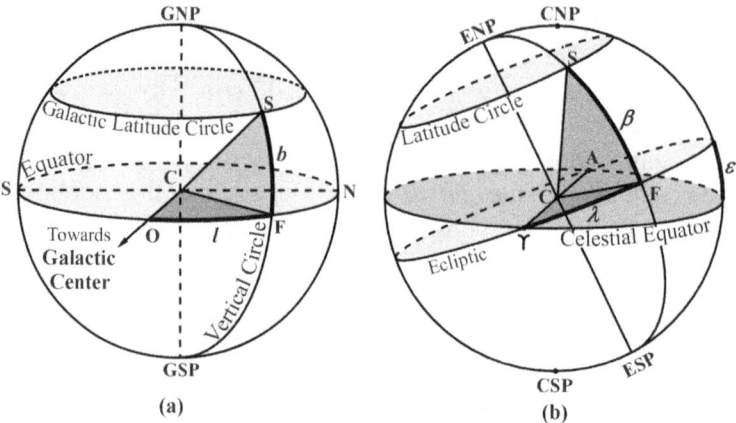

**FIGURE A.3:** Different spherical coordinate systems. (a) The galactic coordinate system $(l,b)$. (b) The ecliptic coordinate system $(\lambda,\beta)$.

## A.7   THE ECLIPTIC COORDINATE SYSTEM

The ecliptic system of coordinates is shown in Fig. A.3b. Here the ecliptic is considered as the fundamental great circle with the vernal equinox $\Upsilon$ as the zero point. The poles formed in the northern and southern hemispheres with respect to the celestial equator are known as the *ecliptic north pole* (ENP) and the *ecliptic south pole* (ESP), respectively. The secondary half-circles joining the ENP and ESP form the celestial longitudes. Small circles parallel to the ecliptic with fixed celestial latitudes are known as *latitude circles*.

Consider a star at a point S on the celestial sphere. As shown in the figure, the longitude of the star in ecliptic coordinates is $\lambda$, which is $\angle \Upsilon CF$. Starting from $\Upsilon$, it is measured as positive towards the east from 0° to 360°. Similarly, the latitude of the star in ecliptic coordinates is $\beta$, which is $\angle FCS$. With the ecliptic as the zero point, $\beta$ is measured as positive towards the north and negative towards the south. The range of $\beta$ is within ±90°.

## A.8   THE $X,Y,Z$ COORDINATE SYSTEM

The $X$, $Y$, $Z$ coordinate system is used for determining the relative positions of the antennas within an antenna array of a radio telescope. This is shown in Fig. A.4a, where *CNP* and *CSP*, respectively, represent the celestial north and south poles. The north and south of the ground plane are marked as N and S, respectively. The zenith and nadir are marked as $Z'$ and $N'$, respectively. A reference antenna is chosen to form the origin of the system at C. The $Z$-axis is directed towards the celestial pole (usually CNP). The $X$-axis lies in the meridian plane of the reference antenna[2]. The $Y$-axis is directed towards the east. Positions of all other antennas relative to the origin are determined as lengths of $X$, $Y$, and $Z$. The position vector $\overline{d_b}$ is a vector from the reference antenna to an antenna whose coordinates are to be measured. It is also called the *baseline vector*. The components of $\overline{d_b}$ are shown along the $X$, $Y$, and $Z$ directions in the figure, and their magnitude relationship is $d_b = \sqrt{X^2 + Y^2 + Z^2}$. The position vector $\overline{d_b}$ makes an hour angle $h$ and declination $d$ with respect to the reference antenna. The values of $X$, $Y$, and $Z$ can be obtained by using Eq. (A.3).

$$\begin{bmatrix} X \\ Y \\ Z \end{bmatrix} = d_b \begin{bmatrix} \cos d \cos h \\ -\cos d \sin h \\ \sin d \end{bmatrix} \tag{A.3}$$

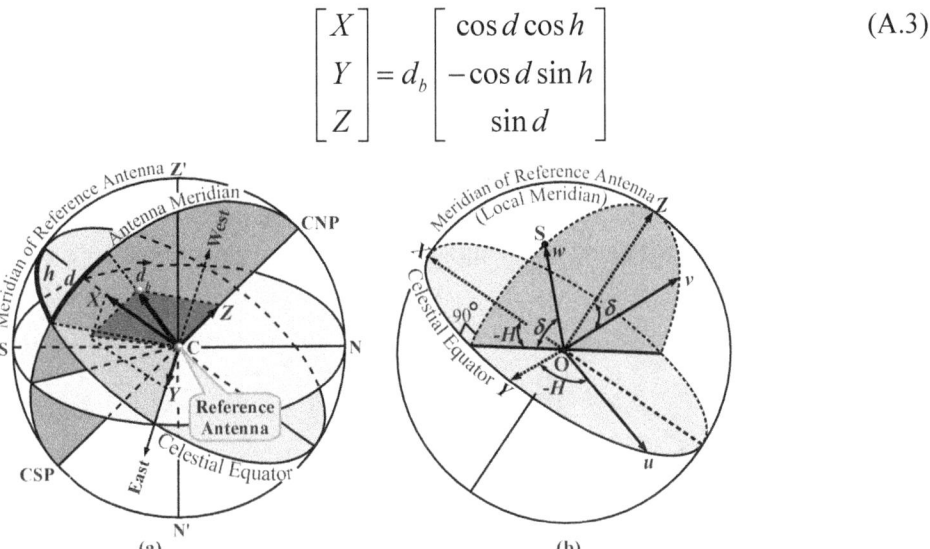

(a)                    (b)

**FIGURE A.4:** Coordinate systems for obtaining relative antenna positions in an array and imaging. (a) The $X$, $Y$, $Z$ coordinate system. (b) The $u$, $v$, $w$ coordinate system with respect to the celestial sphere, and its geometrical relation with the $X$, $Y$, $Z$ coordinate system.

## A.9   THE *U,V,W* COORDINATE SYSTEM

The $u$, $v$, $w$ coordinate system is used for super-synthesis in radio astronomy. For an astronomical radio source located at S, the $w$-coordinate is directed towards it

---

[2] In VLBI (very long baseline interferometry) the local meridian of the reference antenna is generally replaced by the Greenwich meridian.

(along the line of observation), as shown in Fig. A.4b. Thus the $u$-$v$ plane appears perpendicular to the direction of the source. It is the aperture plane of the super-synthesis antenna (formed by an antenna array aided by Earth's rotation). The geometrical relations among the $u$, $v$, $w$ coordinates, local equatorial coordinates, and $X$, $Y$, $Z$ coordinates are shown.

The radio source S under observation is at a declination $\delta$. Let $H$ be the local hour angle of the phase reference position on the radio source with respect to the local meridian. In the figure, the point S lies in the eastern half of the hemisphere, and hence $H$ is negative. The $X$-$Y$ plane sits on the plane of the celestial equator, such that the $X$-axis lies on the meridian plane, the $Y$-axis points towards the east, and the $Z$-axis points towards the north.

## A.10   THE $L,M,N$ COORDINATE SYSTEM AND ITS APPROXIMATIONS

Radio maps constructed of astronomical objects correspond to electomagnetic field distributions produced by objects on the celestial sphere. These maps are intensity contours plotted using astronomical coordinates.

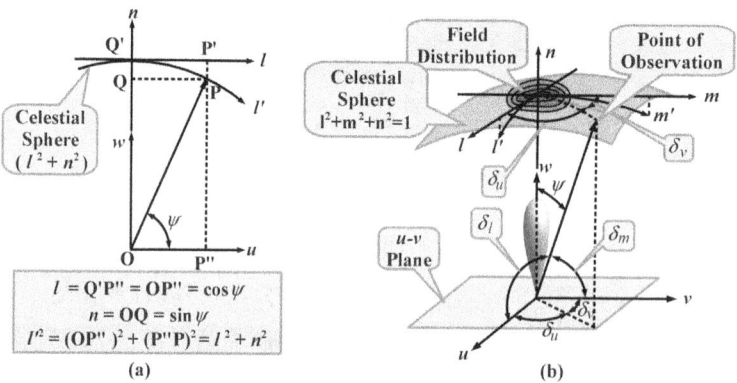

FIGURE A.5: The $l$, $m$, $n$ coordinate system and approximations. (a) An extended radio source is located at Q' on the celestial sphere. The source geometry is measured in two dimensions using $l$ and $n$ coordinates. (b) An antenna scans an extended radio source located on the celestial sphere using $l$, $m$ and $n$ coordinates.

Consider the two-dimensional geometry shown in Fig. A.5a. The celestial sphere is denoted by an arc $l'$. Let the point Q' on the celestial sphere be observed such that the source extent on the right side is P and the angle subtended is $\psi$. A tangent on the celestial sphere at Q' is the coordinate $l$. The point P is now represented by direction cosines. Hence, $l = Q'P' = OP'' = \cos\psi$ and $n = OQ = \sin\psi$. Note that the origin of the $n$-axis is O and not Q'. Therefore, $l'^2 = (OP'')^2 + (P''P)^2 = l^2 + n^2$.

Hence any point on the celestial sphere is represented by $l'^2 + l^2 + n^2$. If the source extent is small, $n \approx 0$ and hence $l' \approx l$.

A radio telescope may be thought of as a single highly directional antenna with a beamwidth much smaller than the source extent, and which scans the source to receive and estimate the spatial intensity distribution of the source. We now consider the three-dimensional geometry shown in Fig. A.5b. The beam-scanning is now a function of two angles, $\delta_l$ and $\delta_m$. The direction cosines are now $l$, $m$, and $n$, as shown. The arcs are represented by $l'$ and $m'$. Any point on the celestial sphere can be represented by $l$, $m$, and $n$, as expressed in Eq. (A.4). The geometrical relationships between $l$, $m$, $n$ are such that their squares always add to unity, as shown in Eq. (A.5).

$$\left.\begin{array}{l} n = \cos\psi \\ l = \sin\psi\cos\delta_u = \cos\delta_l \\ m = \sin\psi\cos\delta_v = \cos\delta_m \end{array}\right\} \tag{A.4}$$

$$l^2 + m^2 + n^2 = 1 \tag{A.5}$$

The surface of the celestial sphere is spherical. A very small surface of the celestial sphere may be approximated as a flat surface. This assumes $n \approx 0$. Under such conditions, radio astronomers use the approximation shown in Eq. (A.6).

$$\left.\begin{array}{l} l = l' \\ m = m' \end{array}\right\} \text{when } n \approx 0 \tag{A.6}$$

## A.11  CONVERSION BETWEEN COORDINATE SYSTEMS

Often it is required to convert one system of astronomical coordinates into another. Various formula used are listed below.

### A.11.1 Conversion Between Local and Universal Equatorial Systems

The declination $\delta$ is the same in both systems. The hour angle $H$ and right ascension $\alpha$ are related as in Eq. (A.7).

$$H = \text{Sidereal Time} - \alpha \tag{A.7}$$

### A.11.2 Conversion Between Local Equatorial and Altazimuth Systems

Eq. (A.8) gives the conversion relation from local equatorial coordinates ($\delta$, $H$) to altazimuth coordinates ($\mathcal{E}$, $\mathcal{A}$), where $\mathcal{L}$ is the latitude of the location. Eq. (A.9) gives the conversion of altazimuth coordinates ($\mathcal{E}$, $\mathcal{A}$) to the local equatorial coordinates ($\delta$, $H$).

$$\cos \mathcal{E} \sin \mathcal{A} = -\cos \delta \sin H$$
$$\cos \mathcal{E} \cos \mathcal{A} = \sin \delta \cos \mathcal{L} - \cos \delta \sin \mathcal{L} \cos H$$
$$\sin \mathcal{E} = \sin \delta \sin \mathcal{L} + \cos \delta \cos \mathcal{L} \cos H$$

(A.8)

$$\cos \delta \sin H = -\cos \mathcal{E} \sin \mathcal{A}$$
$$\cos \delta \cos H = \sin \mathcal{E} \cos \mathcal{L} - \cos \mathcal{E} \sin \mathcal{L} \cos \mathcal{A}$$
$$\sin \delta = \sin \mathcal{E} \sin \mathcal{L} + \cos \mathcal{E} \cos \mathcal{L} \cos \mathcal{A}$$

(A.9)

### A.11.3 Conversion Between Universal Equatorial and Galactic Coordinate Systems

Let $(\alpha_c, \delta_c)$ be the equatorial coordinates of the galactic center, and let $(\alpha_g, \delta_g)$ be the same for the galactic pole. The relationship between the universal equatorial coordinates $(\alpha, \delta)$ and the galactic coordinates $(l, b)$ is given in Eq. (A.10), where $l_\Omega$ is given in Eq. (A.11).

$$\cos b \cos(l - l_\Omega) = \cos \delta \cos(\alpha - \alpha_g)$$
$$\cos b \sin(l - l_\Omega) = \sin \delta \cos \delta_g - \cos \delta \sin \delta_g \cos(\alpha - \alpha_g)$$
$$\sin b = \sin \delta \sin \delta_g + \cos \delta \cos \delta_g \cos(\alpha - \alpha_g)$$

(A.10)

$$\sin l_\Omega = -\sin \delta_c \sec \delta_g$$

(A.11)

### A.11.4 Conversion Between Universal Equatorial and Ecliptic Systems

The universal equatorial coordinates $(\alpha, \delta)$ can be converted to ecliptic coordinates $(\lambda, \beta)$ using Eq. (A.12), where $\epsilon$ represents the obliquity of the ecliptic. For the reverse conversion, use Eq. (A.13).

$$\cos \beta \cos \lambda = \cos \delta \cos \alpha$$
$$\cos \beta \sin \lambda = \sin \delta \sin \varepsilon + \cos \delta \cos \varepsilon \sin \alpha$$
$$\sin \beta = \sin \delta \cos \varepsilon - \cos \delta \sin \varepsilon \sin \alpha$$

(A.12)

$$\cos \delta \cos \alpha = \cos \beta \cos \lambda$$
$$\cos \delta \sin \alpha = -\sin \beta \sin \varepsilon + \cos \beta \cos \varepsilon \sin \lambda$$
$$\sin \delta = \sin \beta \cos \varepsilon + \cos \beta \sin \varepsilon \sin \lambda$$

(A.13)

### A.11.5 Conversion Relations Among X,Y,Z and u,v,w Coordinates

The transformation of the $X, Y, Z$ coordinate system to $u, v, w$ is given in Eq. (A.14), where $X_\lambda$, $Y_\lambda$, and $Z_\lambda$ are, respectively, the values of $X, Y,$ and $Z$ in wavelengths. Mathematically, $Z_\lambda = X/\lambda$, $Y_\lambda = Y/\lambda$, and $Z_\lambda = Z/\lambda$, where $\lambda$ is the wavelength of the radio observation. The $u$, $v$, $w$ coordinates can also be obtained from the baseline length $d_\lambda$ (in wavelengths) using Eq. (A.15), where $d_\lambda = d_b/\lambda$, and $d_b = \sqrt{X^2 + Y^2 + Z^2}$, which is the baseline length.

$$\begin{bmatrix} u \\ v \\ w \end{bmatrix} = \begin{bmatrix} \sin H & \cos H & 0 \\ -\sin\delta\cos H & \sin\delta\sin H & \cos\delta \\ \cos\delta\cos H & -\cos\delta\sin H & \sin\delta \end{bmatrix} \begin{bmatrix} X_\lambda \\ Y_\lambda \\ Z_\lambda \end{bmatrix} \tag{A.14}$$

$$\begin{bmatrix} u \\ v \\ w \end{bmatrix} = d_\lambda \begin{bmatrix} \cos d \sin(H-h) \\ \sin d \cos\delta - \cos d \sin\delta\cos(H-h) \\ \sin d \sin\delta + \cos d \cos\delta\cos(H-h) \end{bmatrix} \tag{A.15}$$

$$\begin{bmatrix} X \\ Y \\ Z \end{bmatrix} = d_b \begin{bmatrix} \cos\mathcal{L}\sin\mathcal{E} - \sin\mathcal{L}\cos\mathcal{E}\cos\mathcal{A} \\ \cos\mathcal{E}\sin\mathcal{A} \\ \sin\mathcal{L}\sin\mathcal{E} + \cos\mathcal{L}\cos\mathcal{E}\cos\mathcal{A} \end{bmatrix} \tag{A.16}$$

## REVIEW QUESTIONS

1. What are spherical coordinates? How are they used to represent celestial objects?
2. List the different astronomical coordinate systems, and give their applications.
3. Describe the altazimuth system and its applications.
4. What is the difference between a local equatorial system and a universal equatorial system?
5. In which direction are the galactic longitude and latitude zero? Name the constellation.                    [*Hint:* Sagittarius]
6. For what kinds of problems are galactic coordinates suitable?
7. With a suitable diagram, explain the ecliptic coordinate system.
8. Write down the conversion relationships between (*i*) local and universal equatorial systems, (*ii*) local equatorial and altazimuth systems, (*iii*) universal equatorial and ecliptic system.
9. Express the relations used for converting universal equatorial to galactic coordinates.

10. Write down the conversion relations from the $x, y, z$ to the $u, v, w$ coordinate system, and vice versa.

11. Using a diagram to describe the $l, m, n$ coordinate system.

12. The radio maps are made two-dimensional with what approximations? [Ans: $n = 0$]

# B

# *ANTENNA TERMINOLOGY FOR RADIO TELESCOPES*

## B.1  INTRODUCTION

The principle of radio telescope observation is similar to that of optical telescopes, which is known as *optical seeing*, where the observation of the astronomical object is made though a combination of lenses. Radio telescopes use antennas as the viewing instrument instead of lenses. The wavelengths used are in the radio spectrum. However, direct image viewing is not possible using current technology. Radio telescopes thus use the principle of *indirect seeing*. The antenna receives signals from the radio source, forming the frontend of the radio telescope. These signals are then processed using electronic circuits and computers and recorded. Finally an image is developed from the data. Pseudocolors that are visible to the human eye are used in the image. The antenna is the prime instrument in a radio telescope. The quality of the image is highly dependent on the performance of the antenna. The basic concepts of antenna theory are explained in this appendix purely from an engineering point of view. These concepts will help one to understand the technical aspects of radio telescopes.

## B.2  RADIATION AND RECEPTION

The antenna receives power from a signal source and radiates it into free space in the form of electromagnetic waves. Conversely, electromagnetic radiation falling on the antenna gets converted to power, which is then available at its terminals and is delivered to a load connected to it. This fundamental property comes from the *reciprocity principle* of an antenna. It may simply be stated as *the properties of an antenna are unchanged when used as a radiator or a receiver*. Understanding the radiating properties is equivalent to knowing its receiving properties, and vice versa.

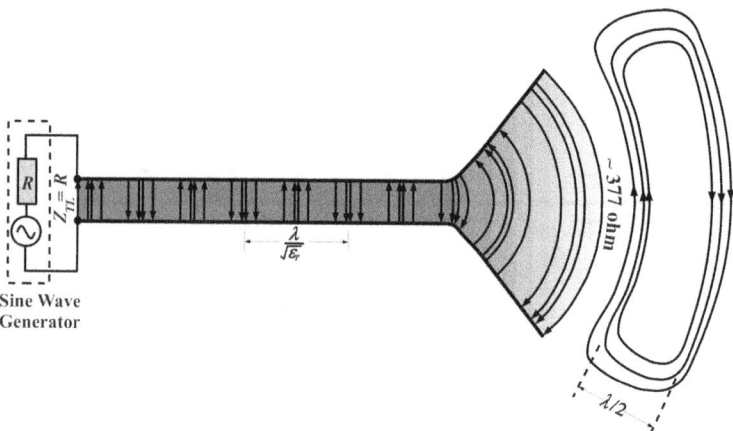

**FIGURE B.1:** Radiation from a gradually flared transmission line.

Fig. B.1 shows one end of a balanced transmission line having characteristic impedance $Z_{TL} = R$ Ohms, which is connected to a sinusoidal voltage source operating at a frequency $v$ and having an internal impedance $R$ Ohms. Let the other end of the transmission line be gradually opened out in free space, such that the impedance at this end is the same as the characteristic impedance of free space (377 ohms, approximately). Hence, the power from the source gets transmitted into free space. Let us have a closer look. Let the dielectric of the transmission line have a relative permittivity $\varepsilon_r$. If $\lambda_{TL}$ is the wavelength inside the transmission line, it can be expressed as in Eq. (B.1), where $\lambda$ is the free-space wavelength obtained as $\lambda = c/v$, and $c$ is the speed of light. The gray color gradient on the gradually opened end shows the variation of characteristic admittance and decreasing electric-field intensity. The distance between the two wires of the transmission line increases, resulting in an increase of the characteristic impedance. The electric fields are shown using arrows. The propagating wave towards the open end sees a gradual increase in the characteristic impedance. At the end, finally it sees an open space with intrinsic impedance of 377 ohms. Because the conductors of the transmission line are absent here, the field lines join together over a distance of $\lambda/2$ due to opposite polarity. This generates a propagating electromagnetic wave with an approximate circular wave front.

$$\lambda_{TL} = \frac{\lambda}{\sqrt{\varepsilon_r}}$$

(B.1)

Let's have another view. From classical electromagnetism, it is known that an electromagnetic wave is generated if there is any acceleration or deceleration of electricly charged particles. In this case, the charges move along the electric field lines. The accelerations and decelerations occur due to a sinusoidal signal from the generator. Hence, electromagnetic waves are generated.

Conversely, if the signal generator is replaced by a resistor of $R$ Ohms, and if any externally generated electromagnetic waves reach the open end of the transmission line, then by applying the reciprocity principle, we find that the antenna converts these moving charges (electromagnetic waves) into a signal power that gets dissipated into the resistor.

## B.3  ISOTROPIC ANTENNA AND INVERSE-SQUARE LAW OF RADIATION

The isotropic antenna or isotropic radiator is an ideal concept. In reality, it does not exist. It is defined as a dimensionless (point) antenna radiating identical power in all directions. It is also termed an *isotropic source* or *unipole*. It is also lossless. The performance of any practical radiating antenna is determined by camparing it with an isotropic antenna, which is thus called the theoretical lossless reference antenna[1].

Let an isotropic antenna be placed at the center of a sphere having a radius of $r$ meter. This is shown in Fig. B.2. The power radiated by it covers the surface area $4\pi r^2$ of the sphere. If $W_r$ is the total power radiated in Watts, then the power radiated per unit area $S$ can be written as in Eq. (B.2), where $ds$ is an infinitesimal area on the surface of the sphere. Thus, $S$ is the flux density[2] in Watt/m$^2$ at a distance of $r$ meter. Because $S$ is inversely proportional to the square of $r$, it is also known as the *inverse-square law of radiation*.

$$S = \frac{W_r}{\iint_\Omega ds} = \frac{W_r}{\int_0^{2\pi}\int_0^{\pi} r^2 \sin\theta\, d\theta\, d\phi} = \frac{W_r}{4\pi r^2} \tag{B.2}$$

---

[1] Note that certain applications may use a half-wave dipole antenna as a reference antenna for specific reasons, but use of the isotropic radiator gives a better understanding of the distribution of radiation in three-dimensional space, and thus it is preferable in a majority of cases.

[2] The term flux density is used for describing the power available over a unit area and unit bandwidth in radio astronomy. It is assumed that the flux density has been integrated here over the bandwidth in use, and hence a different notation has been used.

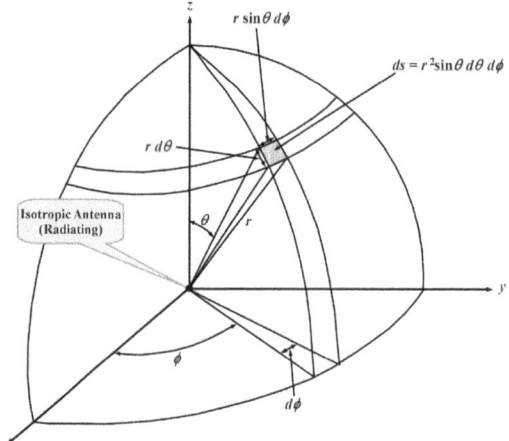

**FIGURE B.2:** Radiation from an isotropic radiator (isotropic antenna).

## B.4 NEAR- AND FAR-FIELD RADIATING REGIONS

The field patterns of a radiating antenna vary with distance. These are primarily associated with two energy types: (*i*) *radiating,* and (*ii*) *reactive*. The space surrounding the antenna can be broadly divided into three regions: (*i*) *reactive field* region, (*ii*) *radiating near-field* region, and (*iii*) *radiating far-field* region. This is shown in Fig. B.3. The boundaries are not precisely defined but only approximated.

The reactive near-field region exists very close to the antenna, where the reactive part of the field is much larger than the radiating part. The reactive energy is carried back and forth from the antenna to the reactive near-field region by electromagnetic radiation of the type that slowly changes electrostatic and magnetostatic effects. For example, current flowing in the antenna creates a magnetic component in the near field, which then attenuates as the antenna current begins to reverse, causing transfer of the field's magnetic energy back to electrons in the antenna as the changing magnetic field causes a self-inductive effect on the antenna that generated it. This returns energy to the antenna in a regenerative way, so that it is not lost. This field component decays with the square and/or cube of the distance from the antenna. For distances greater than $R_1$ (meters), these are considerably weaker than the radiating field and can be neglected. This is expressed in Eq. (B.3), where $D$ is the largest dimension of the antenna (meters) and $\lambda$ is the wavelength (meters).

$$R_1 = 0.62\sqrt{\frac{D^3}{\lambda}}$$

(B.3)

Beyond $R_1$, the field is radiative by nature. This region may further be divided into two subregions: (*i*) *radiating near-field* region, also called the *Fresnel zone,* and (*ii*) *radiating far-field* region, also known as the *Fraunhofer zone*. Not all antennas produce a radiating near-field region. Examples are the electrically small antennas. But all antennas do produce a radiating far-field region. The radiation pattern changes as a function of distance and observation angle within the near-field radiating region. This is because the distance from different sections of the antenna to the observation point are not the same for any practical antenna, because they are not point sources and have arbitrary shapes. Thus notable amplitude and phase changes are produced at the observing points. The radiating near-field exists between $R_1$ (m) and $R_2$ (m). $R_2$ is given by Eq. (B.4), where $D$ is the largest dimension of the antenna (m) and $\lambda$ is the wavelength (m).

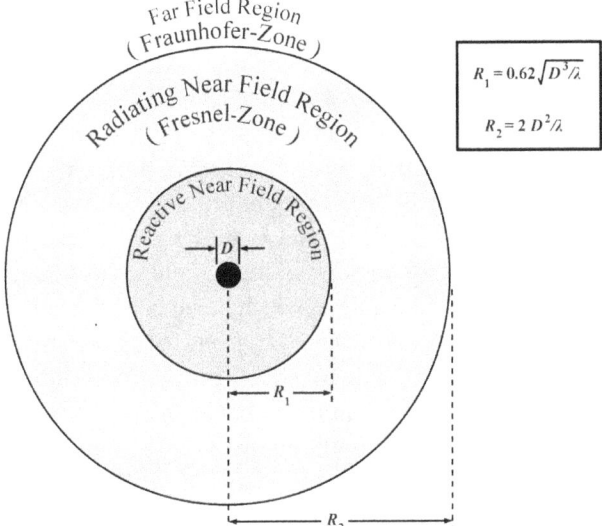

**FIGURE B.3:** Field regions surrounding a radiating antenna.

$$R_2 = \frac{2D^2}{\lambda} \tag{B.4}$$

Beyond $R_2$, the relative amplitude and phase components of the radiating field produced from different parts of the antenna become almost distance independent. Only transverse components of the electric and magnetic field appear, which significantly simplifies the mathematics of the angular properties of radiation.

## B.5   RADIATION PATTERNS

The radiated power from a practical antenna does not produce the same field strength in all directions. The strength is greater in some particular directions and less or null in some other directions. The power radiated in a particular direction is measured as flux density at a point that is at a fixed radial distance from the antenna center. This is usually done in the Fraunhofer region. The antenna is placed at the center of an imaginary sphere with a radius greater than $R_2$, and the field or flux-density distribution on the surface of the sphere is measured. The values are then plotted in a three-dimensional coordinate system with the radial distance vector from its center scaled proportionally to the magnitude of the measured flux density or field strength. This is known as the *three-dimensional radiation pattern* of the antenna. An example of such a pattern obtained from a half-wave dipole antenna is shown in Fig. B.4a. Although it is not shown, it is actually a function of $r$, $\theta$, and $\phi$. Depending on the choice of values (whether flux density or field strength), the radiation pattern is called the *power pattern* or *field-strength pattern*.

The three-dimensional patterns gives a clear understanding of the radiation properties, but it is inconvenient to visualize them on a flat surface like a paper. Hence these patterns are sliced using two or three orthogonal planes. The results are in two dimensions. These are shown in Figs. B.4b and B.4c. Using the spherical coordinates shown in Fig. B.2, slicing may be done in the x-y plane ($\theta = 90°$), y-z plane ($\phi = 90°$), and z-x plane ($\phi = 0°$). The antenna at the center of the coordinate system should be carefully aligned so that when preparing the two-dimensional plots, the maximum radiation direction will be sliced by two orthogonal planes. That is to say, the two-dimensional patterns must contain the direction of maximum radiation. For example, to measure the radiation pattern of a half-wave dipole antenna, place it vertically at the center of the coordinates (Fig. B.4a). Here, the electric field at the center is aligned with the z-axis, and the magnetic fields lie in the x-y plane. Hence the x-y plane is also known as the *H-plane*. If the x-y plane is also parallel to the ground, it is also called the *horizontal plane*. With reference to this, both the y-z and z-x planes may be called *vertical planes*.

### B.5.1 Principal Patterns

The performance of an antenna is usually described in terms of the principal E-plane and H-plane radiation patterns. For linearly polarized antennas like half-wave dipoles, the E-plane pattern is said to be *a plane containing the electric field vector and the direction of maximum radiation,* and the H-plane is *a plane containing the magnetic field vector and the direction of maximum radiation.* In general, orient the

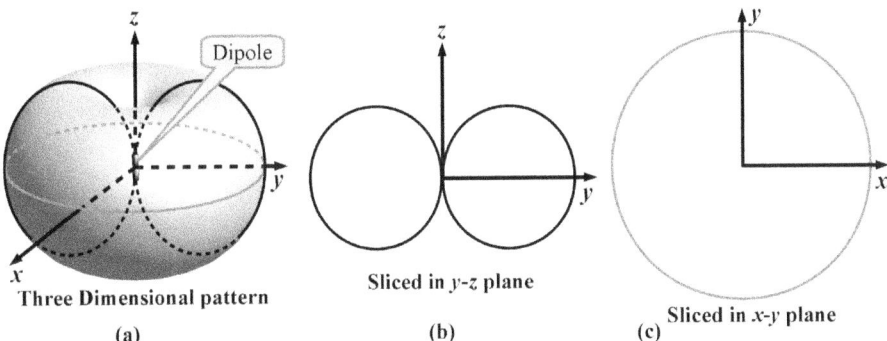

**FIGURE B.4:** Radiation patterns of a half-wave dipole antenna in three and two dimensions.

antennas such that at least one of the principal plane patterns coincides with one geometrical plane. An example of such an alignment is shown in Fig. B.5.

### B.5.2 Lobes in Radiation Pattern and Beam Widths

The magnitude of any radiation pattern is a function of $\theta$ and $\phi$. There exists a very low-radiating region between two adjacent radiating regions, known as a *null*. Regions between two adjacent nulls are known as *lobes*. This is shown in Fig. B.5. The lobe containing the direction of maximum radiation is known as the *main lobe*. As one moves away from the direction of maximum radiation across the main lobe, the radiation power falls until a null is reached, where the radiation is minimum. As shown in Fig. B.5b and B.5c, on a two-dimensional radiation pattern containing the main lobe, the angle between the two points where the radiating power falls to half of the maximum radiation is known as the *half-power beam width,* or *HPBW*. The angle subtended by the major lobe between two adjacent nulls is called the *beam width between first nulls* or *BWFN*. Some amount of radiation also takes place in the opposite direction of the main lobe. This is seen as a small lobe called the *back lobe*. Lobes other than the main and back lobes are known as *side lobes*. In different planes, the beam widths could be different. Hence, they are given separate identities. For example, $\theta_{HP}$ and $\phi_{HP}$, respectively, represent the HPBW in the two orthogonal planes bisecting the major lobe in a spherical coordinate system. Similarly, $\theta_{FN}$ and $\phi_{FN}$, respectively, represent the BWFN in the two orthogonal planes bisecting the major lobe. In these examples, the planes contain the major lobe. Thus, they are also called *principal planes*.

### B.5.3 Front-to-Back Ratio

The ratio of power radiated along a desired direction to the power radiated in the opposite direction is known as the *front-to-back ratio* or *FBR*. A higher *FBR* is thus

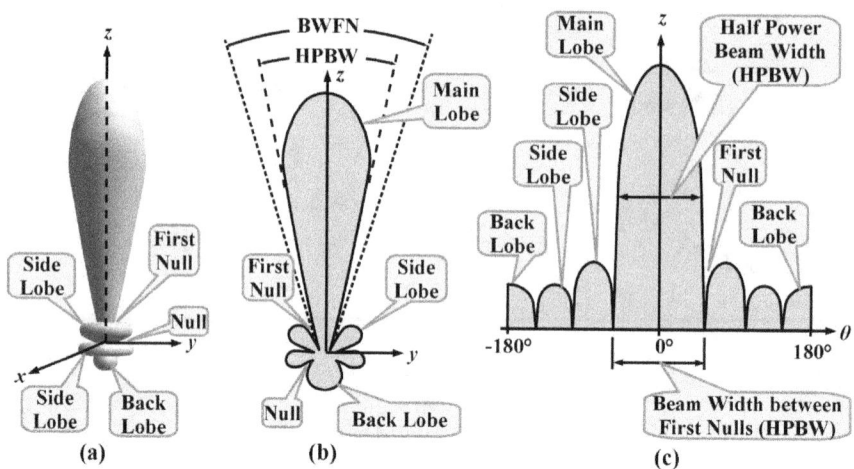

**FIGURE B.5:** Lobes in a radiation pattern. (a) A three-dimensional view. (b) A two-dimensional polar plot of the same radiation pattern in a principal plane. (c) A two-dimensional rectangular plot in one principal plane.

preferable. FBR is also the ratio of the radiated flux density from the center of the major lobe to the radiated flux density from the center of the back lobe. This is expressed in Eq. (B.5). Changes in wavelength causes changes in the electrical length of the antenna, thereby changing the radiation pattern. Hence, *the FBR* changes with the frequency of operation.

$$\text{FBR} = \frac{\text{Radiated flux density from the center of the major lobe}}{\text{Radiated flux density from the center of the back lobe}} \quad \text{(B.5)}$$

### B.5.4 Normalized Radiation Pattern

Consider an antenna at the center of a coordinate system, as shown in Fig. B.6. The electric field has two components $E_\theta(\theta, \phi)$ and $E_\phi(\theta, \phi)$. The radiated power is given by the Poynting vector $\vec{P} = \vec{E} \times \vec{H}$. The normalized electric field pattern $E_\theta(\theta, \phi)_n$ is expressed in Eq. (B.6), where $E_{\theta \text{max}}$ is its maximum value. The normalized power pattern $P_n(\theta, \phi)$ is given in Eq. (B.7), where $S(\theta, \phi)$ is the power pattern and $S_{\text{max}}$ is its maximum value. In terms of the electric field pattern and Poynting vector, $S(\theta, \phi)$ can be expressed as in Eq. (B.8). Here, $Z_0$ is the intrinsic impedance of free space (377 Ohms).

$$E_\theta(\theta, \phi)_n = \frac{E_\theta(\theta, \phi)}{E_{\theta \text{max}}} \quad \text{(B.6)}$$

$$P_n(\theta, \phi) = \frac{S(\theta, \phi)}{S_{\text{max}}} \quad \text{(B.7)}$$

$$S(\theta,\phi) = \left[ E_\theta^2(\theta,\phi) + E_\phi^2(\theta,\phi) \right] / Z_0 \qquad \text{(B.8)}$$

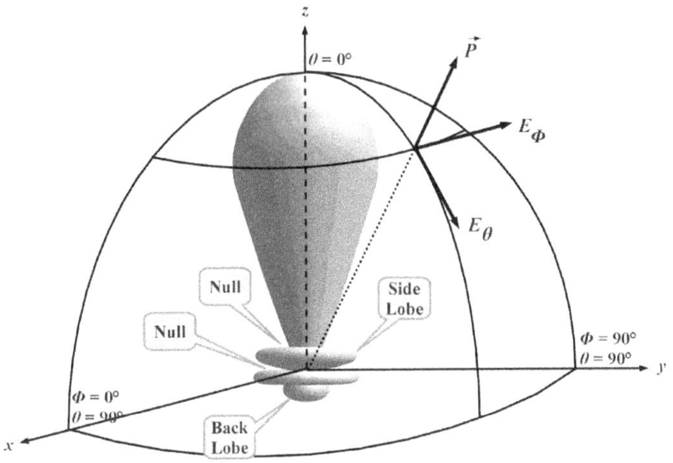

**FIGURE B.6:** A radiation pattern with its direction of maximum radiation aligned to the z-axis. Electric field components and Poynting vector are shown.

$$E_\theta(\theta,\phi)_n = \frac{E_\theta(\theta,\phi)}{E_{\theta\max}} \qquad \text{(B.9)}$$

$$P_n(\theta,\phi) = \frac{S(\theta,\phi)}{S_{\max}} \qquad \text{(B.10)}$$

$$S(\theta,\phi) = \left[ E_\theta^2(\theta,\phi) + E_\phi^2(\theta,\phi) \right] / Z_0 \qquad \text{(B.11)}$$

## B.5.5 Beam Solid Angle

The antenna patterns are not uniform over the angle of radiation. For example, see the radiation pattern in Fig. B.6a. Note that the power is radiated in various directions with different intensities. Now assume a fictitious radiation pattern shown in Fig. B.6b. It has a solid angle $\Omega_A$ and radiates uniformly over this solid angle. If power radiated from both of these patterns are identical, then $\Omega_A$ is defined as the *beam solid angle*. A two-dimensional comparison of both the patterns are shown in Fig. B.7c. Note that the fictitious pattern radiates at the peak intensity of the actual pattern.

Mathematically, the beam solid angle $\Omega_A$ can be expressed as the integration of the normalized radiation pattern over a sphere. This is shown in Eq. (B.12). The beam solid angle roughly equals the product of the half-power beam-width angles $\theta_{\mathrm{HP}}$ and $\phi_{\mathrm{HP}}$. Assume the power radiated by main beam alone is $\kappa_B$ times the total

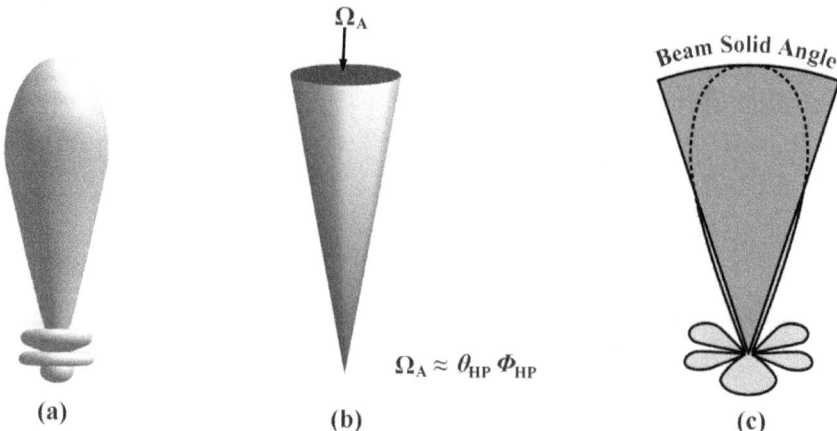

**FIGURE B.7:** (a) A practical radiation pattern. (b) A fictitious beam pattern for calculating the beam solid angle. (c) A comparison between the two patterns in a two-dimensional plane.

power radiated (which includes the side lobes), where $\kappa_B \leq 1$. We define the *main-beam solid angle* as equivalent to the beam solid angle formed from the main beam alone. This is mathematically shown in Eq. (B.13). The limits of integration are now positioned between the first nulls. The main-beam solid angle is usually more than 80% of the beam solid angle, which gives $\kappa_B \leq 0.75$

$$\Omega_A = \iint_{4\pi} P_n(\theta,\phi)\, d\Omega \approx \theta_{\mathrm{HP}}\, \phi_{\mathrm{HP}} \tag{B.12}$$

$$\left. \begin{array}{c} \Omega_M = \iint_{\text{main}} P_n(\theta,\phi)\, d\Omega \approx \kappa_B \theta_{\mathrm{HP}}\, \phi_{\mathrm{HP}} \\[2mm] \text{where,}\ 0.8 \leq \kappa_B \leq 1.0 \end{array} \right\} \tag{B.13}$$

### B.5.6 Beam Efficiency and Stray Factor

As discussed earlier, the radiation distribution over the sphere is not uniform (see Fig. B.7). There seems to be no radiation at certain angles. The antenna beam shape gives a rough estimation of the fractional power radiated along a proper direction. This is given by the *beam efficiency* $\varepsilon_M$. Mathematically, $\varepsilon_M$ is defined as the ratio of the solid angle $\Omega_M$ subtended by the main beam alone to the sum of solid angles $\Omega_A$ subtended by all the lobes. This is shown in Eq. (B.14). The ratio of the total solid angle $\Omega_m$ subtended by the minor lobes to the sum of solid angles $\Omega_A$ subtended by all the lobes is known as the *stray factor* $\varepsilon_M$. It is expressed in Eq. (B.15). Eq. (B.16) shows that the sum of these two factors is unity.

$$\varepsilon_M = \frac{\text{Solid angle subtended by the main beam}}{\text{Sum of solid angles subtended by all the lobes}} = \frac{\Omega_M}{\Omega_A} \tag{B.14}$$

$$\varepsilon_m = \frac{\text{Sum of solid angles subtended by the minor lobes}}{\text{Sum of solid angles subtended by all the lobes}} = \frac{\Omega_m}{\Omega_A} \qquad (\text{B}.15)$$

$$\varepsilon_M + \varepsilon_m = 1 \qquad (\text{B}.16)$$

## B.6   RADIATION POWER DENSITY

The power available per unit area at a large distance from the source is obtained from the mean value $\vec{P}_{av}$ of the instantaneous Poynting vector $\vec{P}$. This is given in Eq. (B.17), where $\vec{E}$ and $\vec{H}$ are the instantaneous electric and magnetic fields. The quantity $\vec{H}^*$ is the complex conjugate of $\vec{H}$. If the units of $\vec{E}$ and $\vec{H}$ are respectively in Volt/m and Amp/m, then $\vec{P}_{av}$ is in Watt/m$^2$.

$$\vec{P}_{av} = \left( \vec{E} \times \vec{H} \right)_{av} = \frac{1}{2} \, \text{Re} \left( \vec{E} \times \vec{H}^* \right) \text{ Watt/m}^2 \qquad (\text{B}.17)$$

We may write $P_{av} = \left| \vec{P}_{av} \right|$.

## B.7   RADIATION INTENSITY

The radiation intensity $U$ is defined as the *power emitted over a unit solid angle from the antenna*. Note that radiation intensity is independent of distance but varies as a function of $(\theta, \phi)$. It is used in determining certain antenna parameters like antenna gain, directive gain, etc. Radiation intensity is measured in Watt per steradian.

We now refer to coordinates shown in Fig. B.2. The infinitesimal area $ds = r^2 \sin\theta d\theta d\phi$ on the surface of the sphere subtends a solid angle $d\Omega = \sin\theta d\theta d\phi$. One can express the total solid angle $\Omega$ of the sphere as given in Eq. (B.18). This evaluates to $4\pi$ steradians. For an isotropic radiator, the radiation intensity $U_i$ is defined as the *total power radiated over $4\pi$ steradian*. This is shown in Eq. (B.19), where $W_r$ is the power radiated by the isotropic antenna. For a general case, the radiation intensity $U$ can be expressed as in Eq. (B.20), where $dW_r$ is the infinitesimal power radiated over the solid angle $d\Omega$. The user's interest is to find the radiation intensity for the direction of maximum radiation. It is generally represented by $U_{max}$. The normalized power pattern $P_n(\theta, \phi)$ may be expressed in terms of radiation intensity as shown in Eq. (B.21).

$$\Omega = \int d\Omega = \int_0^{2\pi} \int_0^{\pi} \sin\theta d\theta d\phi = 4\pi \qquad (\text{B}.18)$$

$$U_i = \frac{W_r}{4\pi} \tag{B.19}$$

$$U = \frac{\text{Infinitesimal power}}{\text{Infinitesimal solid angle}} = \frac{dW_r}{d\Omega} \tag{B.20}$$

$$P_n(\theta,\phi) = \frac{U(\theta,\phi)}{U_{max}} \tag{B.21}$$

## B.8 ANTENNA IMPEDANCE AND RADIATION RESISTANCE

The antenna shows an impedance at its terminals. It consists of two parts: (*i*) resistive, and (*ii*) reactive. The real part is responsible mainly for radiation and to some extent power loss. For transferring maximum power from the generator to the antenna, the reactive part must be removed by impedance that matches the source. Let us assume the antenna impedance as purely resistive. Most of the power fed to it is radiated into space, and the remaining part gets dissipated as heat. Let $W_r$ and $W_1$, respectively, represent the radiated and dissipated powers. The total power $W_t$ consumed by the antenna is shown in Eq. (B.22). If $I$ is the current flowing at the antenna terminals, then $W_t$ can be expressed as in Eq. (B.23). Here $R_r$ is a fictitious resistance that would consume the same amount of power as what is actually radiated. Similarly, $R_1$ is another fictitious resistance that would consume the amount of power lost as heat. Hence, $R_r$ is known as the *antenna radiation resistance*, and $R_1$ is known as the *antenna loss resistance*. For an ideal antenna, $R_1 = 0$.

$$W_t = W_r + W_1 \tag{B.22}$$

$$W_t = \frac{1}{2}I^2\left(R_r + R_1\right) \tag{B.23}$$

## B.9 ANTENNA EFFICIENCY

The ratio of actual power radiated by an antenna to total power consumed by it is known as *antenna efficiency* and denoted by $\eta_a$. If the radiation resistance $R_r$ and the loss resistance $R_1$ are known, the antenna efficiency can be calculated using Eq. (B.24), where $I$ is the current flowing through the antenna terminals. Multiplying $\eta_a$ by 100, one obtains the antenna efficiency as a percentage.

$$\eta_a = \frac{\text{Power radited by the antenna}}{\text{Power input to the antenna}} = \frac{I^2 R_r}{I^2\left(R_r + R_1\right)} = \frac{R_r}{R_r + R_1} \tag{B.24}$$

## B.10   DIRECTIVITY

Practical antennas are designed to radiate maximum power in one specific direction. The *directivity* may be seen as to what extent a lossless antenna concentrates the radiated power as compared to an isotropic radiator. The directivity $G_{D(max)}$ is defined as the *ratio of maximum radiated power density* $P_{max}$ *to its average value* $P_{av}$. It is also defined as the *ratio of maximum radiation intensity to radiation intensity of an isotropic radiator*. The relations are shown in Eq. (B.25), where $\Omega_M$ is the main beam solid angle. Directivity is a dimensionless quantity.

$$G_{D(max)} = \frac{P_{max}}{P_{av}} = \frac{U_{max}}{U_i} = \frac{4\pi}{\Omega_M} \tag{B.25}$$

### B.10.1 Relation Between Directivity and HPBW

From the HPBW of the major lobe in two principal panes, one can estimate the directivity $D$ using Eq. (B.26). Here, $\theta_{HP}$ and $\phi_{HP}$, respectively, are the HPBW measured in two principal planes and expressed in degrees.

$$G_{D(max)} = \frac{40000}{\theta_{HP}\,\phi_{HP}} \tag{B.26}$$

### B.10.2 Directive Gain

Unlike directivity, which is specified for the direction of maximum radiation, the directive gain $G_D$ is a directional quantity. It is given in Eq. (B.27), where $P(\theta, \phi)$ is the radiated power density in any direction, $P_{av}$ is the mean of the power density in all directions, $U(\theta, \phi)$ is the radiation intensity along any direction, and $U_i$ is the mean of the radiation intensities in all directions.

$$G_D = \frac{P(\theta,\phi)}{P_{av}} = \frac{U(\theta,\phi)}{U_i} \tag{B.27}$$

## B.11   GAIN (POWER GAIN)

A concept similar to directive gain is the *gain G* or $G_P$. It is also known as *power gain*. The power gain $G_P$ is equal to the antenna efficiency times the directive gain $G_D$. It is expressed in Eq. (B.28), where $\eta_a$ is the antenna efficiency. Note that $G_P$ is always less than $G_D$ because losses exist in all practical antennas.

$$G_P = \eta_a\,G_D \tag{B.28}$$

## B.12  EFFECTIVE APERTURE AREA

The concept of *effective aperture area* $A_e$ was developed based on a receiving antenna. Let a device convert electromagnetic radiation into electrical power delivered at its terminals. The electromagnetic wave energy collected is proportional to the size of the collecting area. This follows from the fact that the electromagnetic energy is measured in terms of energy per unit time per unit area over a bandwidth, which is none other than the flux density. Hence, the greater the collecting area, the more power one receives. The power $P_{ant}$ received by an antenna is the product of $A_e$ with the flux density $S$. This is expressed in Eq. (B.29). It is assumed that the wave fronts of the incoming radiation are parallel to $A_e$.

$$P_{ant} = SA_e \qquad (B.29)$$

### B.12.1 Aperture Efficiency

The effective aperture area $A_e$ is specific for the type of antenna. For example, a lossless paraboloid dish has an $A_e$ equal to its physical cross-sectional area, whereas for a half-wave dipole $A_e$ is larger than its physical cross-sectional area. Effective usage of the physical aperture depends on the aperture efficiency $\eta_A$, which is defined as the *ratio of effective aperture area $A_e$ to the physical aperture area $A_p$*. This is expressed in Eq. (B.30).

$$\eta_A = \frac{A_e}{A_p} \qquad (B.30)$$

### B.12.2 Wavelength, Beam Angle, and Directivity Relationships

The wavelength $\lambda$ is related to effective aperture area $A_e$ and the main-beam solid angle $\Omega_M$ as given in Eq. (B.31). The directivity $G_{D(max)}$ may be obtained from $A_e$ using Eq. (B.32).

$$\lambda^2 = A_e \Omega_M \qquad (B.31)$$

$$G_{D(max)} = 4\pi \frac{A_e}{\lambda^2} \qquad (B.32)$$

## B.13  EFFECTIVE HEIGHT (EFFECTIVE LENGTH)

The *effective height h* (also called *effective length*) of an antenna is a concept similar to the effective aperture $A_e$. It is used for calculating the potential developed across the receiving antenna terminals produced by an electromagnetic wave. The antenna output voltage $V$ (in Volts) is a product of the electric field $E$ (in Volts per meter) with the effective height $h$ of the antenna (in meters). This is expressed in Eq. (B.33). It is used for wire antennas whose physical aperture area is negligible.

$$V = h\,E \qquad\qquad (B.33)$$

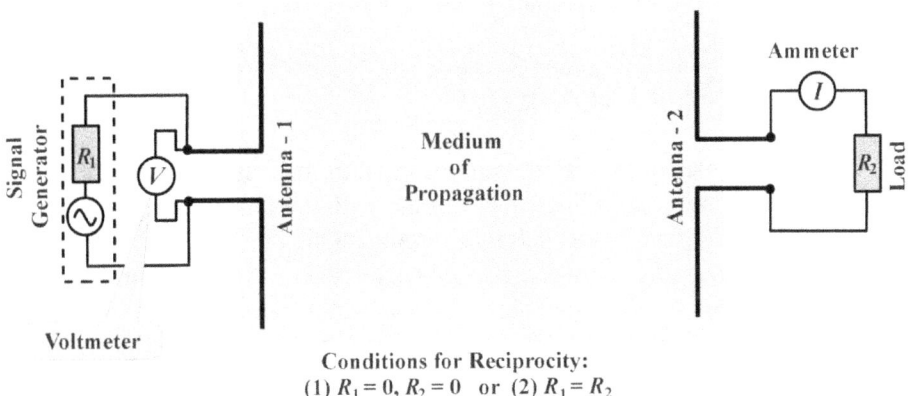

**Conditions for Reciprocity:**
(1) $R_1 = 0$, $R_2 = 0$  or  (2) $R_1 = R_2$

**FIGURE B.8:** Test setup for verification of the antenna reciprocity theorem.

## B.14   ANTENNA RECIPROCITY THEOREM

Consider the test setup shown in Fig. B.8. The antenna reciprocity theorem states that *if an emf is applied to the terminals of an antenna 1 and the current is measured at the terminals of an antenna 2, then an equal current in both amplitude and phase will be obtained at the terminals of antenna 1 if the same emf is applied to the terminals of antenna 2.* The theorem is valid if impedance $R_1$ of the signal generator and the load $R_2$ across the antenna 2 are zero. In practice, all signal generators have nonzero impedance. It is therefore necessary to have $R_1 = R_2$. The impedance of the voltmeter and the ammeter are assumed to be infinity and zero, respectively.

### B.14.1 Use of Reciprocity Theorem in Radio Astronomy

The important applications of the reciprocity theorem on radio-telescope antennas are as follows:

  (*i*) The radiation pattern of an antenna in receiving and transmission modes remains the same. Hence, radiation patterns of the antenna feeds can be measured in the transmitting mode before implementing them on the telescope dishes.

  (*ii*) The impedance of an antenna in receiving and transmission modes remains the same. The impedance of the antenna feeds can be measured in the transmitting mode by measuring the voltage and current at its terminals.

### B.14.2 Practical Limitations of Reciprocity Theorem

The theorem is not valid when the propagating radio waves are noticeably effected by the presence of the Earth's magnetic field or ionospheric disturbances.

## B.15   ANTENNA BANDWIDTH

There is no unique definition for antenna bandwidth, because properties like radiation pattern, radiation resistance, etc. change with the frequency of operation. Hence, the bandwidth is defined such that certain antenna properties remain more or less constant within this range of frequencies. In general, the bandwidth is measured in three categories: (*i*) bandwidth $\Delta v$ over which the directivity of the antenna is higher than some acceptable value; (*ii*) bandwidth $\Delta v$ over which at least a specified front-to-back ratio is met; (*iii*) bandwidth $\Delta v$ over which the antenna matches the load. The last one is expressed in Eq. (B.34), where $v_r$ is the center frequency, and $Q$ is the quality factor of antenna resonance shown in Eq. (B.35). Hence, the lower the $Q$, the higher is its bandwidth, and vice versa.

$$\Delta v = \frac{v_r}{Q} \tag{B.34}$$

$$Q = \frac{\text{Total energy stored by the antenna}}{\text{Energy dissipated or radiated per cycle}} \tag{B.35}$$

## B.16   HALF-WAVE DIPOLE ANTENNA

The half-wave dipole is one of the most common antennas. It is made using two quarter-wavelength conductors placed back to back such that the total length is $\lambda/2$, where $\lambda$ is the dipole wavelength. This is shown in Fig. B.9a. The dipole is excited at a frequency $c/\lambda$, where $c$ is the speed of light. A standing-wave pattern is formed across the dipole. The current distribution across the antenna length is shown. Consider $l$ as an axis lying along the dipole such that, at $l = 0$, the current $I$ is maximum ($I = I_0$), and at $l = \pm\lambda/2$, $I = 0$. If the signal's angular frequency is $\omega$, the current distribution $I(t, l)$ across the dipole as a function of time $t$ may be expressed as in Eq. (B.36). Here, $\beta = 2\pi/\lambda$. At a very large distance $r$ from the dipole center, the electric field $E_\theta$ is given as in Eq. (B.37), where $\varepsilon_0$ is the permittivity of free space and

$c$ is the speed of light. The quantity within "[ ]" behaves more or less like sin$\theta$, which is shown in Fig. B.9b using dotted lines. The electric field distributions are shown in the E- and H-planes. The E-plane can be rotated along the dipole axis. The radiation resistance is given in Eq. (B.38). The dipole impedance also consists of an imaginary part. Hence the impedance is preferably measured. Eq. (B.39) expresses the directivity, which evaluates to approximately 2.15 dBi[3].

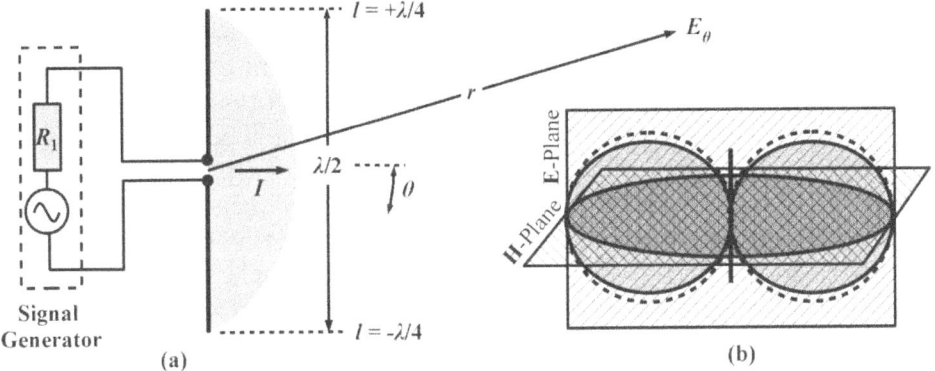

**FIGURE B.9:** (a) Current distribution across a half-wave dipole due to excitation from a source. (b) Radiation patterns in the E- and H-planes.

$$I(t,l) = I_0 \exp\left(j\omega t\right)\cos\left(\beta l\right) \tag{B.36}$$

$$E_\theta = \frac{-jI_0}{2\pi\varepsilon_0 cr}\left[\frac{\cos\left(\dfrac{\pi}{2}\cos\theta\right)}{\sin\theta}\right]e^{j(\omega t-\beta r)} \tag{B.37}$$

$$R_r = 120\int_0^{\pi/2}\frac{\cos^2\left(\dfrac{\pi}{2}\cos\theta\right)}{\sin\theta}\,d\theta = 73.13\ \text{Ohm} \tag{B.38}$$

$$D = \frac{1}{\displaystyle\int_0^{\pi/2}\frac{\cos^2\left(\dfrac{\pi}{2}\cos\theta\right)}{\sin\theta}\,d\theta} \approx 1.641 = 2.15\ \text{dBi} \tag{B.39}$$

---

[3] The unit *dBi* is the decibel conversion of the linear gain of an antenna. The subscript *i* indicates that the gain is measured with respect to an isotropic antenna. Gain in dBi = $10\log_{10}$[linear gain w.r.t isotropic antenna.]

## B.17   ANGULAR SPECTRUM AND RADIATION PATTERN

The electric field $E$ on the antenna aperture can be visualized as the result of interference from a continuum of plane and evanescent waves propagating in various angular directions. A plane wave frequency $v$ may be expressed as $P(\varphi)\,e^{-j(\omega t - kr_\varphi)}$, where $\omega = 2\pi v$ is the angular frequency, $r_\varphi$ is the distance along the propagating direction, and $t$ is time. The function $P(\varphi)$ is the wave amplitude, which is a function of the direction angle $\varphi$. Then $E$ may be expressed as a sum of the continuum of infinite spatial frequencies. Thus $E$ is the Fourier transform of the function $P(\varphi)$. Conversely, $P$ can be expressed as the inverse Fourier transform of $E$. If this second relationship is established in the aperture plane using direction cosines of the waves, one obtains the *angular spectrum* of the antenna.

Fig. B.10a shows an $E$-field radiation pattern in one plane. Let $\vec{E}\left(x/\lambda\right)$ be the component of the electric field vector in the aperture plane (along the $x$-axis), where $\lambda$ is the wavelength. The radiation pattern is an angular function of $\varphi$. The quantity $P$, which is a function of $\sin\varphi$, is the angular spectrum. It is related to $E(x/\lambda)$ by a Fourier transformation, as shown in Eq. (B.40). Its inverse relationship is given in Eq. (B.41).

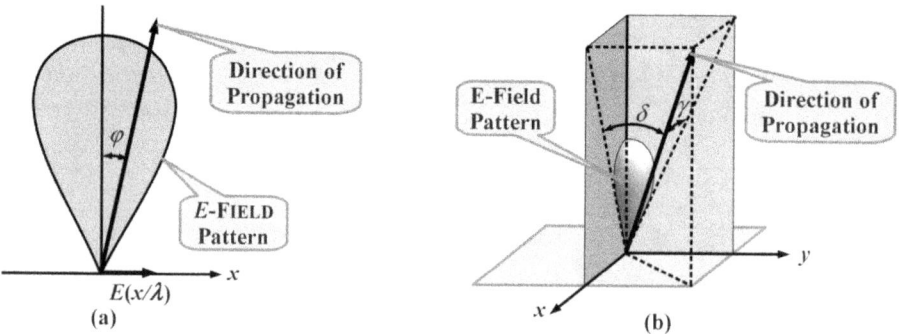

FIGURE B.10: Geometry for studying the angular spectrum in (a) one dimension, (b) two dimensions.

$$P(\sin\varphi) = \int_{-\infty}^{\infty} E\left(\frac{x}{\lambda}\right) e^{j2\pi\left(\frac{x}{\lambda}\right)\sin\varphi}\, d\left(\frac{x}{\lambda}\right) \tag{B.40}$$

$$E\left(\frac{x}{\lambda}\right) = \int_{-\infty}^{\infty} P(\sin\varphi) e^{-j2\pi\left(\frac{x}{\lambda}\right)\sin\varphi}\, d\left(\sin\varphi\right) \tag{B.41}$$

Fig. B.10b shows a two-dimensional case. From this geometry, two-dimensional relations between the angular spectrum and the two-dimensional electric field are shown, respectively, in Eqs. (B.42) and (B.43).

$$P(\sin\gamma,\sin\delta) = \int_{-\infty}^{\infty}\int_{-\infty}^{\infty} E\left(\frac{x}{\lambda},\frac{y}{\lambda}\right) e^{j2\pi\left\{\left(\frac{x}{\lambda}\right)\sin\gamma+\left(\frac{y}{\lambda}\right)\sin\delta\right\}} d\left(\frac{x}{\lambda}\right)d\left(\frac{y}{\lambda}\right) \qquad \text{(B.42)}$$

$$E\left(\frac{x}{\lambda},\frac{y}{\lambda}\right) = \int_{-\infty}^{\infty}\int_{-\infty}^{\infty} P(\sin\gamma,\sin\delta) e^{-j2\pi\left\{\left(\frac{x}{\lambda}\right)\sin\gamma+\left(\frac{y}{\lambda}\right)\sin\delta\right\}} d(\sin\gamma)d(\sin\delta) \quad \text{(B.43)}$$

Note that if the observation point is at infinity, the angular spectrum turns out to be the far-field radiation pattern of the antenna.

## REVIEW QUESTIONS

1. How does an antenna radiate and receive? Explain with a diagram.
   [*Hint:* See Fig. B.1]

2. What is an isotropic antenna?

3. Using a diagram of an antenna, explain the (*i*) reactive near-field region, (*ii*) radiative near-field region, and (*iii*) radiative far-field region. Use equations to specify them. [*Hint:* See Fig. B.3]

4. Explain the difference between the Fraunhofer region and the Fresnel zone?

5. What do you understand concerning the radiation pattern of an antenna?

6. Using a diagram, illustrate the various lobes of an antenna.
   [*Hint:* See Fig. B.5]

7. With the help of a diagram, explain the concepts of (*i*) beam width between first nulls, and (*ii*) half-power beam width.

8. What is meant by the front-to-back ratio of an antenna?

9. What is meant by the normalized radiation pattern of an antenna?

10. Find the directivity of an isotropic antenna. [*Ans:* $G_{D(max)} = 1$]

11. The radiation resistance of an antenna is 73 Ohms, and its dissipative resistance is 2 Ohms. Calculate the antenna efficiency. [*Hint:* Use Eq. (B.21).]

12. For the same antenna efficiency as above, find the gain of an antenna if its directivity is 10. [*Hint:* Use Eq. (B.25).]

13. A paraboloid dish antenna has a cross-sectional area of 100 m³. The effective aperture area is 60 m². What is the aperture efficiency?
    [*Hint:* Use Eq. (B.27).]

14. The directivity of an antenna is 900 at 1 GHz. Find its effective aperture area. [*Hint:* Find the wavelength using $c = f\lambda$, where $c$ is the speed of light, $f$ is the frequency, and $\lambda$ is the wavelength. Then use Eq. (B.29).]

15. An antenna has an effective aperture area of 10 m². If the flux density across the aperture is 1 $\mu$W/m², calculate the power delivered by the antenna to a matched load. Assume the antenna as 100% efficient.

   [*Hint:* Use Eq. (B.26).]

16. A receiving wire antenna with an effective length of 1 m is exposed to an electromagnetic field of 1 mV/m. Find the e.m.f. produced by the antenna at its terminals. [*Hint:* Use Eq. (B.30).]

17. Find the antenna bandwidth for a center frequency of 5 MHz and a $Q$ of 100. [*Hint:* Use Eq. (B.31).]

18. Using a diagram, explain the antenna reciprocity theorem. State its usefulness in radio astronomy. [*Hint:* See section 14.]

19. Draw the radiation patterns of a half-wave dipole. [*Hint:* See Fig. B.4.]

20. Why should the radiation resistance of an antenna be measured?

21. With respect to an isotropic radiator, the directivity of an antenna is 10. Convert this value to dBi.

22. Explain the concept of the angular spectrum of an antenna.

   [*Hint:* See section B.17.]

# RADIATION POTENTIAL FORMULATION

## C.1 INTRODUCTION

Electromagnetic wave propagation can be analyzed by application of Maxwell's equations. For more complex radiation fields, the use of electrodynamic potentials often yields simpler solutions than would be obtainable directly from Maxwell's equations. We first consider the potential formulation of static sources and then outline a generalization for the treatment of sources with arbitrary time dependence. The inhomogeneous wave equations for potentials are developed. Time-harmonic sources are then considered with application to radiation fields with the simplest example of the Hertz dipole antenna. The Hertz dipole solution is then extended to the half-wave dipole considered in Appendix B.

## C.2   STATIC SOURCES

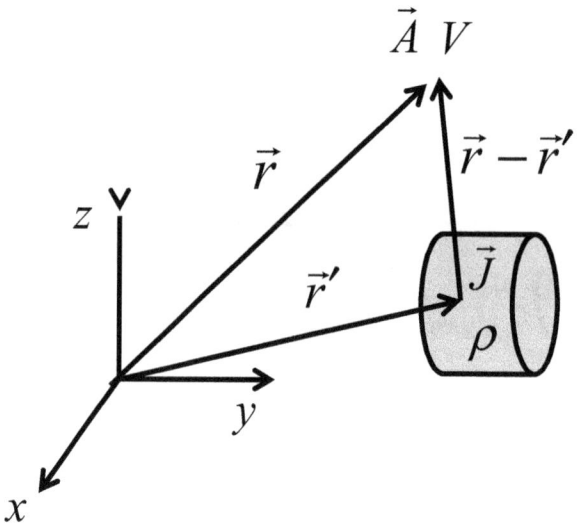

**FIGURE C.1:** Vectors $\vec{r}$ and $\vec{r}'$ locating source and field positions.

In electrostatics the magnetic and electric fields are described by the curl and divergence of the vector and scalar potentials $\vec{A}$ and $V$

$$\vec{B}(\vec{r}) = \nabla \times \vec{A}(\vec{r}) \tag{C.1}$$

$$\vec{E}(\vec{r}) = -\nabla V(\vec{r}) \tag{C.2}$$

Because divergence of the curl of any vector is zero, we see that (C.1) satisfies the Maxwell relation $\nabla \cdot \vec{B} = 0$. The curl of the gradient is always zero so that electrostatic fields are curl free as seen from (C.2). The vector and scalar potentials at the position $\vec{r}$ are obtained from integrals over the source current $\vec{J}$ and charge distribution $\rho$ located at $\vec{r}$ in Fig. (C.1)

$$\vec{A}(\vec{r}) = \frac{\mu_0}{4\pi} \int \frac{\vec{J}(\vec{r}')}{|\vec{r} - \vec{r}'|} dv' \tag{C.3}$$

$$V(\vec{r}) = \frac{1}{4\pi\varepsilon_0} \int \frac{\rho(\vec{r}')}{|\vec{r} - \vec{r}'|} dv' \tag{C.4}$$

The magnetic and electric fields are then obtained from equations (C.1) and (C.2) giving

$$\vec{B}(\vec{r}) = \frac{\mu_0}{4\pi} \int \frac{\vec{J}(\vec{r}') \times (\vec{r} - \vec{r}')}{|\vec{r} - \vec{r}'|^3} dv' \tag{C.5}$$

$$\vec{E}(\vec{r}) = \frac{1}{4\pi\varepsilon_0} \int \frac{\rho(\vec{r}')(\vec{r} - \vec{r}')}{|\vec{r} - \vec{r}'|^3} dv'. \tag{C.6}$$

For time-dependent currents and charge distributions, simple integral relations such as (C.5) and (C.6) can no longer be used to calculate the electric and magnetic fields.

## C.3   ELECTRODYNAMICS

In electrodynamics, the time-dependent magnetic field is still given by the curl of the vector potential,

$$\vec{B}(\vec{r},t) = \nabla \times \vec{A}(\vec{r},t) \tag{C.7}$$

whereas the electric field is also generated by a time-changing vector potential

$$\vec{E}(\vec{r},t) = -\frac{\partial}{\partial t} \vec{A}(\vec{r},t) - \nabla V(\vec{r},t) \tag{C.8}$$

Maxwell's equation $\nabla \times \vec{E} = -\partial \vec{B} / \partial t$ follows directly from these equations.

The vector and scalar potentials at a given location $\vec{r}$ now depend on the source current and charge distribution at $\vec{r}'$ at some time in the past because of the finite speed of light. The retarded time is thus

$$t_r = t - \frac{|\vec{r} - \vec{r}'|}{c}, \tag{C.9}$$

where the vector and scalar potentials become

$$\vec{A}(\vec{r},t) = \frac{\mu_0}{4\pi} \int \frac{\vec{J}(\vec{r}',t_r)}{|\vec{r} - \vec{r}'|} dv' \tag{C.10}$$

$$V(\vec{r},t) = \frac{1}{4\pi\varepsilon_0} \int \frac{\rho(\vec{r}',t_r)}{|\vec{r} - \vec{r}'|} dv' \tag{C.11}$$

Note that we cannot write simple expressions analogous to (C.5) and (C.6) for $\vec{E}$ and $\vec{B}$ because the retarded time is a function of $\vec{r}$.

## C.4   WAVE EQUATIONS

The vector and scalar potentials obey inhomogeneous wave equations that can be derived from the Maxwell equations:

$$\nabla \times \vec{B} = \mu_0 \vec{J} + \mu_0 \varepsilon_0 \frac{\partial \vec{E}}{\partial t} \tag{C.12}$$

$$\nabla \cdot \vec{E} = \frac{\rho}{\varepsilon_0} \tag{C.13}$$

We substitute the relations (C.7) and (C.8) for $\vec{B}$ and $\vec{E}$ into (C.12) and (C.13), respectively:

$$\nabla \times \left( \nabla \times \vec{A} \right) = \mu_0 \vec{J} + \mu_0 \varepsilon_0 \frac{\partial}{\partial t}\left( -\frac{\partial}{\partial t}\vec{A} - \nabla V \right) \tag{C.14}$$

$$\nabla \cdot \left( -\frac{\partial}{\partial t}\vec{A} - \nabla V \right) = \frac{\rho}{\varepsilon_0} \tag{C.15}$$

Using the vector identity

$$\nabla \times \left( \nabla \times \vec{A} \right) = \nabla \left( \nabla \cdot \vec{A} \right) - \nabla^2 \vec{A} \tag{C.16}$$

and rearranging it, gives

$$\nabla^2 \vec{A} - \mu_0 \varepsilon_0 \frac{\partial^2}{\partial t^2} \vec{A} = -\mu_0 \vec{J} + \nabla\left( \mu_0 \varepsilon_0 \frac{\partial}{\partial t}V + \nabla \cdot \vec{A} \right) \tag{C.17}$$

$$\left( -\frac{\partial}{\partial t}\nabla \cdot \vec{A} - \nabla^2 V \right) = \frac{\rho}{\varepsilon_0} \tag{C.18}$$

Equations (C.17) and (C.18) are simplified by the choice of the Lorentz gauge:

$$\nabla \cdot \vec{A} = -\mu_0 \varepsilon_0 \frac{\partial}{\partial t}V \tag{C.19}$$

This gauge choice leaves Maxwell's equations unchanged while giving the inhomogeneous wave equations symmetric forms in $\vec{A}$ and $V$

$$\nabla^2 \vec{A} - \mu_0 \varepsilon_0 \frac{\partial^2}{\partial t^2} \vec{A} = -\mu_0 \vec{J} \tag{C.20}$$

$$\nabla^2 V - \mu_0 \varepsilon_0 \frac{\partial^2}{\partial t^2} V = -\frac{\rho}{\varepsilon_0} \qquad (C.21)$$

These may be expressed succinctly as

$$\Box^2 \vec{A} = -\mu_0 \vec{J} \qquad (C.22)$$

$$\Box^2 V = -\frac{\rho}{\varepsilon_0}, \qquad (C.23)$$

where the d'Alembertian operator is given by

$$\Box^2 = \nabla^2 - \mu_0 \varepsilon_0 \frac{\partial^2}{\partial t^2} \qquad (C.24)$$

In most standard texts on electromagnetics, the integral expressions (C.10) and (C.11) for $\vec{A}$ and $V$ are shown to satisfy the inhomogeneous wave equations by the Green's function method.

## C.5   TIME-HARMONIC RADIATION FIELDS

Time-harmonic currents and charge densities are expressed as

$$\vec{J}(\vec{r}', t_r) = \vec{J}(\vec{r}') e^{-j\omega t_r} \qquad (C.25)$$

$$\rho(\vec{r}', t_r) = \rho(\vec{r}') e^{-j\omega t_r} \qquad (C.26)$$

The vector and scalar potentials also have harmonic time dependence:

$$\vec{A}(\vec{r}, t) = \vec{A}(\vec{r}) e^{-j\omega t} \qquad (C.27)$$

$$V(\vec{r}, t) = V(\vec{r}) e^{-j\omega t} \qquad (C.28)$$

Substituting (C.25) and (C.26) into (C.10) and (C.11), the spatial part of the potentials become

$$\vec{A}(\vec{r}) = \frac{\mu_0}{4\pi} \int \vec{J}(\vec{r}') \frac{e^{j\beta|\vec{r}-\vec{r}'|}}{|\vec{r}-\vec{r}'|} dv' \qquad (C.29)$$

$$V(\vec{r}) = \frac{1}{4\pi\varepsilon_0} \int \rho(\vec{r}') \frac{e^{j\beta|\vec{r}-\vec{r}'|}}{|\vec{r}-\vec{r}'|} dv', \qquad (C.30)$$

where $\beta = \omega/c = 2\pi/\lambda$. These expressions may be evaluated analytically for a few simple antenna configurations where we are interested in the far-field solution.

## C.6  HERTZ DIPOLE ANTENNA

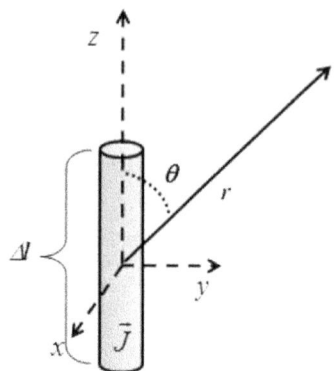

**FIGURE C.2:** Geometry of a Hertz dipole of length $\Delta l$ carrying a uniform ac current in the z-direction

As an application of the potential formalism, we consider the radiation field of a very short dipole antenna of length $\Delta l \ll \lambda$ with cross-section $S$ carrying an AC current $I$ in the z-direction, as shown in Fig. C.2. The radiation field of this antenna was first calculated by H. Hertz before his subsequent discovery of radio waves in 1888. The time-harmonic current density

$$\vec{J} = \frac{I}{S}e^{-j\omega t_r}\hat{z} \tag{C.31}$$

is taken to be uniform over the entire length. We approximate $\left|\vec{r} - \vec{r}'\right| \approx r$ for $L < \lambda/50$ so that the vector potential becomes

$$\vec{A}(\vec{r},t) = \frac{\mu_0}{4\pi}\frac{I}{S}\frac{e^{j\beta r}}{r}\underbrace{\int da \int_{-\Delta l/2}^{\Delta l/2} dz\ e^{j\omega t}\hat{z}}_{S\Delta l}, \tag{C.32}$$

where the integral over the source coordinates simply gives the volume of the antenna. The term $e^{j\beta r}/r$ is called the spherical propagation factor. Expressing the vector potential in spherical coordinates, we write the Cartesian unit vector $\hat{z}$ in terms of its spherical components:

$$\hat{z} = \cos\theta\hat{r} - \sin\theta\hat{\theta} \tag{C.33}$$

The vector potential has both radial and angular components given by

$$\vec{A}(\vec{r},t) = \frac{\mu_0 I \Delta l}{4\pi} \frac{e^{j\beta r}}{r}\left(\cos\theta\hat{r} - \sin\theta\hat{\theta}\right)e^{-j\omega t}. \tag{C.34}$$

We now take the curl in spherical coordinates

$$\nabla \times \vec{A} = \left(\frac{1}{r}\frac{\partial}{\partial r}rA_\theta - \frac{1}{r}\frac{\partial}{\partial\theta}A_r\right)\hat{\phi} \tag{C.35}$$

to obtain a magnetic field that has only a $\phi$-component, given by

$$\vec{B}(\vec{r},t) = \frac{\mu_0 I \Delta l}{4\pi}\frac{e^{j\beta r}}{r}\left(\frac{1}{r} - j\beta\right)\sin\theta e^{-j\omega t}\hat{\phi}. \tag{C.36}$$

The electric field may be obtained from the Maxwell relation Eq. (C.12). We now choose the ac form of the Lorentz gauge in Eq. (C.19)

$$\nabla\cdot\vec{A} = j\omega\mu_0\varepsilon_0 V \tag{C.37}$$

to first obtain the electric potential where we must take the divergence in spherical coordinates

$$\nabla\cdot\vec{A} = \frac{1}{r^2}\frac{\partial}{\partial r}r^2 A_r + \frac{1}{r\sin\theta}\frac{\partial}{\partial\theta}\left(\sin\theta A_\theta\right). \tag{C.38}$$

This becomes

$$\nabla\cdot\vec{A} = \frac{\mu_0}{4\pi}I\Delta l\frac{e^{j\beta r}}{r}\cos\theta\left(j\beta - \frac{1}{r}\right)e^{-j\omega t} \tag{C.39}$$

so that the electric potential is

$$V = \frac{1}{4\pi\varepsilon_0}\frac{I\Delta l}{\omega}\frac{e^{j\beta r}}{r}\cos\theta\left(\beta + \frac{j}{r}\right)e^{-j\omega t}. \tag{C.40}$$

To obtain the electric field, we take the gradient in spherical coordinates

$$\nabla V = \hat{r}\frac{\partial V}{\partial r} + \hat{\theta}\frac{1}{r}\frac{\partial V}{\partial\theta}, \tag{C.41}$$

which has radial and angular components,

$$\nabla V = \frac{1}{4\pi\varepsilon_0}\frac{I\Delta l}{\omega}\frac{e^{j\beta r}}{r}\left[\cos\theta\left(j\beta^2 - \frac{2\beta}{r} - \frac{2}{r^2}\right)\hat{r} - \sin\theta\left(\frac{\beta}{r} + \frac{j}{r^2}\right)\hat{\theta}\right]e^{-j\omega t}. \tag{C.42}$$

Factoring $\beta^2$, we have

$$\nabla V = \frac{1}{4\pi} \underbrace{\frac{\beta^2}{\varepsilon_0 \omega^2}}_{\mu_0} \omega I \Delta l \frac{e^{j\beta r}}{r} \left[ \cos\theta \left( j - \frac{2}{\beta r} - \frac{2}{(\beta r)^2} \right) \hat{r} \right. \tag{C.43}$$

$$\left. - \sin\theta \left( \frac{1}{\beta r} + \frac{j}{(\beta r)^2} \right) \hat{\theta} \right] e^{-j\omega t} .$$

The time derivative of the vector potential is

$$\frac{\partial \vec{A}}{\partial t} = \frac{\mu_0}{4\pi} \omega I \Delta l \frac{e^{j\beta r}}{r} \left( -j\cos\theta \hat{r} + j\sin\theta \hat{\theta} \right) e^{-j\omega t} . \tag{C.44}$$

The electric field is the negative of the sum of (C.43) and (C.44), giving

$$\vec{E} = \frac{\mu_0}{4\pi} \omega I \Delta l \frac{e^{j\beta r}}{r} \left[ \cos\theta \left( \frac{2}{\beta r} + \frac{2}{(\beta r)^2} \right) \hat{r} + \right.$$

$$\left. \sin\theta \left( -j + \frac{1}{\beta r} + \frac{j}{(\beta r)^2} \right) \hat{\theta} \right] e^{-j\omega t} \tag{C.45}$$

In the radiation zone, both the electric and magnetic fields fall off as l/r:

$$\vec{E} = -j\frac{\mu_0}{4\pi} \omega I \Delta l \frac{e^{j\beta r}}{r} \sin\theta e^{-j\omega t} \hat{\theta} . \tag{C.46}$$

$$\vec{B} = -j\frac{\mu_0}{4\pi} \beta I \Delta l \frac{e^{j\beta r}}{r} \sin\theta e^{-j\omega t} \hat{\phi} . \tag{C.47}$$

The time-averaged Poynting vector

$$\vec{P}_{av} = \frac{1}{2} \text{Re}\left( \vec{E} \times \vec{H}^* \right) = \frac{1}{2} \text{Re}\left( \frac{1}{\mu_0} \vec{E} \times \vec{B}^* \right) = \frac{1}{2} \mu_0 \beta \omega \left( \frac{I\Delta l}{4\pi} \right)^2 \frac{\sin^2\theta}{r^2} \hat{r} \tag{C.48}$$

provides the radiated Watts/m². The total power radiated in Watts is thus

$$W_r = \int \vec{P}_{av} \cdot d\vec{a} = \frac{1}{2} \mu_0 \beta \omega \left( \frac{I\Delta l}{4\pi} \right)^2 \underbrace{\int_0^\pi \frac{\sin^2\theta}{r^2} r^2 \sin\theta d\theta \int_0^{2\pi} d\phi}_{\frac{4}{3} \cdot 2\pi} . \tag{C.49}$$

Numerically we have

$$W_r = \frac{\mu_0 \beta \omega}{12\pi}(I\Delta l)^2 = \frac{\pi\mu_0 c}{3}\left(\frac{I\Delta l}{\lambda}\right)^2 = 395 \text{ Ohm } \left(\frac{I\Delta l}{\lambda}\right)^2. \quad (C.50)$$

From the expression

$$W_r = \frac{1}{2}I^2 R_r \quad (C.51)$$

we identify the radiation resistance of the Hertz dipole as $R_r = 790(\Delta l/\lambda)^2$ Ohm. For a short dipole, $\Delta l < \lambda / 50$, we have $R_r < 0.32$ Ohm, which is much smaller than the radiation resistance of the half-wave dipole in Appendix B, where we found $R_r = 0.32$ Ohm. The efficiency $\eta_a$ of the Hertz dipole may be computed by Eq. B.21 of Appendix B, giving $\eta_a = R_r/(R_r + R_1)$ for any loss resistance $R_1$. We see that $\eta_a$ of the Hertz dipole would be much smaller than that of the half-wave dipole for comparable $R_1$ values.

## C.7 HALF-WAVE DIPOLE ANTENNA REVISITED

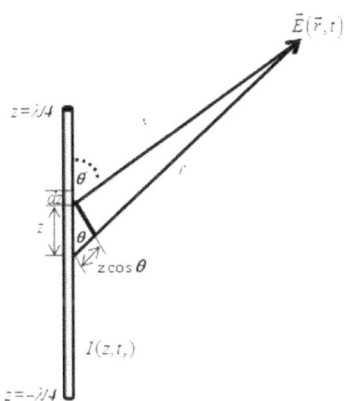

**FIGURE C.3:** Geometry for calculating the electric field of the half-wave dipole antenna.

Although the infinitesimal Hertz dipole has limited practical value with such a small efficiency, we may calculate the radiation emitted by longer antennas with more realistic current configurations by stacking Hertz dipoles and integrating over the current distribution. As an example, we consider the half-wave dipole antenna discussed in Appendix B with current along the z-axis given by Eqn. B.33 $I(z,t_r) = I_0 \exp(-j\omega t_r)\cos(\beta z)$ with $l \to z$ for $-\lambda/4 \le z \le \lambda/4$. In this case, the current is properly zero at the ends of the antenna. The far-field contribution to the electric field $d\vec{E}$ due to an infinitesimal element of the half-wave dipole is given by Eq. (C.44) after making the replacements $\Delta l \rightarrow dz$ and $I_0 \rightarrow I_0 \cos(\beta z)$:

$$d\vec{E} = -j\frac{\mu_0}{4\pi}\omega I_0 \cos(\beta z)\, dz\, \frac{e^{j\beta s}}{s}\sin\theta e^{-j\omega t}\hat{\theta}, \qquad (C.52)$$

where $s = r - z\cos\theta$, as shown in Fig C.3. Approximating the spherical propagation factor in the far-field region

$$\frac{e^{j\beta s}}{s} \approx \frac{e^{j\beta(r-z\cos\theta)}}{r} \qquad (C.53)$$

We now have

$$dE = -j\frac{\mu_0}{4\pi}\omega I_0 \cos(\beta z)\, dz\, \frac{e^{j\beta(r-z\cos\theta)}}{r}\sin\theta e^{-j\omega t}\hat{\theta}. \qquad (C.54)$$

Integrating over the length of the antenna, we obtain the electric field,

$$\vec{E} = -j\frac{\mu_0}{4\pi}\omega I_0 \frac{e^{j\beta r}}{r}\sin\theta e^{-j\omega t}\hat{\theta}\int_{-\lambda/4}^{\lambda/4}\cos(\beta z)e^{-jz\beta\cos\theta}dz, \qquad (C.55)$$

where we are taking $\theta \approx \theta'$ in the far-field region. Integrate by parts twice to obtain a term that contains the original integral, then factor to obtain

$$\int_{-\lambda/4}^{\lambda/4}\cos(\beta z)e^{-jz\beta\cos\theta}dz = \frac{2}{\beta}\frac{\cos\left(\dfrac{\pi}{2}\cos\theta\right)}{\sin^2\theta}. \qquad (C.56)$$

The electric field of the half-wave dipole is then

$$\vec{E} = -j\frac{\mu_0}{4\pi}\omega I_0 \frac{2}{\beta}\frac{e^{j\beta r}}{r}\frac{\cos\left(\dfrac{\pi}{2}\cos\theta\right)}{\sin\theta}e^{-j\omega t}\hat{\theta}, \qquad (C.57)$$

which is equivalent to Eq. B.34 of Appendix B, and

$$E_\theta = \frac{-jI_0}{2\pi\varepsilon_0 c}\frac{e^{j\beta r}}{r}\left[\frac{\cos\left(\dfrac{\pi}{2}\cos\theta\right)}{\sin\theta}\right]e^{-j\omega t}, \qquad (C.58)$$

where we have chosen $e^{-j\omega t}$ time dependence.

## C.8   GENERAL ANTENNA RECIPROCITY

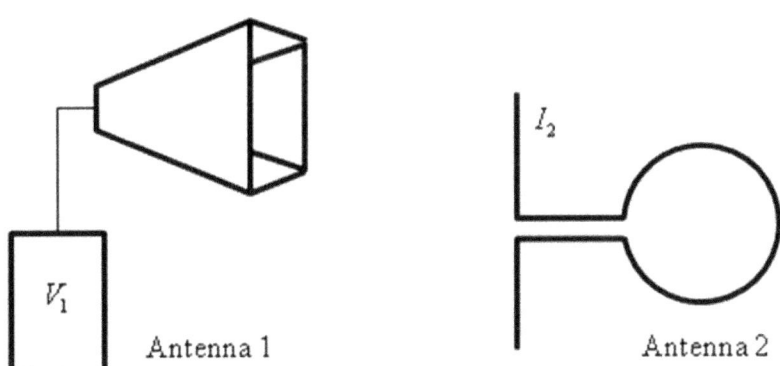

Antenna 1                                    Antenna 2

*FIGURE C.4:* Configuration with antenna 1 transmitting and antenna 2 receiving

The antenna reciprocity theorem and its application to radio astronomy with practical limitations is discussed in Section B.14 of Appendix B. Below we outline a proof of the theorem involving the electric and magnetic fields generated by two antennas alternately in transmitting and receiving configurations. Consider the two antennas shown in Figure C.4. An emf $V_1$ applied to antenna 1 will induce a current $I_2$ in antenna 2. Likewise an emf $V_2$ applied to antenna 2 will induce a current $I_1$ in antenna 1. The reciprocity theorem may be expressed as

$$V_2 I_1 = V_1 I_2 \tag{C.59}$$

An AC current density $\vec{J}_1$ in antenna 1 will produce electric $\vec{E}_1$ and magnetic $\vec{H}_1$ fields related by Maxwell's equations for time-harmonic fields

$$\nabla \times \vec{H}_1 = \vec{J}_1 + j\omega\varepsilon\vec{E}_1 \qquad \nabla \times \vec{E}_1 = -j\omega\mu\vec{H}_1. \tag{C.60}$$

Likewise, an AC current density $\vec{J}_2$ in antenna 2 produces electric $\vec{E}_2$ and magnetic $\vec{H}_2$ fields

$$\nabla \times \vec{H}_2 = \vec{J}_2 + j\omega\varepsilon\vec{E}_2 \qquad \nabla \times \vec{E}_2 = -j\omega\mu\vec{H}_2. \tag{C.61}$$

Taking the divergence of the cross product of the electric and magnetic fields

$$\nabla \cdot \left( \vec{E}_1 \times \vec{H}_2 \right) \text{ and } \nabla \cdot \left( \vec{E}_2 \times \vec{H}_1 \right) \tag{C.62}$$

while using the vector identity

$$\nabla \cdot \left( \vec{E} \times \vec{H} \right) = \vec{H} \cdot \left( \nabla \times \vec{E} \right) - \vec{E} \cdot \left( \nabla \times \vec{H} \right) \tag{C.63}$$

and subtracting the two divergence terms in Eq. (C.62) gives

$$\nabla \cdot \left( \vec{E}_1 \times \vec{H}_2 - \vec{E}_2 \times \vec{H}_1 \right) = \vec{E}_2 \cdot \vec{J}_1 - \vec{E}_1 \cdot \vec{J}_2 . \qquad (C.64)$$

Calculate the volume integral of Eq. (C.64), bounded by a surface enclosing both antennas:

$$\int_{\text{vol}} \nabla \cdot \left( \vec{E}_1 \times \vec{H}_2 - \vec{E}_2 \times \vec{H}_1 \right) dv = \int_{\text{vol}} \left( \vec{E}_2 \cdot \vec{J}_1 - \vec{E}_1 \cdot \vec{J}_2 \right) dv . \qquad (C.65)$$

The left-hand side may be transformed using Gauss's divergence theorem, giving

$$\int_{\text{surf}} \left( \vec{E}_1 \times \vec{H}_2 - \vec{E}_2 \times \vec{H}_1 \right) \cdot d\vec{a} . \qquad (C.66)$$

This integral vanishes over surfaces with areas $\sim r^2$ sufficiently far removed, where $\left| \vec{E}_1 \times \vec{H}_2 - \vec{E}_2 \times \vec{H}_1 \right| \rightarrow 0$ faster than $1/r^2$, so that the volume integral

$$\int_{\substack{\text{all} \\ \text{space}}} \left( \vec{E}_2 \cdot \vec{J}_1 - \vec{E}_1 \cdot \vec{J}_2 \right) dv = 0 . \qquad (C.67)$$

If we let volume 1 contain antenna 1 and volume 2 contain antenna 2, then

$$\int_{\text{vol 1}} \vec{E}_2 \cdot \vec{J}_1 dv_1 = \int_{\text{vol 2}} \vec{E}_1 \cdot \vec{J}_2 dv_2 \qquad (C.68)$$

because $\vec{J}_1 = 0$ outside of volume 1, and $\vec{J}_2 = 0$ outside of volume 2. If we take the infinitesimal volume of antenna 1, $dv_1 = dS_1 d\ell_1$, with current density $J_1 = I_1 / S_1$ and voltage $V_1 = E_1 d\ell_1$; and for antenna 2, $dv_2 = dS_2 d\ell_2$, $J_2 = I_2 / S_2$, and $V_2 = E_2 d\ell_2$, we arrive at $V_2 I_1 = V_1 I_2$ from Eq. (C.68).

## REVIEW QUESTIONS

1. Show that gauge transformation performed by the replacements

$$A \rightarrow A + \nabla \lambda$$

$$V \rightarrow V - \frac{\partial \lambda}{\partial t}$$

leaves $\vec{E}$ and $\vec{B}$ unchanged, where $\lambda$ is any scalar function.

2. Show that $\vec{E} \cdot \vec{B} = 0$ by using (C.36) and (C.45) for the electric and magnetic fields of the Hertz dipole.

3. Calculate the electric field of the Hertz dipole directly from the magnetic field using Maxwell's equation relating the time variation of the electric field to the curl of the magnetic field.

[***Hint:*** use Eq. C.12].

4. Write an expression for the power radiated (in Watts/m$^2$) by a linear array of N Hertz dipoles, each separated by a distance $d$, and each having the same amplitude and phase

[***Hint:*** use Eq. (4.81) in Chapter 4 and Eq. (C.50)].

# *RADIO SPECTRAL LINES*

## D.1 INTRODUCTION

Details of spectral-line observation are discussed in Chapter 8. Radio spectral lines are detected in cold molecular clouds with temperatures ~10 K, in cool neutral HI gas ~$10^2$ K, and in warmer ionized HII regions ~$10^4$ K. Radio lines are produced by various physical mechanisms, including the hyperfine transition in hydrogen, by atomic transitions between energy levels with large principle quantum number (recombination lines), and by transitions between molecular rotation states. Absorption processes followed by spontaneous and stimulated emission can result in the enhancement of radio lines in astrophysical masers. We first discuss the profile of radio lines and how line shapes are affected by the velocity of radiating atoms and molecules in the ISM.

## D.2 RADIO SPECTRAL-LINE PROFILES

Radio spectral lines are often approximately Gaussian in shape, although large-scale turbulence, pressure broadening, and large optical depths can distort line shapes, resulting in asymmetric profiles. An ideal Gaussian line profile $\varphi(v)$ with half-intensity width $\Delta v$ centered at frequency $v_0$ is shown in Fig. D.1 and described by Eq. (D.1):

$$\varphi(v) = A \exp\left[ -4\ln 2 \left( \frac{v_0 - v}{\Delta v} \right)^2 \right], \tag{D.1}$$

with normalization constant

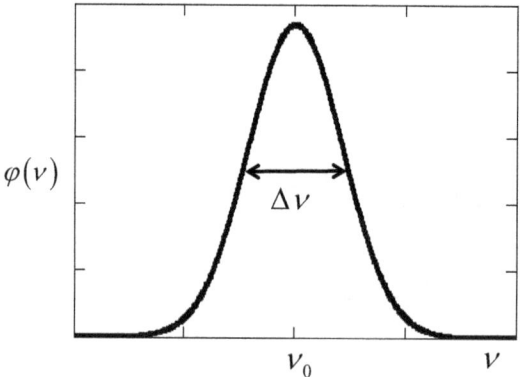

**FIGURE D.1:** Gaussian line profile with center frequency $v_0$ and full width at half-maximum $\Delta v$.

$$A = \left(\frac{\ln 2}{\pi}\right)^{1/2} \frac{2}{\Delta v},$$ (D.2)

such that

$$\int_0^\infty \varphi(v)\,dv = 1.$$ (D.3)

The average line intensity is then

$$I_{\text{avg}} = \int_0^\infty I\varphi(v)\,dv,$$ (D.4)

where $I$ is the radiance or intensity.

### D.2.1 Doppler Line Broadening

If we consider radiating particles of mass $M$ in a gas cloud at temperature $T$ to have a Maxwell-Boltzmann velocity distribution, the number of particles with line-of-site velocity between $v_x$ and $v_x + dv_x$ is given by the 1D distribution:

$$N(v_x)\,dv_x = N\left(\frac{M}{2\pi k_B T}\right)^{1/2} \exp\left[-\frac{Mv_x^2}{2k_B T}\right] dv_x,$$ (D.5)

where $N$ is the total number of radiating particles contributing to the radio line, and $k_B$ is the Boltzmann constant. The observed frequency $v$ is related to the source frequency $v_0$ and the line-of-sight velocity of each particle by the classical Doppler formula

$$v = v_0 \left(1 - \frac{v_x}{c}\right), \tag{D.6}$$

where $c$ is the speed of light. We may therefore transform the velocity distribution of Eq. (D.5) into a frequency distribution. Taking the differentials of Eq. (D.6), we obtain

$$dv = -\frac{v_0}{c} dv_x . \tag{D.7}$$

Substituting $v_x$ from Eq. (D.6) into Eq. (D.5), the frequency distribution becomes

$$N(v)\,dv = N\left(\frac{M}{2\pi k_B T}\right)^{1/2} \exp\left[-\frac{Mc^2 (v_0 - v)^2}{2 k_B T v_0^2}\right] dv . \tag{D.8}$$

We now equate the exponential arguments of the Gaussian line profile in Eq. (D.1) and the Maxwell–Boltzmann distribution in Eq. (D.8) to obtain the half-intensity width:

$$\Delta v = \left(8 \ln 2 \frac{k_B T}{Mc^2}\right)^{1/2} v_0 . \tag{D.9}$$

The Doppler-broadened line width can therefore be used to infer the temperature of a gas in the ISM. Additional line broadening may be due to nonthermal microturbulent velocity components.

## D.3   EINSTEIN COEFFICIENTS

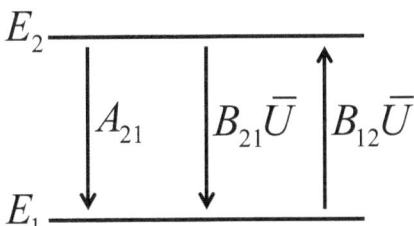

**FIGURE D.2:** Diagram illustrating transitions between states with energies $E_1$ and $E_2$ with Einstein coefficients corresponding to spontaneous emission ($A_{21}$), stimulated emission ($B_{21}$), and absorption ($B_{12}$) .

The Einstein coefficients relate emission and absorption transition rates, where atoms and molecules may radiate by spontaneous as well as stimulated emission.

Consider a two-state system shown schematically in Fig. D.2, where $N_2$ and $N_1$ are the numbers of atoms in higher energy $E_2$ and lower energy $E_1$ states, respectively. The number of atoms transitioning from states $1 \to 2$ per unit time is proportional to $N_1$ and the average energy density $\bar{U}$. We express this transition rate to the higher energy level as

$$R_{1 \to 2} = N_1 B_{12} \bar{U} \qquad \text{(D.10)}$$

where $B_{12}$ is the absorption coefficient. The number of atoms per unit time that transition from $2 \to 1$ by stimulated emission is similarly $N_2 B_{21} \bar{U}$, where $B_{21}$ is the stimulated emission coefficient. Photons induced by stimulated emission have the same frequency and phase as the incident photons. The number of atoms transitioning from $2 \to 1$ per unit time by spontaneous emission is $N_2 A_{21}$, which is not proportional to the energy density. The total transition rate from $1 \to 2$ is a sum of spontaneous and stimulated emission

$$R_{2 \to 1} = N_2 A_{21} + N_2 B_{21} \bar{U} \qquad \text{(D.11)}$$

Because $R_{1 \to 2} = R_{2 \to 1}$ we have

$$N_2 A_{21} + N_2 B_{21} \bar{U} = N_1 B_{12} \bar{U} \qquad \text{(D.12)}$$

Solving for the average energy density $\bar{U}$,

$$\bar{U} = \frac{A_{21}}{\left( \dfrac{N_1}{N_2} B_{12} - B_{21} \right)} . \qquad \text{(D.13)}$$

For a two state system in thermodynamic equilibrium with statistical weights $g_1$ and $g_2$

$$\frac{N_2}{N_1} = \left( \frac{g_2}{g_1} \right) \exp\left( -\frac{(E_2 - E_1)}{k_B T} \right) = \left( \frac{g_2}{g_1} \right) \exp\left( -\frac{h\nu}{k_B T} \right) . \qquad \text{(D.14)}$$

Substituting the Boltzmann relation for $N_1 / N_2$ into the average energy given by Eq. (D.13), we obtain

$$\bar{U} = \frac{A_{21}}{\dfrac{g_1}{g_2} \exp\left( \dfrac{h\nu}{k_B T} \right) B_{12} - B_{21}} . \qquad \text{(D.15)}$$

This must be equal to the average energy density given by the Planck distribution function

$$\bar{U} = \frac{8\pi v^2}{c^3} \frac{hv}{\exp\left(\dfrac{hv}{k_B T}\right) - 1}$$ (D.16)

so that the absorption and stimulated emission coefficients are related by

$$g_1 B_{12} = g_2 B_{21}$$ (D.17)

and the spontaneous and stimulated emission coefficients are

$$A_{21} = \frac{8\pi h v^3}{c^3} B_{21}$$ (D.18)

From this expression, we see that stimulated emission dominates at lower frequencies, while radiation at higher frequencies is mostly due to spontaneous emission. The characteristic lifetime before spontaneous emission is given by $1/A_{21}$.

## D.4  HYPERFINE TRANSITION IN HYDROGEN

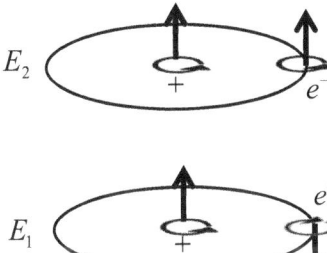

$E_2$

$E_1$

**FIGURE D.3:** Electron spin flip in hydrogen resulting in 21-cm radiation: spins are parallel with energy $E_2$ (top), spins are opposite with energy $E_1$ (bottom).

Hyperfine transitions in neutral hydrogen (H1) result from the spin flip of the electron orbiting a proton, which results in a spectral line at 1420 MHz, or 21 cm. In 1944, van de Hulst predicted that 21 cm radiation could be observed in interstellar neutral hydrogen. This radiation was first detected by H. Ewen and E. M. Purcell in 1951 at Harvard University. Later maps of neutral hydrogen in the Milky Way revealed the spiral structure of our galaxy. Doppler shifts in the 21 cm line have been used to determine the speed of the Milky Way's spiral arms and to plot the rotation curve of the Milky Way and other galaxies.

The electron spin in hydrogen can either be parallel to or opposite from the proton spin, as shown in Fig. D.3. When the spins are parallel (triplet state), the total energy $E_2$ is higher than the energy $E_1$ when the spins are opposite (singlet state). The energy difference $\Delta E = E_2 - E_1$ is calculated from perturbation theory, giving

$$\Delta E = \frac{4 g_p \hbar^4}{3 m_p m_e^2 c^2 a_0^4}, \tag{D.19}$$

where the g-factor of the proton is $g_p = 5.59$, $a_0$ is the Bohr radius, and $\hbar = 1.055 \times 10^{-34}$ Js is Planck's constant divided by $2\pi$. Numerically, $\Delta E = 5.88 \times 10^{-6}$ eV, which corresponds to an emission frequency $\nu = \Delta E / h = 1420$ MHz with a wavelength $\lambda = 21.106$ cm.

The Einstein coefficient giving the probability per unit time for the transition between the triplet and singlet states is $A_{21} = 2.85 \times 10^{-15}$ s$^{-1}$, with a characteristic lifetime of the triplet state $\tau = 1 / A_{21} \approx 1.11 \times 10^7$ yr. Collisional spin flips occur once every 400 years on average, and there are roughly equal numbers of atoms in each spin configuration.

### D.4.1 Spin Temperature and H1 Column Density

The spin temperature $T$ is related to the ratio of the number of hydrogen atoms $N_2$ and $N_1$ in states with respective energies $E_2$ and $E_1$,

$$\frac{N_2}{N_1} = \left( \frac{g_2}{g_1} \right) \exp\left( -\frac{h\nu}{k_B T_s} \right), \tag{D.20}$$

where the ratio of statistical weights $g_2 / g_1 = 3$. Here the spin temperature $T_s$ is equal to the kinetic temperature if collisions dominate.

The H1 column density $N_H$ is the number of atoms/cm$^2$ along a line of sight. $N_H$ is proportional to $T_s$.

$$\frac{N_H}{\text{cm}^{-2}} = 1.82 \times 10^{18} \left( \frac{T_s}{\text{K}} \right) \int_{-\infty}^{\infty} \tau(v_x) \, d\left( \frac{v_x}{\text{km/s}} \right), \tag{D.21}$$

assuming a constant spin temperature along the line of sight, where $\tau(v_x)$ is the velocity-dependent optical depth. This integral may be approximated by a Gaussian curve with area $\sqrt{\pi / 4 \ln 2} \, \tau(v_{x0}) \Delta v_x = 1.064 \tau(v_{x0}) \Delta v_x$, so that

$$\frac{N_H}{cm^{-2}} = 1.94 \times 10^{18} \left(\frac{T_s}{K}\right) \tau\left(v_{x0}\right) \frac{\Delta v_x}{km/s}, \qquad \text{(D.22)}$$

where $\Delta v_x$ is the half-intensity width centered at $v_{x0}$.

## D.5   RECOMBINATION LINES

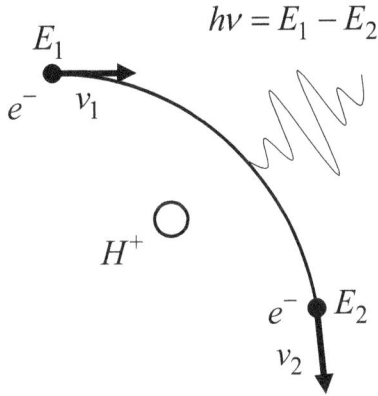

FIGURE D.4: Free-free radiation emitted during the acceleration of an electron in the field of an ion.

Bremsstrahlung radiation, also known as free-free radiation, results from noncapture events where an electron passes near an ion, as discussed in Section 1.3 of Chapter 1. An electron moving in the Coulomb field of an ion will experience a centripetal acceleration with a corresponding deviation in path, as show in Fig. D.4. Both particles are unbound before and after scattering. Deviated electrons will radiate over a wide band of frequencies, which can range from X-rays and gamma rays for head-on collisions to radio waves for smaller deviations with large-impact parameters. Slower moving electrons will experience larger path deviations and more power will be radiated.

In capture events, where the electron and ion become bound, the electron most often descends immediately to the ground state. The electron may also cascade down in a series of steps from very high quantum numbers, giving off radiation known as recombination lines. One such step is depicted in Fig. D.5.

Radio recombination lines are observed in HII regions consisting of low-density gas partially ionized by UV radiation from hot O and B stars. Here, nearly all atoms are either ionized or in the ground state. Although the density of HII regions is $\sim 10^4$ atoms /cm$^3$, radio lines are observable because of very long line-of-sight ~ 0.5 pc path lengths. Thermal widths and the ratio of recombination line strengths to the continuum background can be used to learn the temperature and composition of the interstellar medium in these regions. It turns out that the ratio of bound to free electrons in HII regions is proportional to the ratio of recombination line strength to the continuous free-free background radiation:

$$\frac{\text{line}}{\text{continuum}} \propto \frac{v^{2.1}}{nT_e^{1.15}}, \tag{D.23}$$

where $v$ is the line frequency and $n$ is the principle quantum number. From Eq. (D.23), typical HII-region electron temperatures are calculated to be $T_e \approx 10^4$ K.

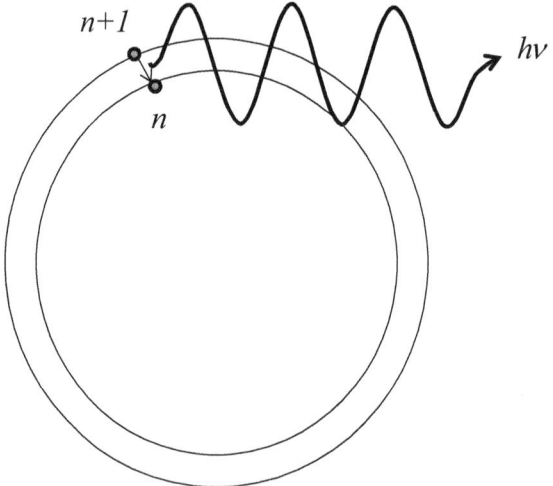

**FIGURE D.5:** Transition from energy level $n+1$ to $n$ with emission of a photon.

## D.5.1 Recombination Line Frequencies

Neils Bohr first suggested in 1914 that atoms with very large principle quantum numbers might be found in space. The atomic-energy levels are given by the Bohr formula

$$E_n = -\frac{2\pi^2 m_e e^4}{h^2} \frac{Z^2}{n^2} \tag{D.24}$$

Transitions between nearby energy levels with $n_f \approx 100-1000$ will produce radio frequency emission. The first radio recombination line in the Orion and Omega nebulae, measured by Z. Dravskikh and A. Dravskikh in 1963, had a frequency of $v = 5.76 \times 10^9$ Hz. The corresponding transition from $n_i = 105$

to $n_f = 104$ is known as the H104$\alpha$, where transitions with $\Delta n = 1, 2, \ldots$ are denoted by $\alpha$, $\beta$, $\ldots$, respectively. The recombination line corresponding to the $n_i = 110$ to $n_f = 109$ transition (H109$\alpha$) was first detected by P. Mezger in 1965 with a frequency of $\nu = 5.0089 \cdot 10^9$ Hz.

Transitions between very large principle quantum numbers in multielectron atoms are very hydrogen-like because of the large separation between the electron and the nucleus, and so the Bohr model is applicable. In neutral atoms, the distant electron sees a screened nuclear charge of $Z = 1$. Atoms in space have been detected with very large principle quantum numbers, including H$n\alpha$ (with $n = 486$, 538, 603, 611, 631, 640, 686, and 732) and C$n\alpha$ (with $n = 382, 486, 530, 538, 552$, 603, 611, 631, 640, 686, and 732). The orbital radii of these large atoms is

$$r_n = \frac{a_0}{Z} n^2 \tag{D.25}$$

Atoms with principle quantum number as high as 766 have been reported with effective diameters $\approx 0.1$ mm (about the thickness of a sheet of paper). It is believed that the limit of detectability corresponds to $n \approx 1000$ because of nonthermal galactic background radiation.

The emission frequency is calculated from Eq. (D.24) with $E_i - E_f = h\nu$, giving

$$\nu = R_M c Z^2 \left( \frac{1}{n_f^2} - \frac{1}{n_i^2} \right) \tag{D.26}$$

Accounting for the electron and nuclear masses $m_e$ and $M$

$$R_M = R_\infty \left( \frac{M}{M + m_e} \right), \tag{D.27}$$

where the Rydberg constant $R_\infty = \dfrac{2\pi^2 m_e e^4}{h^3 c} = 1.097373 \times 10^7 \, \text{m}^{-1}$. For heavier multielectron atoms, $M$ is replaced by the nuclear mass plus the mass of the inner electrons.

For $n_f = n$ and $n_i = n + \Delta n$, we write Eq. (D.26) as

$$\nu = R_M c Z^2 \frac{1}{n^2} \left[ 1 - \left( 1 + \frac{\Delta n}{n} \right)^{-2} \right]. \tag{D.28}$$

For $n \gg \Delta n$, we use the binomial theorem to approximate the term

$$1-\left(1+\frac{\Delta n}{n}\right)^{-2} \approx 1-\left(1-2\frac{\Delta n}{n}\right),\qquad\text{(D.29)}$$

and the frequency becomes

$$\nu = R_M c Z^2 \frac{2\Delta n}{n^3}\qquad\text{(D.30)}$$

The frequency separation between adjacent lines is

$$\nu_n - \nu_{n+1} \approx \frac{3}{n}\nu_n,\qquad\text{(D.31)}$$

so that the interline separation is nearly evenly spaced for large values of $n$. We can compare the interline separation to the thermal line width evaluated from Eq. (D.9) with $T$ in Kelvins, $M$ in atomic mass units, and centerline frequency $\nu_0 = \nu_n$:

$$\Delta\nu \approx 7.16 \times 10^{-7}\left(\frac{T}{M}\right)^{1/2}\nu_n\qquad\text{(D.32)}$$

Also, an approximation for the natural line width due to uncertainty broadening is given by

$$\Delta\nu \approx 1.2 \times 10^{-6}\frac{\ln n}{n^2}\nu_n\qquad\text{(D.33)}$$

It turns out that the thermal line width is five orders of magnitude greater than the natural line width of H109$\alpha$ at 8000 K.

## D.6   MOLECULAR LINE SPECTRA

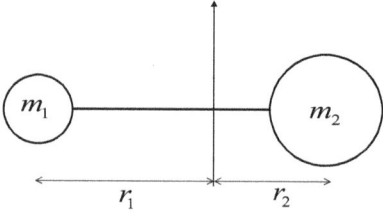

**FIGURE D.6:** A diatomic molecule rotating about an axis passing through its center of mass.

Molecular lines are found in the ISM at temperatures ranging between $10 \sim 100$ K. These lines can be used to determine the temperature, density, composition, and radial velocity of molecular clouds. Molecules cannot exist in regions where ionizing radiation and highly energetic collisions break molecular bonds.

The first molecular line was discovered in 1963 by Weinreb, et al., at 1665/7 MHz from OH. The most common molecule in the ISM is $H_2$. Molecular line spectra are not found in this and other symmetric molecules with no net dipole moment, such as $N_2$, $CH_4$, or $CO_2$. Because $H_2$ is closely associated with CO in the ISM, CO lines can be used to trace molecular hydrogen.

Molecules exhibit rotational and vibrational energies that are also quantized. Transitions between rotational and vibrational states result in emissions at radio and infrared frequencies, respectively. We first consider rotational transitions in a diatomic molecule. The rotational energy of a molecule is half the moment of inertia $I$ times the square of the angular frequency $\omega$:

$$E_{\text{rot}} = \frac{1}{2} I \omega^2 . \qquad (D.34)$$

The moment of inertia about a given axis of rotation is the sum of each mass times the square of its distance to the axis of rotation:

$$I = \sum_i m_i r_i^2 \qquad (D.35)$$

For a diatomic molecule, such as that shown in Fig. D.6, the moment of inertia has two terms,

$$I = m_1 r_1^2 + m_2 r_2^2 , \qquad (D.36)$$

where the equilibrium bond length is give by $r_e = r_1 + r_2$. For CO the moment of inertia about the center of mass of the molecule has the numerical value of $1.46 \times 10^{-46}$ kg m$^2$. The angular momentum $L$ is given by the product of the moment of inertia times the angular frequency,

$$L = I \omega . \qquad (D.37)$$

The angular momentum is quantized as

$$L = \hbar \sqrt{J(J+1)} , \qquad (D.38)$$

where $J$ is an integer ( $J$ = 0, 1, 2, 3 ...). The rotational kinetic energy $E_{\text{rot}} = L^2 / 2I$ is also quantized:

$$E_{\text{rot}} = \frac{J(J+1)\hbar^2}{2I} . \qquad (D.39)$$

Angular momentum must be conserved in rotational transitions with the corresponding selection rule

$$\Delta J = \pm 1 . \tag{D.40}$$

The change in rotational energy corresponding to the $J \rightarrow J-1$ transition is

$$\Delta E_{\text{rot}} = \frac{\hbar^2}{2I}\left[ J(J+1) - (J-1)J \right] = \frac{\hbar^2 J}{I}, \tag{D.41}$$

which is proportional to the emission frequency of the photon $h\nu = \Delta E_{\text{rot}}$, or

$$\nu = \frac{\hbar J}{2\pi I} \tag{D.42}$$

so that the emission spectra will have a ladder appearance with steps of ~115 GHz for CO, as shown in Fig. D.7. From Eq. (D.42), we see that the frequency is inversely proportional to the moment of inertia. Hence larger molecules with greater mass will emit lower rotational transition frequencies. The spectra of larger molecules with more than one principle moment of inertia, such as ammonia ($NH_3$), will be more complex, composed of a superposition of ladders with each ladder corresponding to a moment of inertia about a given axis of rotation.

Molecular transitions may result from radiative or collisional excitations where $E_{\text{rot}} \approx k_B T$. Thus we define a minimum temperature for a given energy $T_{\text{min}} \approx E_{\text{rot}} / k_B$ from Eq. (D.39):

$$T_{\text{min}} = \frac{J(J+1)\hbar^2}{2k_B I} \tag{D.43}$$

Writing the minimum temperature required for the $J \rightarrow J-1$ transition in terms of $\nu$ from Eq. (D.42), we have

$$T_{\text{min}} = \frac{\nu h(J+1)}{2k_B} \tag{D.44}$$

Using this equation, we find the minimum gas temperature $T_{\text{min}} \approx 8.3$ K for the $J = 1 \rightarrow 0$ transition in CO. Molecular excitations resulting from collisions are most likely due to interactions with hydrogen atoms that have the greatest relative abundance in the ISM.

Molecules in higher angular-momentum states will experience greater centripetal accelerations resulting in increased bond lengths. The corresponding increase in moment of inertia will result in a slightly lower frequency than predicted by the static moment of inertia. This frequency difference may be used to determine the bond strength that determines the vibrational frequency $\omega_{\text{vib}}$. The quantized vibrational-energy levels in the simple harmonic oscillator approximation are

$$E_{\text{vib}} = \left(n + \frac{1}{2}\right)\hbar\omega_{\text{vib}}, \tag{D.45}$$

where $n = 0, 1, 2...$ is an integer. The total energy is given by a sum of vibrational and rotational components:

$$E_{\text{rot-vib}} = \frac{J(J+1)\hbar^2}{2I} + \left(n + \frac{1}{2}\right)\hbar\omega_{\text{vib}}, \tag{D.46}$$

where vibrational transitions also obey a selection rule, $\Delta n = \pm 1$. As mentioned before, vibrational transitions emit infrared photons, such as from the $n = 1 \rightarrow 0$ transition in CO at $6.42 \times 10^{13}$ Hz. The $n = 1 \rightarrow 0$ transition in other diatomic molecules, such as NO, HI, HBr, HCl, and HF emit in the range of $\left(5.63 \times 10^{13} - 8.72 \times 10^{13}\right)$ Hz.

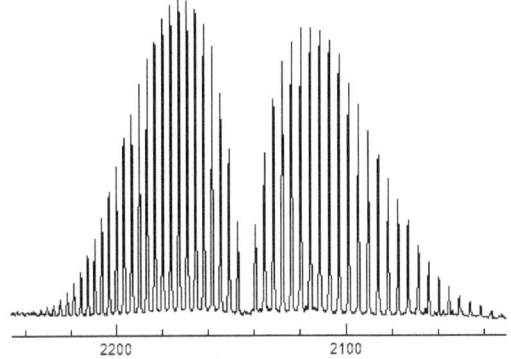

**FIGURE D.7:** CO rotational–vibrational spectrum (from FTIR) plotted as absorbance vs. wave number showing the presence of P ($<2140$ cm$^{-1}$) and R ($>2140$ cm$^{-1}$) branches. Image credit: Wikimedia Commons.

## D.7 ASTROPHYSICAL MASERS

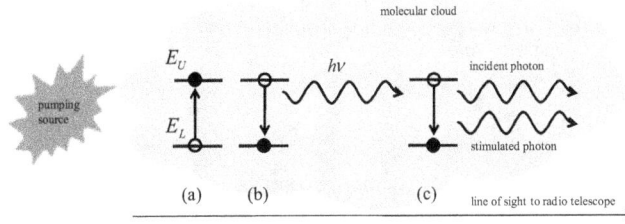

**FIGURE D.8:** Schematic of astrophysical maser emission. (a) Population inversion in energy levels is sustained by a pumping source; (b) spontaneous transitions to a lower energy level with photon emission; (c) stimulated emission with beam intensity increasing with path length.

Masers (Microwave Amplification by the Stimulated Emission of Radiation) are the microwave equivalent of lasers, with population inversions in the molecular rotational or vibrational energy levels. Laboratory masers (and lasers) require multiple reflections within a resonant cavity where the beam intensity increases exponentially from repeated transversals along the length of the cavity. Amplification in astrophysical masers occurs over very long path lengths in the ISM. The first such naturally occurring maser was discovered by Weaver in 1965. Masers are common in star-forming regions and in extended atmospheres of red giant stars. CO masers have been observed in star-forming galaxies. The amplification of molecular lines by maser action enables the study of gas dynamics in more distant stellar objects and active galactic nuclei. Keplerian orbits of water masers have been used to determine the mass of super-massive black holes at the center of distant galaxies.

A schematic of maser action is shown in Fig. D.8. Astrophysical masers are energized by a nearby star or other source of intense radiation that serves as a pump to sustain a non-Boltzmann distribution of the energy levels. One pumping mechanism occurs as stellar UV radiation heats gas and dust that re-emits in the IR. Molecules such as OH, SiO, and $H_2O$ are excited by IR radiation, resulting in a population inversion with subsequent spontaneous and stimulated emission. As an example, we consider an ensemble of two-state systems with the population ratio of upper $N_U$ and lower $N_L$ energy levels given by

$$\frac{N_U}{N_L} = \left(\frac{g_U}{g_L}\right) \exp\left(-\frac{h\nu}{k_B T_s}\right), \tag{D.47}$$

with a population inversion for $N_U > N_L\left(g_U / g_L\right)$. When $N_U > N_L\left(g_U/g_L\right)$, we have a negative state temperature $T_s$, where a resonant photon beam will be amplified as it propagates through the ISM. The photon beam is attenuated when we have a positive state temperature and $N_U < N_L\left(g_U / g_L\right)$.

The pumping action resulting in a negative state temperature may be collisional or radiative. Unsaturated masers occur with an exponential increase in specific intensity $I$ along a path length $s$:

$$I(s) = I_0 \exp\left(\alpha_0 s\right), \tag{D.48}$$

where $\alpha_0 = h\nu_0 B_{21} R / C$, $\nu_0$ is the center-line frequency, $B_{21}$ is the Einstein coefficient corresponding to stimulated emission, $R$ is the rate at which the upper level is populated, and $C$ is the transition rate due to collisions. Maser saturation occurs when transitions are predominately induced by radiation with intensity proportional to the path length:

$$I(s) = \frac{\alpha_0}{B_{21}} s \tag{D.49}$$

## REVIEW QUESTIONS

1. Show that the line strength at $v_0$ of a normalized Gaussian line is inversely proportional to the half-intensity line width

$$\varphi(v_0) = \left(\frac{\ln 2}{\pi}\right)^{1/2} \frac{2}{\Delta v}.$$

2. Show that $\Delta E$ in Eq. (D.19) can be written

$$\Delta E = \frac{8}{3} g_p \frac{m_e}{m_p} \alpha^2 \frac{e^2}{a_0}$$

in terms of the fine-structure constant $\alpha = \dfrac{\hbar}{m_e c a_0}$ and the Bohr radius $a_0 = \dfrac{\hbar^2}{m_e e^2}$.

For a quantum mechanical derivation of $\Delta E$ from first-order perturbation theory, see D. J. Griffiths *Introduction to Quantum Mechanics*, pp. 250–251.

3. The H109$\alpha$ line width at half-maximum in an HII region with $T \approx 10^4$ K is $3.6 \times 10^5$ Hz. What fraction of the interline separation does this correspond to?

4. For $\Delta n = 1$ Eq. (D.30) becomes

$$v = \frac{2 R_M c Z^2}{n^3}.$$

Calculate the error in using this approximation for H109$\alpha$.

5. The uncertainty estimate of the natural line width is

$$\Delta E \Delta t \geq \frac{\hbar}{2} \quad \Delta E = h\Delta v \text{ and } \Delta t = A_{n+1,n}^{-1},$$

where the Einstein coefficient for transition $n + 1$ to $n$ is approximated by

$$A_{n+1,n} \approx \frac{64\pi^6 m_e e^{10}}{3c^3 h^6 n^5} = 5.3 \times 10^9 \; \frac{1}{n^5} s^{-1}.$$

Compare the uncertainty estimate of $\Delta v$ above with the natural line width calculated from Eq. (D.33) for H109$\alpha$.

# E

# *RADIO TELESCOPES*

The following tables list a majority of the U. S. and non-U.S. radio observatories. Several historical telescopes may no longer be in commission. Radio telescopes are sometimes renamed, upgraded, moved, or decommissioned.

## E.1  U.S. RADIO OBSERVATORIES

| Observatory/ Location | Telescope Diameter | | Observations/Notes |
|---|---|---|---|
| Allen Telescope Array (ATA) Hat Creek Radio Observatory: Hat Creek, California | Array of 42 6.1 × 7.0 m offset Gregorian telescopes, each with a 2.4 m subreflector | 0.5–11.2 GHz | Radio spectra of starburst galaxies, spectropolarimetry of radio galaxies |
| Arecibo Observatory National Astronomy and Ionosphere Center Arecibo, Puerto Rico | 305 m spherical reflector | 1–10 GHz | Planetary science, asteroid imaging, binary pulsar, spectroscopy, SETI |
| Caltech Submillimeter Observatory Mauna Kea, Hawaii | Steerable 10.4 m paraboloid | 250 GHz 350 GHz 450 GHz 650 GHz 850 GHz | Galactic, extragalactic, spectral line, continuum |
| Combined Array for Research in Millimeter-wave Astronomy (CARMA) California | Dish array: 6 × 10.4 m 9 × 6.1 m 8 × 3.5 m | 30–100 GHz | CO tracing of molecular hydrogen, proto planetary disk studies imaging cold dust |
| Crawford Hill Holmdel, NJ (horn antenna) | 7 m horn antenna | 1400 MHz (tunable up to 40 GHz) | HI line survey (not in use) |

*(Contd.)*

| Observatory/ Location | Telescope Diameter | | Observations/Notes |
|---|---|---|---|
| Crawford Hill Holmdel, NJ (mm telescope) | 7 m offset Cassegrain | 70–250 GHz (tunable) | mm-wavelength spectral line and continuum |
| Deep Space Network: Goldstone Deep Space Communications Complex (DSS 13, 14, 15, 16, 26, 27 and 28) (NASA Jet Propulsion Laboratory, California Institute of Technology) | | | |
| DSS 14 Goldstone, California (Mars Antenna) | 70 m parabolic Cassegrain | 1670 MHz 2290 MHz 8.45 GHz 22 GHz | Spacecraft communication and navigation, radio astronomy |
| DSS 15 Goldstone, California (Uranus Antenna) | 34 m parabolic High-Efficiency (HEF) Cassegrain | 2250 MHz 8.45 GHz | Spacecraft communications and radio astronomy |
| DSS 16 Goldstone, California | 26 m antenna | 2025–2120 MHz | Support near Earth missions |
| DSS 26 Goldstone, California | 34 m beam waveguide (BWG) Cassegrain antenna | 2250 MHz 8.42 GHz | Spacecraft communications and radio astronomy |
| DSS 27 Goldstone, California (Gemini Antenna) | 34 m beam waveguide (BWG) Cassegrain antenna | 2250 MHz 8.42 GHz | Spacecraft communications and radio astronomy |
| DSS 28 Goldstone, California (Gemini Antenna) | 34 m beam waveguide (BWG) Cassegrain antenna | 2250 MHz 8.42 GHz | Spacecraft communications and radio astronomy |
| Deep Space Network: Canberra Deep Space Communications Complex (DSS 34, 35, 36, 43, 45 and 46) (NASA Jet Propulsion Laboratory, California Institute of Technology) | | | |
| DSS 34 Tidbinbilla, Australia | 34 m beam waveguide (BWG) Cassegrain antenna | 2250 MHz 8.42 GHz | Spacecraft communications and radio astronomy |
| DSS 35 Tidbinbilla, Australia (under construction) | 34 m beam waveguide (BWG) Cassegrain antenna | 2250 MHz 8.42 GHz | Spacecraft communications and radio astronomy |
| DSS 36 Tidbinbilla, Australia (under construction) | 34 m beam waveguide (BWG) Cassegrain antenna | 2250 MHz 8.42 GHz | Spacecraft communications and radio astronomy |
| DSS 43 Tidbinbilla, Australia | 70 m parabolic Cassegrain antenna | 1670 MHz 2290 MHz 8.45 GHz 22 GHz | Spacecraft communication and navigation, radio astronomy |

| Observatory/ Location | Telescope Diameter | | Observations/Notes |
|---|---|---|---|
| DSS 45<br>Tidbinbilla, Australia | 34 m parabolic<br>High-Efficiency<br>(HEF)<br>Cassegrain | 2250 GHz<br>8.45 GHz | Spacecraft communications and radio astronomy |
| DSS 46<br>Tidbinbilla, Australia | 26 m antenna | 2025–2120 MHz | Support near Earth missions |
| Deep Space Network: Madrid Deep Space Communications Complex<br>(DSS 54, 55, 63, 65 and 66)<br>(NASA Jet Propulsion Laboratory, California Institute of Technology) | | | |
| DSS 54<br>Robledo, Spain | 34 m beam waveguide<br>(BWG) Cassegrain<br>antenna | 2250 MHz<br>8.42 GHz | Spacecraft communications and radio astronomy |
| DSS 55<br>Robledo, Spain | 34 m beam waveguide<br>(BWG) Cassegrain<br>antenna | 2250 MHz<br>8.42 GHz | Spacecraft communications and radio astronomy |
| DSS 63<br>Robledo, Spain | 70 m parabolic<br>Cassegrain antenna | 1670 MHz<br>2290 MHz<br>8.45 GHz<br>22 GHz | Spacecraft communication and navigation, radio astronomy |
| DSS 65<br>Robledo, Spain | 34 m parabolic<br>High-Efficiency<br>(HEF)<br>Cassegrain | 2250 MHz<br>8.45 GHz | Spacecraft communications and radio astronomy |
| DSS-66<br>Robledo, Spain | 26 m antenna | 2025–2120 MHz | Support near Earth missions |
| Five College Radio<br>Astronomy Observatory<br>New Salem, MA | 13.7 m steerable<br>Cassegrain paraboloid<br>(radome, altazimuth<br>mount) | 86–115 GHz | Planetary, cometary, galactic, and extra-galactic spectroscopy |
| Combined Array for<br>Research in Millimeter-<br>wave Astronomy<br>(CARMA)<br>Hat Creek Radio<br>Observatory:<br>Hat Creek, California | Six 10.4 m<br>Nine 6.1 m<br>Eight 3.5 m<br>Cassegrains | 27–35 GHz<br>80–115 GHz<br>210–270 GHz | Molecular gas, cold dust in planet forming regions |
| Haystack Observatory<br>Tyngsboro, MA | 36 m Cassegrain<br>paraboloid<br>(radome altazimuth<br>Mount) | 2240–2340 MHz<br>7.5–8.7 GHz<br>21.2–25.4 GHz<br>35–49 GHz<br>85–115 GHz | Spectral line, continuum, and VLBI at all frequencies |

*(Contd.)*

| Observatory/ Location | Telescope Diameter | | Observations/Notes |
|---|---|---|---|
| Morehead Radio Telescope<br>Morehead, KY | 13.25 × 3.35 m full-motion partial paraboloidal section | 1.38–1.42 GHz (planned)<br>1.6–1.7 GHz<br>2.4–4.2 GHz<br>22–24 GHz | Microvariability in active galactic nuclei, radio afterglow of gamma ray bursts, HI studies |
| National Radio Astronomy Observatory: Green Bank<br>Green Bank, WV | 42.7 m paraboloid, equatorial mount | 50–88 MHz<br>110–240 MHz<br>450–500 MHz<br>280–350 MHz<br>350–410 MHz<br>450–500 MHz<br>500–750 MHz<br>750–1000 MHz<br>1.00–1.45 GHz<br>4.47–5.05 GHz<br>1.3–1.5 GHz<br>1.30–1.80 GHz<br>2.64–2.75 GHz<br>2.9–3.4 GHz<br>4.6–5.0 GHz<br>3.12–3.37 GHz<br>4.6–5.1 GHz<br>4.7–7.2 GHz<br>7.6–11.2 GHz<br>12–16.2 GHz<br>18.2–25.2 GHz | Continuum, spectral line, and VLBI |
| National Radio Astronomy Observatory: Green Bank<br>Green Bank, WV<br>(Two-Element Interferometer) | Two 25.9 m paraboloids, equatorial mounts | 2.1–2.3 GHz<br>8.2–8.6 GHz | Continuum flux density monitoring for extreme scattering events |
| National Radio Astronomy Observatory: Green Bank<br>Green Bank, WV | 25.9 m paraboloid, equatorial mount | 2.1–2.3 GHz<br>8.2–8.6 GHz | VLBI monitoring of UT1 and polar motion |
| National Radio Astronomy Observatory: Green Bank<br>Green Bank, WV | 100 m parabolic off-axis reflector | 0.1–116 GHz | HI distribution between galaxies, pulsar, planetary and asteroid studies |
| Steward Observatory: Kitt Peak<br>Kitt Peak, AZ | 12 m Paraboloid, altazimuth mount | 68–300 GHz | Continuum, spectral line, and VLBI |

| Observatory/ Location | Telescope Diameter | | Observations/Notes |
|---|---|---|---|
| Steward Observatory: SMT Telescope Mt. Graham, AZ | 10 m Paraboloid, altazimuth mount | 125–1100 GHz | Continuum, spectral line |
| National Radio Astronomy Observatory: Very Large Array (VLA) Plains of San Agustin, New Mexico | 27 shaped 25 m paraboloids, altazimuth mounts, 4 standard 3-armed Wye configurations, baselines 1–36 km | 73.0–74.6 MHz 306–340 MHz 1340–1730 MHz 4500–5000 MHz 8000–8800 MHz 14.4–15.4 GHz 22.0–24.0 GHz | Continuum, spectral line, and VLBI |
| National Radio Astronomy Observatory: Very Long Baseline Array (VLBA) Antenna sites: Pie Town, NM Kitt Peak, AZ Los Alamos, NM Fort Davis, TX North Liberty, IA Brewster, WA Owens Valley, CA St. Croix, VI Mauna Kea, HI Hancock, NH Operations center and correlator: Socorro, NM | 10 25 m paraboloids | 312–342 MHz 580–640 MHz 1.35–1.75 GHz 2.15–2.35 GHz 4.6–5.1 GHz 8.0–8.8 GHz 10.2–11.2 GHz 14.4–15.4 GHz 21.7–24.1 GHz 42.3–43.5 GHz | Maser emissions, pulsar, 3D structure of the Milky Way |
| Ohio State University Radio Observatory Near Delaware, OH | $100 \times 30$ m Kraus-type paraboloid | 1575 MHz | SETI, OH and H lines, continuum |
| Owens Valley Radio Observatory Facilities Big Pine, California | 39.6 m steerable paraboloid | 319–339 MHz 580–620 MHz 1250–1490 MHz 1575–1765 MHz 2250–2330 MHz 4.75–5.15 GHz 2220–2370 MHz 8.085 GHz 10.695 GHz 19–24 GHz 42.6–43.4 GHz | |

*(Contd.)*

| Observatory/ Location | Telescope Diameter | | Observations/Notes |
|---|---|---|---|
| Owens Valley Radio Observatory Facilities Big Pine, California | 6-element interferometer, 10.4 m steerable paraboloids | 26–36 GHz (HEMT) 80–115 GHz (SIS) 210–270 GHz (SIS) | Galactic, extragalactic, planetary, spectral lines, continuum, VLBI |
| Owens Valley Radio Observatory Facilities Big Pine, California | Two-element interferometer, steerable 27.4 m equatorial mounted paraboloid | 500 MHz– 18 GHz 2.4–8.2 GHz | Solar, galactic, continuum |
| Pisgah Astronomical Research Institute Pisgah National Forest near Balsam Grove, NC | West 26 m telescope | 1.4 GHz 1.6 GHz 4.829 GHz 6.668 GHz 12.178 GHz 327 MHz | HI/OH measurements, formaldehyde and methanol detection, pulsar timing |
| Pisgah Astronomical Research Institute Pisgah National Forest near Balsam Grove, NC | East 26 m telescope | 1.4 GHz 3.4–4.4 GHz 11.7–12.2 GHz 327 MHz 4.829 GHz | |
| Pisgah Astronomical Research Institute Pisgah National Forest near Balsam Grove, NC | 12.2 m antenna With removable Cassegrain subreflector and feed assembly | 3.3–4.4 GHz 10.9–12.75 GHz 6.6–7.7 GHz 21.7–24.1 GHz 36.0–36.5 GHz 41.0–45.0 GHz 48.0–49.0 GHz | |
| Pisgah Astronomical Research Institute Pisgah National Forest near Balsam Grove, NC | 4.6 m antenna | 1.42 GHz 4.829 GHz 6.668 GHz 11.178 GHz | 21-cm radiation, early NASA ATS satellite experiments |
| Pisgah Astronomical Research Institute Pisgah National Forest near Balsam Grove, NC | Two large 7-element log periodic Yagis phased for LHCP and RHCP reception | 17–30 MHz | A NASA JOVE project: feeds Jupiter and Solar flare data to the Internet. |
| Radio Solar Telescope Network (RSTN): Palehua Solar Observatory Palehua, Hawaii | Semi-bicone antenna; 3 tracking 8.5, 2.4, and 0.9 m paraboloid antennas | 50 MHz for Semi-bicone; 246.8 MHz 400.21 MHz 610.21 MHz 1415.92 MHz 2695.67 MHz | Solar |

| Observatory/ Location | Telescope Diameter | | Observations/Notes |
|---|---|---|---|
| | | 4962.07 MHz<br>8829.8 MHz<br>15441.85 MHz<br>for others | |
| Radio Solar Telescope Network (RSTN): Learmonth Solar Observatory Learmonth, Western Australia | Semi-bicone antenna; 3 tracking 8.5, 2.4, and 0.9 m paraboloid antennas | 55 MHz for Semi-bicone;<br>245.7 MHz<br>410.4 MHz<br>609.0 MHz<br>1415.5 MHz<br>2695.0 MHz<br>4965.5 MHz<br>8829.7 MHz<br>15421.0 MHz<br>for others | Solar |
| Radio Solar Telescope Network (RSTN): San Vito Solar Observatory San Vito de Normanni, Italy | Semi-bicone antenna; 3 tracking 8.5, 2.4, and 0.9 mparaboloid antennas | 50 MHz for Semi-bicone;<br>245.32 MHz<br>389.84 MHz<br>609.62 MHz<br>1415.72 MHz<br>2695.67 MHz<br>4967.74 MHz<br>8834.9 MHz<br>15427.76 MHz<br>for others | Solar |
| Radio Solar Telescope Network (RSTN): Sagamore Hill Radio Observatory Hamilton, Massachusetts | Semi-bicone and log-periodic antennas; 3 tracking 8.5, 2.4, and 0.9 m paraboloid antennas | 25–75 MHz for Semi-bicone;<br>75–180 MHz for Log-periodic;<br>245 MHz<br>410 MHz<br>610 MHz<br>1415 MHz<br>2695 MHz<br>4995 MHz<br>8800 MHz<br>15.4 GHz<br>for others | Solar |
| Smithsonian Astrophysical Observatory: Center for Astrophysics Cambridge, MA | 1.2 m steerable Cassegrain | 105–116 GHz | Spectral line |

*(Contd.)*

## E.2 NON-U.S. RADIO OBSERVATORIES

| Country Name | Telescope–Diameter | Frequency Bands | Observations/Notes |
|---|---|---|---|
| Antarctica: South Pole Telescope (SPT) Amundsen-Scott South Pole Station | 10 m off-axis Gregorian (Superconducting bolometer array cameras) | 95 GHz 150 GHz 220 GHz | Distortions of CMB radiation by galaxy clusters, Sunyaev-Zeldovich effect |
| Argentina: Instituto Argentino de Radioas-tronomia | Two 30 m steerable paraboloids | antenna one: 1420 MHz 1605 MHz 1612 MHz 1667 MHz 1720 MHz antenna two: 1415 MHz | HI, recombination lines, OH, continuum |
| Australia: Ceduna Radio Astronomy Observatory | 30 m dish | 2–25 GHz | Sources with variability timescales on the order of a few hours |
| Australia: Molonglo Observatory Synthesis Telescope (MOST) | Two 778 × 12 m cylindrical paraboloids with 15 m separation | 843 MHz | Continuum, galactic, extragalactic, solar |
| Australia: Mopra Radio Telescope (near Coonabarabran) | 22 m dish | 75–115 GHz | Spectroscopy and VLBI experiments. Remote observing can be conducted from anywhere in the world. |
| Australia: Murchison Widefield Array (MWA) | 128 × 16-element dual-polarization antennas | 80–300 MHz | H1 studies of the Epoch of Reionization (EoR), solar and transient phenomena |
| Australian National Radio Astronomy Observatory, Parkes (Parkes Radio Telescope) | 64 m steerable paraboloid | 418–454 MHz 642–678 MHz 1250–1750 MHz 2200–2500 MHz 4.5–5.0 GHz | Pulsar timing and surveys (440, 660 and 1420 MHz) |
| | | 5.95–6.25 GHz 6.45–6.75 GHz 8.15–8.65 GHz 12.0–12.4 GHz 12.5–15.5 GHz 21.5–23.5 GHz | Atomic hydrogen in galaxies (1420 MHz) Methanol and water masers (6, 12 and 23 GHz) |

| Country Name | Telescope–Diameter | Frequency Bands | Observations/Notes |
|---|---|---|---|
| Australian Square Kilometer Array Pathfinder (ASKAP) | 36 × 12 m diameter antennas | 0.7 – 1.8 GHz | Polarization surveys, transient sources, continuum surveys, study of galaxy formation, ten major science projects |
| Australia Telescope Compact Array, Narrabri | 22 m steerable paraboloid, altazimuth mounted | 1250–1760 MHz<br>2200–2500 MHz<br>4.4–6.1 GHz<br>8.0–9.2 GHz<br>20–25.5 GHz<br>42–50 GHz<br>84–98.5 GHz<br>105–116 GHz | Earth rotation aperture synthesis, VLBI, continuum, spectral line, pulsar gating, frequency switching, band switching, and mosaicing |
| Australia: Wiruna Telescope Project | 5 m dish | 1420 MHz | 21–cm radiation |
| Belgium: Humain Radio Astronomical Station | 7.5 m steerable paraboloid | 600 MHz | Solar |
| Belgium: Humain Radio Astronomical Station Interferometer | 48-antenna (4 m) equatorial-mount interferometer (E-W, N-S) | 408 MHz | Solar |
| Brazil: Brazilian Decimetric Array (BDA) | 38-element | 1.2–6.0 GHz | Solar and nonsolar observations and space weather phenomena |
| Brazil: Fortaleza Observatory | 14.2 m dish | 2.6 GHz<br>8.8 GHz | Dedicated to geodetic VLBI |
| Brazil: Itapetinga Radio Observatory | 13.7 m steerable paraboloid, radome-enclosed, Cassegrain, altazimuth mount | 10.7 GHz–22 GHz<br>30 GHz<br>44 GHz<br>48 GHz<br>94 GHz | Solar, galactic, extragalactic, continuum and line works, and VLBI astrophysics and geodetics, at 48 GHz multiple-beam solar observations only |
| Canada: Dominion Radio Astrophysical Observatory | 26 m steerable paraboloid | 1500 MHz<br>1420 MHz<br>1610 MHz<br>1660 MHz<br>1665 MHz | Spectroscopy of HI and OH, 1.6–GHz VLBI |
| Canada: Dominion Radio Astrophysical Observatory Synthesis Telescope | Seven steerable paraboloids, Earth-rotation synthesis, | Operates simultaneously at 408 MHz 1420 MHz (256 21-cm channels) | Continuum and HI spectroscopy |

*(Contd.)*

| Country Name | Telescope–Diameter | Frequency Bands | Observations/Notes |
|---|---|---|---|
| Chile: Atacama Desert Llano de Chajnantor Observatory Atacama Large Millimeter Array (ALMA) | 54 dishes with 12 m diameter and 12 dishes with 7 m diameter | sensitive to wavelengths between radio and infrared (0.3 – 9.6 mm) | Proposed investigation of star birth in the early universe |
| Chile: Atacama Cosmology Telescope | 6 m off-axis Gregorian, bolometer array camera | 145 GHz 215 GHz 280 GHz | Measurements of CMB, galaxy clusters via Sunyaev-Zeldovich effect |
| China: Five hundred meter Aperture Spherical Telescope (FAST) | 500 m dish under construction | 0.3–5.1 GHz | Sky mapping using an adaptive surface |
| China: High Time Resolving Solar Radio Telescope, Nanjing | 2 m steerable paraboloid | 9.375 GHz | Solar |
| China: Primeval Structure Telescope (PaST) | 10,000 log periodic antennas under construction | 50–200 MHz | Detecting 21-cm radiation from earliest objects in the universe |
| China: Yunnan Solar Observatory | 2.5, 3.0, 3.2, and 10 m steerable paraboloids | 2902 MHz 1420 MHz 3653 MHz 1426 MHz 230–300 MHz | Solar |
| China: Purple Mountain | 0.4 m steerable paraboloid | 3500 MHz | Solar radio emission (total intensity) |
| China: Purple Mountain | 13.7 m steerable paraboloid | 85–115 GHz | Continuum and spectral line |
| China: Purple Mountain | 1.5 m steerable paraboloid | 2700 MHz 9375 MHz | Solar radio emission (total intensity) |
| China: Purple Mountain | 2 m steerable paraboloid | 5000 MHz | Solar radio emission (total intensity) |
| Czech Republic: Ondrejov Astronomical Observatory | 3 m dish | 3 GHz 2–4 GHz | Solar |
| Czech Republic: Ondrejov Astronomical Observatory | 7.5 m steerable paraboloid | 260 MHz 536 MHz 808 MHz | Solar |
| Czech Republic: Ondrejov Astronomical Observatory (with spectrograph) | 7.5 m steerable paraboloid | 100–1200 MHz | Solar |

| Country Name | Telescope–Diameter | Frequency Bands | Observations/Notes |
|---|---|---|---|
| Finland: Metsähovi Radio Research Station | 13.7 m steerable dish | 22.2 GHz<br>37.6 GHz<br>75–115 GHz | Solar and extragalactic continuum on all three frequencies, galactic lines at 22.2<br>and<br>75–115 GHz. |
| France: Instituto de Radio Astronomia Millimetrica (IRAM) | 30 m steerable paraboloid | 74–115 GHz<br>130–170 GHz<br>203–270 GHz<br>250 GHz<br>320–360 GHz | Millimeter wavelength spectroscopy and photometry |
| France: Observatoire de l'Université de Bordeaux, POM 1 Telescope | 2.5 m steerable paraboloid | 100 GHz<br>(75 to 120 GHz) | Spectral-line observations: in our galaxy (dark clouds, HII regions, large-scale surveys), in the terrestrial atmosphere (ozonosphere, mesosphere) |
| France: Observatoire de Grenoble, C.E.R.M.O., POM 2 Telescope | 2.5 m Cassegrain paraboloid | 220 GHz | Mainly CO isotopes |
| France: Observatoire du Plateau de Bure | 3 × 15 m (3-element interferometer) | 80–115 GHz<br>230 GHz | Millimeter wavelength spectroscopy, radiometry, and interferometry |
| France: Station de Radioastronomie de Nancay (16-element network) Observatoire de Paris-Meudon | Network of 16 1 m paraboloids distributed over 23 m, E-W | 9.400 GHz | Solar |
| France: Station de Radioastronomie de Nancay (18-element network) Observatoire de Paris-Meudon | Network of 16 4 m paraboloids and two 10 m paraboloids distributed over 3.2 km, E-W | 150–450 MHz (5 simultaneously observed frequencies) | Solar and calibration radio sources (Cygnus A, Cassiopeia, Hydra) |
| France: Station de Radioastronomie de Nancay (24-paraboloid network) Observatoire de Paris-Meudon | Network of 24 5 m paraboloids, distributed over 1.2 km, north to south | 150–450 MHz (5 simultaneously observed frequencies) | Solar and calibration radio sources |

*(Contd.)*

| Country Name | Telescope–Diameter | Frequency Bands | Observations/Notes |
|---|---|---|---|
| France: Station de Radioastronomie de Nancay (dual reflecting antenna) Observatoire de Paris-Meudon | 40 × 200 m Dual reflecting antenna: tiltable flat reflector and standing spherical reflector | 1400 MHz 1650 MHz 3300 MHz | Planetary, cometary, galactic, extragalactic; continuum and spectral lines |
| France: Station de Radioastronomie de Nancay (log-periodic antennas) Observatoire de Paris-Meudon | 144 (2 × 72) conical log periodic antennas | 15 MHz < $f_0$ < 80 MHz | Solar and Jovian bursts |
| Georgia: Abastumani Astrophysical Observatory | 6.6 m × 3.1 m flat synphase antenna | 221 MHz | Solar radio flares |
| Germany: Radioobser-vatorium Effelsberg | 100 m fully steerable altazimuth | 408 MHz – 86 GHz | Pulsars, star forming regions, active galactic nuclei |
| Germany: Observato-rium für Solare Radioastronomie | Three paraboloids: 1.5, 4, and 10.5 m, and Yagi antenna | 9.500 GHz 775 MHz 1470 MHz 2920 MHz 64 MHz 113 MHz 234 MHz 30 MHz 40 MHz | Total solar flux and polarization |
| India: Gauribidanur Low-Frequency Array | T-shaped array: E-W 1.5 km (25 m), S array 0.5 km (15 m), N array 0.34 km (25 m) | 34.5 MHz 45 MHz 55 MHz 80 MHz 150 MHz | Solar, planetary, galactic, extragalactic, and spectral line |
| India: Giant Me-trewave Radio Telescope(GMRT) | 30 45 m parabolic dishes, 14 of which are in a 1 sq km array; the rest are along 3 arms of a Y (max baseline 25 km) | 38 MHz 151 MHz 235 MHz 327 MHz 610 MHz 1420 MHz | Primordial hydrogen in protogalaxies, millisec-ond and binary pulsars, galactic and extragalactic radio sources |
| India: Mauritius Radio Telescope | Aperture synthesis in-strument: T array with a 2 km long E-W arm with 1024 helices and 880 m long S arm | 151.6 MHz | Continuum imaging of the southern sky |

| Country Name | Telescope–Diameter | Frequency Bands | Observations/Notes |
|---|---|---|---|
| India: Ooty Radio telescope and Synthesis Radio telescope | Equatorially-mounted parabolic cylinder, steered mechanically in E-W and N-S | 326.5 MHz | Lunar occultation, IPS, spectral line, synthesis mapping, pulsar search, VLBI, galactic, and extragalactic. |
| Italy: Stazione Radioas-tronomica di Medicina | Two arms (T-shaped), transit instrument, parabolic cylinder | 408 MHz | Continuum galactic and extragalactic observations |
| Italy: Stazione Radioas-tronomica di Medicina | 32 m steerable paraboloid | 1400 MHz 1600 MHz 5.0 GHz 10.7 GHz 22.3 GHz 2.2–8.2 GHz | VLBI for astronomy and geodesy, spectral lines |
| Italy: Trieste Astro-nomical Observatory, Basovizza Solar Radio Telescope | 10 m single steerable paraboloid | 237 MHz 327 MHz 408 MHz 610 MHz 1413 MHz 1665 MHz 2695 MHz | Solar |
| Japan: HALCA (Highly Advanced Laboratory for Communication and Astronomy) | Space-based 8 m telescope | 1.6 GHz 5.0 GHz | Space VLBI, studied hydroxyl masers, quasars and pulsars |
| Japan: Hiraiso Solar Terrestrial Research Center | Four paraboloids: 1.1 m 6 m 10 m (two) | 100.5 MHz 201 MHz 501 MHz 9.500 GHz 31.65 GHz | Solar radio at 31.65 GHz, solar patrol at other frequencies |
| Japan: Solar-Terrestrial Environment Labo-ratory UHF Radio Telescope, Nagoya | Cylindrical parabolic antenna 100 m E-W, 20 m N-S | 327 MHz | Solar |
| Japan: Kashima Space Research Center | 26.0 m steerable parabolic | 200–2320 MHz 8160–8800 MHz | Extragalactic and galactic continuum observation, VLBI scheduling for geodesy |
| Japan: Kisarazu College Observatory | 46 m paraboloid dish | 115.27 GHz | CO spectral-line observation |

| Country Name | Telescope–Diameter | Frequency Bands | Observations/Notes |
|---|---|---|---|
| Japan: Nobeyama Radio Observatory | 45 m telescope | 9.8 GHz<br>23.0 GHz<br>36–49 GHz*<br>36–49 GHz*<br>70–90 GHz*<br>85–115 GHz*<br>(* tunable) | Galactic and extragalactic continuum, spectral line, VLBI (except 70–90 GHz) |
| Japan: Nobeyama Radio Observatory | Five-element interferometer array, T baseline is 560 m E-W, 520 m N-S | 22–24 GHz*<br>40–50 GHz**<br>80–120 GHz*<br>(*SIS mixer, FX correlator<br>**HEMT amp, FX correlator) | Aperture synthesis, continuum and spectral lines |
| Japan: Nobeyama Radio Observatory | 16-element interferometer:<br>1.2 m steerable paraboloids | 17 GHz | Solar |
| Japan: Nobeyama Radio Observatory | Two steerable 8- and 6 m paraboloids | 70–1000 MHz | Solar |
| Japan: Nobeyama Radio Observatory (17-element interferometer) | Seventeen 6 m steerable paraboloids (interferometer) | 160.4 MHz | Solar |
| Japan: Nobeyama Radio Observatory (Three paraboloid telescope) | One 0.3 m and two 0.25 m paraboloids on the same equatorial mount | 35 GHz*<br>80 GHz**<br>(0.3 m*)<br>(0.25 m**) | Solar |
| Japan: Toyokawa Observatory | 0.85 m dish equatorial mount | 9.411 GHz | Solar |
| Japan: Toyokawa Observatory | 1.5 m dish equatorial mount | 3750 MHz | Solar |
| Japan: Toyokawa Observatory | 2.0 m dish equatorial mount | 2060 MHz | Solar |
| Japan: Toyokawa Observatory | 3.0 m dish equatorial mount | 1060 MHz | Solar |
| Japan: Toyokawa Observatory | Array:<br>32 × 16 tee, 32 × 2 linear<br>(2 m E-W1.2 m N-S) | 9.407 GHz | Solar |

| Country Name | Telescope–Diameter | Frequency Bands | Observations/Notes |
|---|---|---|---|
| Japan: Toyokawa Observatory | Large Array: 32 × 16 T, 32 × 2 linear, 16 × 2 linear (3 m) | 3748.5 MHz | Solar |
| Mexico: Large Millimeter Telescope | Steerable 50 m paraboloids, two-element interferometer | Any between 85–350 GHz | Continuum and spectroscopic |
| Mexico: Tonantzintla Solar Radio Interferometer | Steerable 1 m paraboloids, two-element interferometer | 9 GHz | Solar |
| Mexico: UNAM Solar Radio Interferometer | Two-element interferometer (1.4 m) | 7.5 GHz | Solar |
| Netherlands: Dwingeloo Radio Observatory | Steerable 25 m paraboloid | 1400 MHz 1652.5 MHz 4895 MHz | Auto-correlation spectrometry |
| Netherlands: Low-Frequency Array | Phased array of ~20,000 dipole antennas in 48 stations (40 in the Netherlands and 8 spanning ~1500 km throughout Europe ) | 10–240 MHz | Epoch of reionization following the dark ages, cosmic magnetic fields in galaxies, space weather |
| Netherlands: Westerbork Synthesis Radio Telescope | Array: 14 25 m steerable paraboloids (4 movable 10 fixed) on an E-W baseline (min-max baseline: 32 m–2769 m) | 310–390 MHz 606–610 MHz 1200–1450 MHz 1590–1750 MHz 2215–2375 MHz 4770–5020 MHz 8150–8650 MHz 250–460 MHz 700–1200 MHz | Continuum synthesis (to 160 MHz), spectral-line synthesis, VLBI (tied-array), pulsar (tied-array) |
| Russia: Ratan-600 Special Astrophysical Observatory | Variable profile antenna, 895 2×7 m reflector elements in a 600 m circle | 1000 MHz 1500 MHz 2500 MHz 3750 MHz 7.600 GHz 9.375 GHz 11.100 GHz 13.000 GHz 15.000 GHz 37.000 GHz 1420 MHz 1670 MHz | Solar, Solar Corona, SETI |

*(Contd.)*

| Country Name | Telescope–Diameter | Frequency Bands | Observations/Notes |
|---|---|---|---|
| | | 4850 MHz | |
| | | 22.200 GHz | |
| | | 960 MHz | |
| | | 2300 MHz | |
| | | 3650 MHz | |
| | | 3900 MHz | |
| | | 7.700 GHz | |
| | | 14.400 GHz | |
| | | 21.800 GHz | |
| | | 3650 MHz | |
| | | 11.500 GHz | |
| | | 3900 MHz | |
| | | 7.700 GHz | |
| | | 14.400 GHz | |
| Russia: Spektr-R | Space-based 10 m parabolic reflector | 0.327 GHz 1.665 GHz 4.830 GHz 18.39–25.11 GHz | Space VLBI |
| South Africa: Hartebeesthoek Radio Astronomy Observatory | 25.9 m equatorially mounted Cassegrain | 4970 MHz 8400 MHz | VLBI, Pulsar |
| South Africa: MeerKAT | 64 × 13.5 m offset Gregorian dishes under construction | 0.58–1.015 GHz 1–1.75 GHz 8–14.5 GHz | Continuum, spectral line, pulsar timing |
| South Korea: Taeduk Radio Astronomy Observatory | Steerable 13.7 m paraboloid | 80–115 GHz | Spectral line and continuum |
| Spain: Observatorio del Ebro | 3.24 m parabolic reflector | 1390–1567 MHz 2540–2720 MHz 1830–5030 MHz | Solar |
| Spain: Instituto de Radio Astronomia Millimetrica (IRAM) | 30 m steerable paraboloid | 74–115 GHz 130–170 GHz 203–270 GHz 250 GHz 320–360 GHz | Millimeter wavelength spectroscopy and photometry |
| Sweden: ESO Submillimeter Telescope | Three paraboloids | 85–117 GHz 220–280 GHz 300–360 GHz | Planetary, galactic, extragalactic, spectral-line and continuum studies |
| Sweden: Onsala Space Observatory | 20 m and 25 m | 116 MHz–8 GHz 7 MHz–1 GHz | Planetary, galactic, extragalactic; continuum and spectral-line observations |

| Country Name | Telescope–Diameter | Frequency Bands | Observations/Notes |
|---|---|---|---|
| Sweden: European Incoherent Scatter Facility (EISCAT) Scientific Association | 4 32 m parabolic dishes | 933 MHz | Incoherent scatter radar and radio astronomy |
| Sweden: European Incoherent Scatter Facility (EISCAT) Scientific Association | 4 parabolic cylinders | 224 MHz | Incoherent scatter radar and radio astronomy |
| Switzerland: Bleien Radio Astronomy Observatory | 5 m steerable paraboloid | 100–1000 MHz (spectrometer) | Solar radio bursts |
| Switzerland: Bleien Radio Astronomy Observatory | 7 m steerable paraboloid | 100–3000 MHz (spectrometer) | Solar radio spectrum of flares |
| Tasmania: Ceduna Radio Astronomy Observatory | 30 m steerable paraboloid | 1.6–24 GHz | VLBI, line ad continuum, galactic, extragalactic |
| Tasmania: Mount Pleasant Radio Observatory | 14 m steerable paraboloid | 600 MHz 900 MHz 1390 MHz | Pulsars (emission properties and timing) |
| Tasmania: Mount Pleasant Radio Observatory | 26 m steerable paraboloid (X-Y mount prime focus) 12 m AuScope | 600 MHz–12 GHz | Galactic and extragalactic; line and continuum, VLBI (AuScope) |
| UK: Chilbolton Observatory | Steerable 25 m paraboloid, prime focus | 3–30 GHz | Tropospheric propagation |
| UK: James Clerk Maxwell Telescope | 15 m altazimuth mounted, steerable paraboloid, Cassegrain | 218–235 GHz 256–270 GHz 320–370 GHz 150–1000 GHz | Molecular line, continuum |
| UK: Jodrell Bank | 13 m steerable paraboloid, az.-el. mounted | 610 MHz 966 MHz 1420 MHz | Pulsar timing |
| UK: Jodrell Bank MERLIN Cambridge | 32 m shaped paraboloid, altazimuth mounted | 151 MHz 408 MHz 1350–1750 MHz 4600–5100 MHz 21.0–24.0 GHz | Continuum and spectral-line synthesis mapping; VLBI; spectral line |
| UK: Jodrell Bank MERLIN Darnhall | 25 m steerable paraboloid, equatorially mounted | 151 MHz 408 MHz 1350–1750 MHz 4600–5100 MHz 21.0–24.0 GHz | Continuum and spectral-line synthesis mapping; VLBI; spectral line |

*(Contd.)*

| Country Name | Telescope–Diameter | Frequency Bands | Observations/Notes |
|---|---|---|---|
| UK: Jodrell Bank MERLIN Defford | 25 m steerable paraboloid equatorially mounted | 151 MHz<br>408 MHz<br>1350–1750 MHz<br>4600–5100 MHz | Continuum and spectral-line synthesis mapping; VLBI; spectral line |
| UK: Jodrell Bank MERLIN Knockin | 2- m steerable paraboloid, altazimuth mounted | 151 MHz<br>408 MHz<br>1350–1750 MHz<br>4600–5100 MHz<br>21.0–24.0 GHz | Continuum and spectral-line synthesis mapping; VLBI; spectral line |
| UK: Jodrell Bank Observatory Lovell telescope | 76 m steerable paraboloid, altazimuth mounted | 151 MHz<br>240 MHz<br>327 MHz<br>408 MHz<br>610 MHz<br>930 MHz<br>966 MHz<br>1350–1750 MHz<br>2300 MHz<br>2700 MHz<br>4600–5000 MHz<br>5.863 GHz<br>6.0–6.7 GHz | Continuum and spectral-line synthesis mapping; VLBI; pulsars; spectral line; continuum |
| UK: Jodrell Bank (Mk II) | 38 × 25.9 m steerable paraboloid, altazimuth mounted | 151 MHz<br>240 MHz<br>327 MHz<br>408 MHz<br>610 MHz<br>930 MHz<br>966 MHz<br>1350–1750 MHz<br>2300 MHz<br>2700 MHz<br>4600–5000 MHz<br>21.0–24.0 GHz | Continuum and spectral-line synthesis mapping; VLBI; pulsars; spectral line; continuum |
| UK: Jodrell Bank MERLIN Pickmere | 25 m steerable paraboloid, altazimuth mounted | 151 MHz<br>408 MHz<br>1350–1750 MHz<br>4600–5100 MHz<br>21.0–24.0 GHz | Continuum and spectral-line synthesis mapping; VLBI; spectral line |
| UK: One-Mile Telescope | 3 × 18 m telescopes | 408 MHz<br>1407 MHz | Deep field surveys, produced the 5C catalogue of radio sources |

| Country Name | Telescope–Diameter | Frequency Bands | Observations/Notes |
|---|---|---|---|
| Ukraine: UTR-2 Array | T-shaped array | 10.0 MHz<br>12.6 MHz<br>14.7 MHz<br>16.7 MHz<br>20 MHz<br>25 MHz | (1) Systematic survey of discrete sources in the northern sky, (2) Search for recombination lines with large quantum numbers, (3) Pulsars, (4) Supernova remnants, (5) H II regions, (6) Fine structure of objects by scintillation method, (7) Jupiter, (8) VLBI, (9) the Sun |
| Ukraine: Yevpatoria RT-70 | 70 m dish | 5–300 GHz | Radar studies of planets and asteroids, transmission for space experiments |

# BIBLIOGRAPHY

[1] Abhyankar, K. D. 2001. *Astrophysics Star and Galaxies*. Hyderabad, India: Universities Press.

[2] Alfven, H. and Falthammar, C. G. 1963. *Cosmic Electrodynamics*. Oxford, UK: Clarendon

[3] Anderson, J., et al. 2011. The LOFAR Magnetism Key Science Project. Proceedings of "*Magnetic*

*Fields in the Universe: From Laboratory and Stars to Primordial Structures*," eds. M. Soida, et al. Zakopane, Poland.

[4] Balanis, C. A., Elliot, S., Perea, J., and Nolan, H. 1992. *Antenna Theory: Analysis and Design*. New York, NY: McGraw-Hill.

[5] Ball, J. A. 1973. The Harvard Microcorrelator. *IEEE Transactions on Instrumentation and Measurement.* IM-22:193.

[6] Birney, D. S. 1991. *Observational Astronomy*. New York, NY: Cambridge University Press.

[7] Bracewell, R. N. 2003. *Fourier Analysis and Imaging*. New York, NY: Kluwer Academic/Plenum.

[8] Bracewell, R. N. 1956. Two-Dimensional Aerial Smoothing in Radio Astronomy. *Australian Journal of Physics* 9:297–314.

[9] Bracewell, R. N. and Riddle, A. C. 1967. Inversion of Fan-Beam Scans in Radio Astronomy. *The Astrophysical Journal* 150:427–434.

[10] Bracewell, R. N. and Thompson, A. R. 1974 Interpolation of Synthesis Data and Some Effects on Ringlobes. Astronomy and Astrophysics Supplement Series 15:453.

[11] Bracewell, R. N. and Thompson, A. R. 1973. The Main Beam and Ringlobes of an East-West Rotation Synthesis Array. *Astrophysical Journal* 182:77–94. 1973.

[12] Bracewell, R. N. and Roberts, J. A. 1954. Aerial Smoothing in Radio Astronomy. *Australian Journal of Physics* 7:615–640.

[13] Briggs, D. 1955. High Fidelity Deconvolution of Moderately Resolved Sources. PhD thesis, The New Mexico Institute of Mining and Technology.

[14] Burke, B. F. and Graham-Smith, F. 2010. *An Introduction to Radio Astronomy*. Cambridge, UK: Cambridge University Press.

[15 ] Carlstrom, J. E., et al. 2011. The 10 Meter South Pole Telescope. *Publications of the Astronomical Society of the Pacific* 123:568-581.

[16] Carrel, R. L. 1961. The Design of Log Periodic Antennas. *IRE International Convention Record* 1:61 – 75.

[17] Clarke, D. and. Roy, A. E. 1993. *Astronomy: Principles and Practice.* Bristol, UK: Adam Hillger.

[18] Christiansen ,W. N. and Högbom, J. A. 1985. *Radio Telescopes 2$^{nd}$ ed.* Cambridge, UK: Cambridge University Press.

[19] Cole, T. 1970. Finite Sample Correlations of Quantized Gaussians. *Australian Journal of Physics* 21:273–282.

[20] Colvin, R. S. and D'Addario, L. R. 1974, The Stanford Five-Element Array: An Instrument for Real-Time Fan Beam Synthesis and Rapid Rotation-Synthesis. *Astronomy and Astrophysics Supplement Series* 15:465–468.

[21] Condon, J. J. and Ransom, S. M. 2010. *Essential Radio Astronomy.* National Radio Astronomy Observatory (online course). Available at http://www.cv.nrao.edu/course/astr534/ERA.shtml

[22] Conway, J. E., Cornwelln, T. J., and Wilkinson, P. N. 1990. Multi-frequency Synthesis: A new Technique in Radio Interferometric Imaging. *Monthly Notices of the Royal Astronomical Society* 246:490–509.

[23] Cooper, B. F. C. 1970. Correlators with Two Bit Quantization. *Australian Journal of Physics, ,* Vol. 23, 521–527, 1970.

[24] Cornwell, T. J. 1998. Radio-interferometric Imaging of Very Larger Objects. *Astronomy & Astrophysics* 202:316–321.

[25] Cornwell, T. J. and Perley, R. A. 1992. Radio-Interferometric Imaging of Very large Fields; The problem of Non-Coplanar Arrays. *Astronomy & Astrophysics* 261:353–364.

[26] Duric, N., Bourneuf, E., and Gregory, P. C. 1988. The Separation of Synchrotron and Bremsstrahlung Radio Emissions in Spiral Galaxies. *The Astronomical Journal* 96: 81–91.

[27] Ebenezer, E., et al. 2007. Gauribidanur Radio Array Solar Spectrograph (GRASS). *Bulletin of the Astronomical Society of India* 35:111–119.

[28] Engargiola, G. 2002. Non-Planar Log-Periodic Antenna Feed for Integration with a Cryogenic Microwave Amplifier. *Antennas and Propagation Society International Symposium- IEEE* 4:140–143.

[29] Esquivel, A., Lazarian, A., Pogosyan, D., and Cho, J. 2003. Velocity Statistics from Spectral Line Data: Effects of Density-Velocity Correlations, Magnetic Field, and Shear. *Monthly Notices of the Royal Astronomical Society* 342:325–336.

[30] Findlay, J. W. 1964. Radio Telescopes. *IEEE Transactions on Antennas and Propagation* 8:187–198.

[31] Fomalont, E. B. 1973. Earth-Rotation Aperture Synthesis. *Proceedings of the IEEE* 61: 1211–1219.

[32] Gardner, F. F. and Whiteoak, J. B. 1963. Polarization of Radio Sources and Faraday Rotation Effects in the Galaxy, *Nature* 197:1162–1164.

[33] Giacconi, R., Gursky, H., and Van Speybroeck, L. P. 1968. Observational Techniques in X-Ray Astronomy. *Annual Review of Astronomy and Astrophysics* 6:373–416.

[34] Graham, M. H. 1966. Radiometer Circuits. *Proceedings of the Institute of Radio Engineers* 46:1966.

[35] Green, R. M. 1985. *Spherical Astronomy.* New York, NY: Cambridge University Press.

[36] Gregson, S., McCormick, J., and Parin, C. 2007. Principles of Planar Near-Field Antenna Measurements, 3rd ed. London, UK: The Institution of Engineering and Technology.

[37] Goldsmith, P. F., ed. 1988. *Instrumentation and Techniques for Radio Astronomy.* New York, NY: IEEE Press.

[38] Goodman, J. W. 2007. *Introduction to Fourier Optics*, 3$^{rd}$ ed. Greenwood Village, CO: Roberts.

[39] Gordon, M. A. and Sorochenko, R. L. 2009. *Radio Recombination Lines.* New York, NY: Springer Science + Business Media.

[40] Griffiths, D. J. 1995. *Introductory Quantum Mechanics.* Upper Saddle River, NJ: Prentice Hall.

[41] Haddock, F. T. 1958. Introduction to Radio Astronomy. *Proceedings of the Institute of Radio Engineers* 46: 3–12

[42] Hollinger, J. P., Mayer, C. H., and Mennella, R. A. 1964. Polarization of Cygnus A and Other Sources at 5 cm. *Astrophysical Journal* 140:656–665.

[43] Högbom, J. A. 1974. Aperture Synthesis with a Non-Regular Distribution of Interferometer Baselines. *Astronomy and Astrophysics Supplement Series* 15:417–426.

[44] Högbom, J. A. and Brouw, W. N. 1974. The Synthesis Radio Telescope at Westerbork. Principles of Operation, Performance and Data Reduction. *Astronomy & Astrophysics* 33:289–301.

[45] Intema, H. T., et al. 2009. Ionospheric Calibration of Low Frequency Radio Interferometric Observations Using the Peeling Scheme. *Astronomy & Astrophysics* 501:1185–1205.

[46] Jansky, K. G. 1932. Directional Studies of Atmospherics at High Frequencies. *Proceedings of the Institute of Radio Engineers* 20:1920–1932.

[47] Jansky, K. G. 1933 Electrical Disturbances Apparently of Extraterrestrial Origin. *Proceedings of the Institute of Radio Engineers* 21:1387–1398.

[48] Jansky, K. G. 1933. Radio Waves from Outside the Solar System. *Nature* 132:66–66.

[49] Jansky, K. G. 1933. Electrical Phenomena that Apparently are of Interstellar Origin. *Popular Astronomy* 41:548.

[50] Jansky, K. G. 1937. Minimum Noise Levels Obtained on Short-Wave Radio Receiving Systems. *Proceedings of the Institute of Radio Engineers* 25:1517–1530.

[51] Joardar, S., et al. 2010. Radio Astronomy and Super-Synthesis: A Survey. *Progress in Electromagnetics Research B* 22:73–102.

[52] Joardar, S., Bhattacharya, A. B. 2007. Uniform Gain Power-Spectrum Antenna-Pattern Theorem. *Progress in Electromagnetics Research* 77:97–110.

[53] Joardar, S. 2008. Design, Development and Applications of Non-Redundant Instruments for Efficient Observations in Low Frequency Radio Astronomy. PhD thesis, University of Kalyani.

[54] Joardar, S. and Bhattacharya, A. B. 2008. A Novel Method for Testing Ultra Wideband Antenna-Feeds on Radio Telescope and Dish Antennas. *Progress in Electromagnetics Research* 81:41–59.

[55] Joardar, S. and Bhattacharya, A. B. 2006. Two New Ultra Wide-band Dual Polarized Antenna-Feeds using Planar Log Periodic Antenna and Innovative Frequency

Independent Reflectors. *Journal of Electromagnetic Waves and Applications* 20:465–1479.

[56] Joardar, S. and Bhattacharya, A.B. 2006. New Designs of Ultra Wideband Dual Polarized Antenna Feeds for Broadband and Ultra Broadband Satellite and Microwave Links. Proceedings of the National Conference on Broad Band Communication Systems, Tata Mc-Graw Hill, 91–96, 2006.

[57] Johnson, R. C. 1993. *Antenna Engineering Handbook 3$^{rd}$ ed.* New York, NY: McGraw-Hill.

[58] Kitchin, C. R. 2003. *Astrophysical Techniques* 4$^{th}$ ed. Boca Raton, FL: Taylor & Francis.

[59] Kreysa, E., et al. 1999. Bolometer Array Development at the Max-Planck-Institute fur Radioastronomie. *Infrared Physics and Technology* 40:191–197.

[60] Kraus, J. D., Tiuri, M. Räisänen, A. V. 1986. *Radio Astronomy*. Powell, Ohio: Cygnus-Quasar.

[61] Kraus, J. D. and Marhefka, R. J. 2002. *Antennas for All Applications*. New York, NY: McGraw-Hill.

[62] Longair, M. S. 2007. *The Cosmic Century: A History of Astrophysics and Cosmology*. Cambridge, UK: Cambridge University Press.

[63] Longair, M. S. 1992. *High Energy Astrophysics*, 2$^{nd}$ ed. Cambridge, UK: Cambridge University Press.

[64] Lorimer, D. and Kramer, M. 2005. *Handbook of Pulsar Astronomy*. Cambridge, UK: Cambridge University Press.

[65] Lyne, A. G. and Graham-Smit, F. *Pulsar Astronomy*, 2$^{nd}$ ed., Cambridge, UK: Cambridge University Press.

[66] Machin, K. E., Ryle, M., and Vonberg, D. D. 1952. The design of an equipment for Measuring Small Radio Frequency Noise Powers. *Proceedings of the IEEE* 99: 127–134.

[67] Mayer, C. H., McCullough, T. P., and Sloanaker, R. M. 1964. Linear Polarization of the Centimeter Radiation of Discrete Sources. *Astrophysical Journal* 139: 248–268.

[68] Minkowski, R. 1960. A New Distant Cluster of Galaxies. *Astrophysical Journal* 132: 908–910.

[69] Mulcahy, D. D., et al. 2011. Probing the Magnetic Fields of Nearby Spiral Galaxies at Low Frequencies with LOFAR. Proceedings of "*Magnetic Fields in the Universe: From Laboratory and Stars to Primordial Structures*," eds. M. Soida, et al. Zakopane, Poland.

[70] Napier, P. J., Thompson, A. R., and Ekers, R. D. 1983. The Very Large Array: Design and Performance of a Modern Synthesis Radio Telescope. *Proceedings of the IEEE* 71:1295–1320.

[71 ] National Academy of Sciences. *Non-US Radio Observatories.* http://www.nas.edu/bpa1/NonUS_Radio_Astronomy_Observatories.htm

[72] National Academy of Sciences. *US Radio Observatories.* http://www.nas.edu/bpa1/US_Radio_Astronomy_Observatories.htm

[73] Nityananda, R. and Narayan, R. 1982. Maximum Entropy Image Reconstruction–A Practical Non-Information-Theoretic Approach. *Journal of Astrophysics and Astronomy* 3:419–450.

[74] Osterbrock, D. E. 1989. *Astrophysics of Gaseous Nebulae and Active Galactic Nuclei*, 2$^{nd}$ ed. Mill Valley, CA: University Science Books.

[75] Pachholczyk, A. B. 1970. *Radio Astrophysics.* San Francisco, CA: Freeman.

[76] Prasad, K. D. 1996. *Antenna and Wave Propagation*, New Delhi, India: Tech India Publications.

[77] Proakis, J. G. and Manolakis, D.G. 2006. *Digital Signal Processing: Principles, Algorithms and Applications*, 4$^{th}$ ed. Upper Saddle River, NJ: Prentice-Hall.

[78 ] RadioAstron Science Operations Group. 2012. *The RadioAstron User Handbook.* Available at http://www.asc.rssi.ru/radioastron

[79] Rau, U., Bhatnagar, S., Voronkov, M.A., and Cornwell, T. J. 2009. Advances in Calibration and Imaging Techniques in Radio Interferometry. *Proceedings of the IEEE* 97:1472–1481.

[80] Reber, G. 1940. Cosmic Static. *Proceedings of the Institute of Radio Engineers* 28: 68–70.

[81] Reber, G. 1942. Cosmic Static. *Proceedings of the Institute of Radio Engineers* 30:367–378.

[82] Reber, G. 1944. Cosmic Static. *Astrophysical Journal* 100:279–287.

[83] Reber, G. 1958. Early Radio Astronomy at Wheaton, Illinois. *Proceedings of the Institute of Radio Engineers* 46:15–23.

[84] Reber, G. 1988. A Play Entitled The Beginning of Radio Astronomy. *Journal of the Royal Astronomical Society of Canada* 82:93–106.

[85] Rohlfs, K. and WilsonT. L. 2006. *Tools of Radio Astronomy 4$^{th}$ ed.* Berlin, Germany: Springer.

[86] Roy, J., et al. 2010. A Real-Time Software Backend for the GMRT. *Experimental Astronomy* 28:25–60.

[87] Rudge, A. W. (ed.), et al. 1983. *The Handbook of Antenna Design Vols. 1&2*. London, UK: Peter Peregrinus.

[88] Ruhl, J. E., et al. 2004 The South Pole Telescope. *Proceedings of the International Society for Optics and Photonics* 5498:11–29.

[89] Runyan, M. C., et al. 2003 ACBAR: The Arcminute Cosmology Bolometer Array Receiver. *Astroph. J. Supp. Series* 149:265–287.

[90] Rybicki, G. B. and Lightman, A. P. 1979. *Radiative Processes in Astrophysics*. New York, NY: Wiley.

[91] Ryle, M. 1952. A New Radio Interferometer and its Application to the Observation of Weak Radio Stars. *Proceedings of the Royal Society of London* 211:351–375.

[92] Ryle, M. and Hewish, A. 1960. The Synthesis of Large Radio Telescopes. *Monthly Notices of the Royal Astronomical Society* 120:220–230.

[93] Sault, R. J. and Osterloo, T. A. 1996. *Imaging Algorithm in Radio Interferometry* in URSI Review of Radio Science 1993–1996, Stone, W. R., ed. New York, NY: Oxford Science Press.

[94] Seidelmann, P. K. 1992. *Explanatory Supplement to the Astronomical Almanac.* Mill Valley, CA: Univ. Science Books.

[95] Shirokoff , E., et al. 2009. The South Pole Telescope SZ-Receiver Detectors. *IEEE Transactions on Applied Superconductivity* 19:517–519.

[96] Smart, W. M. 1977. *Spherical Astronomy 6th ed.* Cambridge, UK: Cambridge University Press.

[97] Smith, G. S. 1997. *An Introduction to Classical Electromagnetic Radiation.* New York, NY: Cambridge University Press.

[98] Staff of NCRA and GMRT. 2007. *Low Frequency Radio Astronomy*, 3rd ed. Pune, India: National Center for Radio Astrophysics, TIFR.

[99] Stanimirovic, S., Altschuler, D. R., Goldsmith, P. F., and Salter, C. J., eds. 2002. *Single-Dish Radio Astronomy: Techniques and Applications.* San Francisco, CA: ASP.

[100] Stanimirovic, S., et al. 1999. The Large-Scale HI Structure of the Small Magellanic Cloud. *Monthly Notices of the Royal Astronomical Society* 302:417–436.

[101] Stutzman, W. L., and Thiele, G. A. 1998. *Antenna Theory and Design 2nd ed.* New York, NY: JohnWiley and Sons.

[102] Sullivan, W. T. 2009. *Cosmic Noise: A History of Early Radio Astronomy.* Cambridge, UK: Cambridge University Press.

[103] Swenson, G.W. and Mathur, N. C. 1968. The Interferometer in Radio Astronomy. *Proceedings of the IEEE* 56:2114–2130.

[104] Taylor, G. B., Carilli, C. L., and Perley R. A., eds.. 1999. *Synthesis Imaging in Radio Astronomy II.* San Francisco, CA: ASP.

[105] Taylor, J. H. 1973. Pulsar Receivers and Data Processing. *Proceedings of the IEEE* 61: 1295–1298.

[106] Thompson, A. R., Moran, J. M., and Swenson, G. W. Jr. 2004. *Interferometer and Synthesis in Radio Astronomy 2nd ed.* Weinheim, Germany: John Wiley-VCH.

[107] Thompson, A. R. and Bracewell, R. N. 1974. Interpolation and Fourier Transformation of Fringe Visibilities. *Astronomical Journal* 79:11–24.

[108] Tiuri, M. E. 1964. Radio Astronomy Receivers. *IEEE Transactions on Antennas and Propagation* AP-12:930–938.

[109] Ulaby, F. T. 2004. *Fundamentals of Applied Electromagnetics.* Upper Saddle River, NJ: Pearson / Prentice Hall.

[120] Verschuur, G. L. 1987. *The Invisible Universe Revealed – The Story of Radio Astronomy.* New York, NY: Springer Verlag.

[121] Verschuur, G. L. and Kellermann, K. I., eds. 1988. *Galactic and Extragalactic Radio Astronomy*, 2nd ed. New York, NY: Springer Verlag.

[122] Weaver, et al. 1965. Observations of a Strong Unidentified Microwave Line and of Emission from the OH Molecule. *Nature* 208: 29–31.

[123] Wellman, P. and Schmid, H. A. 1975. *Radio Astronomy for Amateur Astronomers.* Cambridge, MA: Sky Publishing.

[124] Wiaux, Y., Puy, G., Boursier, Y., and Vandergheynst, P. 2009. Spread Spectrum for Imaging Techniques in Radio Interferometry. *Monthly Notices of the Royal Astronomical Society* 400:1029–1038.

[125] Wilson, T. L., Rohlfs, K. and Hüttemeister, S. 2009. *Tools of Radio Astronomy*, 5th ed. Berlin, Germany: Springer.

[126] Woan, G. and Duffett-Smith, P. J. 1989. Terrestrial Transmittters as Phase Calibrators. *Astronomy & Astrophysics* 208:381–384.

[127] Zeilik, M. and Gregory, S. A. 1998. *Introductory Astronomy Astrophysics.* Orlando, FL: Saunders College Publishers.

# INDEX